I0044553

Photonics: Devices, Systems and Sensors

Photonics: Devices, Systems and Sensors

Editor: Juan Landers

NY RESEARCH
P R E S S

New York

Published by NY Research Press
118-35 Queens Blvd., Suite 400,
Forest Hills, NY 11375, USA
www.nyresearchpress.com

Photonics: Devices, Systems and Sensors
Edited by Juan Landers

© 2019 NY Research Press

International Standard Book Number: 978-1-63238-639-7 (Hardback)

This book contains information obtained from authentic and highly regarded sources. Copyright for all individual chapters remain with the respective authors as indicated. All chapters are published with permission under the Creative Commons Attribution License or equivalent. A wide variety of references are listed. Permission and sources are indicated; for detailed attributions, please refer to the permissions page and list of contributors. Reasonable efforts have been made to publish reliable data and information, but the authors, editors and publisher cannot assume any responsibility for the validity of all materials or the consequences of their use.

Trademark Notice: Registered trademark of products or corporate names are used only for explanation and identification without intent to infringe.

Cataloging-in-Publication Data

Photonics : devices, systems and sensors / edited by Juan Landers.
   p. cm.
Includes bibliographical references and index.
ISBN 978-1-63238-639-7
1. Photonics. 2. Optoelectronic devices. 3. Optical detectors. I. Landers, Juan.
TA1520 .P46 2019
621.36--dc23

Contents

Permissions

List of Contributors

Index

Preface

Photonics is the scientific study of generation, detection and manipulation of light through processes of emission, modulation, amplification, etc. It also studies the transmission and properties of photons. The applications of photonics are in the areas of information processing, telecommunications, spectroscopy, lasers, etc. Important applications of photonic devices are in data recording, laser printing and displays. Sensor is a type of photonic device that detects the change in its immediate environment and sends the signals further to other electronic systems for analysis. There has been rapid progress in this field and its applications are finding their way across multiple industries. This book brings forth some of the most innovative and unexplored aspects of photonics. The various studies that are constantly contributing towards advancing technologies and evolution of this field are examined in detail. For all readers who are interested in photonics, the case studies in this book will serve as an excellent guide to develop a comprehensive understanding.

The researches compiled throughout the book are authentic and of high quality, combining several disciplines and from very diverse regions from around the world. Drawing on the contributions of many researchers from diverse countries, the book's objective is to provide the readers with the latest achievements in the area of research. This book will surely be a source of knowledge to all interested and researching the field.

In the end, I would like to express my deep sense of gratitude to all the authors for meeting the set deadlines in completing and submitting their research chapters. I would also like to thank the publisher for the support offered to us throughout the course of the book. Finally, I extend my sincere thanks to my family for being a constant source of inspiration and encouragement.

Editor

Rapid Multi-Wavelength Optical Assessment of Circulating Blood Volume Without a Priori Data

Ekaterina V. LOGINOVA[1], Tatyana V. ZHIDKOVA[1],
Mikhail A. PROSKURNIN[1,2*], and Vladimir P. ZHAROV[3]

[1]M.V. Lomonosov Moscow State University, Chemistry Department, Moscow, 119991, Russia

[2]Agilent Technologies Partner Laboratory / M.V. Lomonosov Moscow State University, Analytical Center, Moscow, 119991, Russia

[3]Philips Classic Laser Laboratories, University of Arkansas for Medical Sciences, Little Rock, Arkansas, 72205, USA

*Corresponding author: Mikhail A. PROSKURNIN   E-mail: Proskurnin@gmail.com

Abstract: The measurement of circulating blood volume (CBV) is crucial in various medical conditions including surgery, iatrogenic problems, rapid fluid administration, transfusion of red blood cells, or trauma with extensive blood loss including battlefield injuries and other emergencies. Currently, available commercial techniques are invasive and time-consuming for trauma situations. Recently, we have proposed high-speed multi-wavelength photoacoustic/photothermal (PA/PT) flow cytometry for *in vivo* CBV assessment with multiple dyes as PA contrast agents (labels). As the first step, we have characterized the capability of this technique to monitor the clearance of three dyes (indocyanine green, methylene blue, and trypan blue) in an animal model. However, there are strong demands on improvements in PA/PT flow cytometry. As additional verification of our proof-of-concept of this technique, we performed optical photometric CBV measurements *in vitro*. Three label dyes—methylene blue, crystal violet and, partially, brilliant green—were selected for simultaneous photometric determination of the components of their two-dye mixtures in the circulating blood *in vitro* without any extra data (like hemoglobin absorption) known a priori. The tests of single dyes and their mixtures in a flow system simulating a blood transfusion system showed a negligible difference between the sensitivities of the determination of these dyes under batch and flow conditions. For individual dyes, the limits of detection of $3\times10^{-6}\,M - 3\times10^{-6}\,M$ in blood were achieved, which provided their continuous determination at a level of $10^{-5}\,M$ for the CBV assessment without a priori data on the matrix. The CBV assessment with errors no higher than 4% were obtained, and the possibility to apply the developed procedure for optical photometric (flow cytometry) with laser sources was shown.

Keywords: Circulating blood volume assessment, spectrophotometry, methylene blue, crystal violet

1. Introduction

Accurate and rapid assessment of circulating blood volume (CBV) is required in many clinical applications like evaluation of outpatients and inpatients experiencing extensive blood loss [1], rapid fluid administration, and transfusion of packed red blood cells (RBC) [2]; estimation of haemodilution during surgery requiring cardiopulmonary bypass without blood transfusion [3, 4]; monitoring total blood loss during surgery or blood filtration during, e.g. haemodialysis [5]; ascertainment of RBC mass at baseline preoperatively and changing in response to erythropoietic therapies [6−9], and estimation of the requirements of cardiac-assist devices [10]. However, existing techniques and assays require some improvements. Limitations include inaccuracies, high labour requirements, and slow cycle times for initial and repeat measurements.

The mostly used methods for CBV assessment use optical photometric [11−15], fluorescence [16, 17], or isotopic [18, 19] monitoring of dilution curves for contrast labels (dyes, tags, or markers) injected into an unknown blood volume. In photometric CBV assessment, a label dye (the most widespread is Evans blue [11]) is injected into a bloodstream and attaches itself to albumin [12] or in plasma [13, 14] or serum [20] with predetermined haematocrit (Ht) value [15]. The dye concentration in the bloodstream diminishes due to dilution in the whole blood volume. Its average concentration is measured with an optical photometer as a change in the absorbance at a certain wavelength. Assays using fluorescence-labeled albumins or radio-iodinated serum albumins are based on the above-mentioned label-dilution principle, but the dilution curve is measured as a fluorescence signal [16] or ^{131}I pre-tagged to albumin [21, 22]. These assays cannot be used for a rapid CBV assessment without a priori data on haemoglobin (Hb) concentration and are insensitive. Isotopic dilution methods are also not reproducible [12, 23]. Tagged transfusion methods like ^{51}Cr tagging of RBCs require a donor blood infusion [19]. In current practice, most clinicians agree that the transfusion of donor blood should be avoided unless necessary, thus making tagged transfusion methods less practical. Nowadays, clinical CBV assessment is implemented as *in vivo*

pulse dye densitometry (PDD) [3, 4, 24−28]. This method is based on the principles of pulse oximetry and label-dilution techniques at the absorption maximum of indocyanine green (ICG) dye [24, 29−33]. PDD provides a rapid (every 20 min–30 min, after the ICG from the previous injection is safely cleared by the liver [5]), semi-non-invasive, and convenient bedside CBV assessment [28, 34, 35] and is used for diagnostics [3, 4, 36−39] and treatment of blood losses [30, 40, 41]. PDD correlates well with ^{131}I or ^{51}Cr [5, 42, 43] and satisfactorily with other methods [3, 4]. The haemoglobin concentration is measured from a pre-sampled blood before an ICG injection to establish the baseline absorbance for the current patient, otherwise PDD accuracy is degraded significantly. PDD also experiences several clinical problems: (1) it is barely suitable in many liver diseases and low cardiac outputs [38, 44]; (2) signal amplitudes of optical detection are low [3]; (3) the impact of PDD on the mortality and morbidity of the critically ill patients is still under verification [28].

Thus, existing assays provide a clinical justification and examples of CBV assessment; however, no single method is feasible and reliable enough to be widely applied. Label dilution using radioisotopes or dyes is unsuitable for clinical applications, as they do not allow for frequent repeated measurements and require high concentrations of labeled proteins or contrast agents. The state-of-the-art method is PDD, and the use of ICG holds promise as the least invasive technique for measurement. However, the sensitivity and precision are not sufficient, and a priori data on Hb concentration for each patient from an independent method are required.

We have proposed high-speed multi-wavelength photoacoustic/photothermal (PA/PT) flow cytometry (PAFC/PTFC) [45−47] for *in vivo* CBV assessment with multiple dyes as PA and PT contrast labels [48]. PA and PT techniques are based on direct measurements of absorbed energy through its non-radiative transformation into thermal effects [49-55]. Thermal changes in optical and acoustic properties of a sample upon the absorption of laser radiation allows detecting absorbances of 10^{-10} abs − 10^{-6} abs. units, concentrations of 10^{-11} M [56] and to analyze volumes of 10^{-12} L with as low as several absorbing molecules, including biotissues and nanoparticles [50, 51, 54, 57-63]. We have characterized the capability of PAFC/PTFC to monitor the clearance of several dyes (ICG, methylene blue, and trypan blue) in an animal model in vivo and in real time [48]. Strong dynamic PA signal fluctuations were observed, which were associated with interactions of the dyes with circulating blood cells and plasma proteins [48]. PAFC demonstrated enumeration of circulating red and white blood cells labeled with ICG and methylene blue, respectively, and the detection of rare dead cells uptaking trypan blue directly in a bloodstream [48]. PAFC/PTFC offer advanced alternatives to PDD and other CBV assays as they provide much higher sensitivity of direct measurements of absorbed energy compared with light attenuation (as in absorption spectroscopy) or re-emitted light (fluorescent techniques) [49, 64-68]. The PA technique with natural chromophores (Hb or melanin) or lowly toxic nanoparticles as PA contrast agents is currently the fastest-growing area of biomedical imaging, providing higher sensitivity and resolution at a single-cell level in deeper tissues (up to 3 cm − 5 cm) compared with other optical modalities [50, 51, 55, 63, 64]. These techniques offer the highest absorption sensitivity (100-fold − 1000-fold better than PDD/absorption spectroscopy), which allows for noninvasive detection of unlabeled biomolecules at a threshold comparable with that of fluorescence labeling (which is toxic for humans) [54, 69]. Methods are safe; the short-term temperature rise of less than or equal to 0.01 °C − 1 °C at low laser fluences (5 mJ/cm^2 − 20 mJ/cm^2) is well within the laser safety standard of 35 mJ/cm^2 − 100 mJ/cm^2 at 650 nm − 1100 nm [55, 70]. The tremendous clinical potential and safety of PA/PT in vivo have been demonstrated in many clinical trials of other applications [64]. Examples include imaging of breast tumours at centimetre-scale depths [61, 62] or blood microvessels [55], continuous monitoring of blood oxygenation in 15-mm-diameter jugular veins despite light scattering in a 15-mm − 20-mm-thick layer of overlying tissue [71], and measurement of blood [Hb] [58]. PT determination of various Hb species (desoxyhaemoglobin, oxyhaemoglobin, carboxyhaemoglobin, methaemoglobin, cyanohaemoglobin, and hemichrome) with the limits of detection of 10^{-8} M was made [72]. The error of determination of desoxyhaemoglobin vs. oxyhaemoglobin and vice versa was not above 3% − 5% for the ratio of the species of 10:1 [72]. Recently, PAFC has been developed for in vivo detection of circulating tumour cells and bacteria targeted by nanoparticles. As a whole, rapid PA/PT tests can be implemented as the determination of several dyes introduced in the blood as the difference in their absorption spectra can be used for the determination of their dilution. However, there are still strong demands on improvements of PAFC/PTFC.

As an additional step in the verification of the PAFC/PTFC as a CBV assessment platform [48], we focused on photometric CBV assessment with several dyes in vitro. Several clinically relevant absorbing dyes were tested in the presence and absence of blood to select optimum dye/wavelength combination for the simultaneous determination of components of their two-dye mixtures in blood under the conditions of CBV assessment.

2. Experiments

2.1 Reagents, solvents, and solutions

The following dyes were used: methylene blue (MB, CAS no. 61-73-4), brilliant green (BG, CAS no. 633-03-4), crystal violet (CV, CAS no. 548-62-9), indigo carmine (CAS no. 860-22-0), bromsulphalein (CAS no. 71-67-0), and Evans blue (CAS no. 314-13-6); their stock solutions of 0.10% wt. in PBS (20 mM, pH 7.4) were used throughout. Other reagents were high-purity 0.1 M KOH, conc. cp HNO_3, and p.a. acetone. Water from a TW-600RU water purification system (Nomura MicroScience Co., Ltd.; Okada, Atsugi-City, Kanagawa, Japan) was used: pH 6.8; specific resistance 18.2 MΩ·cm, Fe, 2 ppt; dissolved SiO_2, 3 ppb; total ion amount, <0.2 ppb; TOC, <10 ppb. Solutions were made using a Branson 1510 ultrasonic bath (USA), power 1 W (exposure times 10 min − 15 min). The glassware was washed with acetone followed by conc. nitric acid. The blood of rats stabilized with heparin was used at the stages of blood batch and flow tests.

2.2 Flow manifold for *in vitro* measurements

The flow manifold (Fig. 1) consists of a circular flow system pumped with a Watson–Marlow 501U peristaltic pump (UK), a cylindrical flow cell (l = 15 mm; 16 cm^3), and a changeable (volume-adjusting) vessel (volume, 0.25 L to 6 L). All the tubing parts are from a KDM blood-transfusion system (KD Medical GmbH, Germany; length 90 cm, i.d. 0.3 cm). The system is filled with doubly distilled water, PBS, or stabilized rat blood. The dye is injected through a valve before the measurement cell to emulate the injection of the target dye in *in vivo* tests. The flow rate is kept at (35±1) mL/min (linear velocity 2 cm/s). After each measurement, the working liquid is drained to waste and washed with distilled water through a secondary injection valve (Fig. 1).

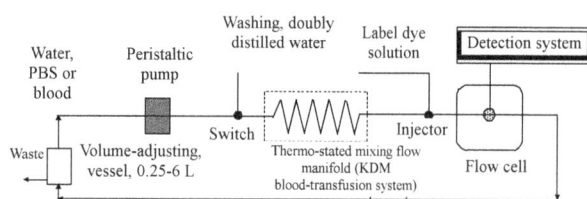

Fig. 1 Flow manifold for circulating blood volume estimation using flow photometry.

2.3 Measurements

A laser-based setup on the basis of a previously described continuous-wave (CW) mode PT-lens spectrometer was used [73]. The schematic (Fig. 2) was based on recording an induced refractive-index heterogeneity (thermal-lens effect, excitation: (IDLS5 diode lasers, Polyus, Moscow; wavelengths 532 nm, 610 nm, 635 nm, 660 nm, and 690 nm; waist diameter, 59.8 μm ± 0.5 μm; power range 1 mW − 20 mW) causing defocusing of a collinear probe beam (diode laser, wavelength, 808 nm; waist diameter, 25.0 μm ± 0.2 μm; (attenuated) power, 1 mW) and, hence, a reduction in the probe beam intensity at its center by a far-field photodiode (sample-to-detector distance 180 cm) supplied with a stained-glass broadband-range (610 nm − 640 nm) bandpass filter and a 2-mm-diameter pinhole (Fig. 2).

The synchronization of the measurements was implemented by in-house written software. The PT spectrometer has a linear dynamic range of the signal of four orders of magnitude (the range of absorption coefficients for 10 mm optical pathways is 1×10^{-6} cm^{-1} to 2×10^{-2} cm^{-1}) and the response time of 0.005 s − 2 s (depending on the selected measurement parameters, namely, on the data throughput rate and time, the number of points to be averaged, etc.). The spectrometer implements a secondary channel (for gathering the scattering signal, if present). The probe beam was reflected by a dichroic mirror; the residual excitation beam was removed with a stained-glass bandpass filter and after a 2-mm pinhole appeared at the primary PT detector. If the photometric or PT channel was not needed, the corresponding detector was switched off.

Spectrophotometric measurements in a batch and flow modes were made using an Agilent Cary 60 spectrophotometer (USA) with l=1 mm, 0.3 cm^3. The pH values were measured by an inoLab pH Level 1 pH-meter (Germany) with a glass pH-selective electrode (precision ±5%). Solutions were mixed with a Biosan MMS 3000 automixer and a micro-stirrer.

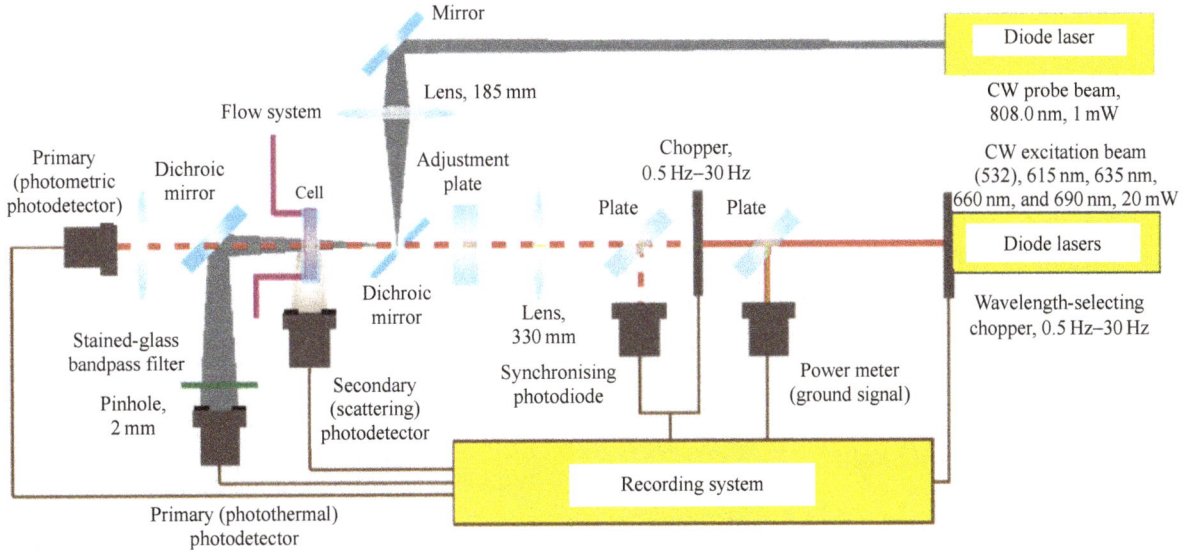

Fig. 2 Schematics of the dual-beam flow photometer / thermal-lens spectrometer.

2.4 CBV assessment

We used dyes for intravenous administration for optical absorbance determination of single dyes or components of two-dye mixtures in circulating blood. CBV was measured as an average of two signals for each dye to decrease the interference of both dyes on one another and thus to improve the accuracy (and determined after their dilution in circulation). The concentration of each dye diminished due to dilution. From a curve of the relative decrease in the concentration, CBV was calculated from the ratio of concentrations of the initial dye solution and the solution of dye in the blood according to the following [2, 5, 15, 74]:

$$CBV = V_0 A_0 / A_x = V_0 c_0 / c_x \qquad (1)$$

where A is absorbance, V_0 and c_0 are initial volume and concentration of the dye solution, and c_x is the dye concentration in the blood after its dilution (Fig. 3). The determination of two components a and b of a dual-component mixture was made using two methods (1) a standard Vierordt's method at two wavelengths λ_1 and λ_2 [48, 75, 76]:

$$\begin{cases} A^{\lambda_1} = l(c_a \varepsilon_a^{\lambda_1} + c_b \varepsilon_b^{\lambda_1}) \\ A^{\lambda_2} = l(c_a \varepsilon_a^{\lambda_2} + c_b \varepsilon_b^{\lambda_2}) \end{cases} \qquad (2)$$

and (2) using an overdetermined Vierordt's system at four wavelengths to decrease the overall error [48]:

$$\begin{cases} \Delta A_a = A^{\lambda_1} - A^{\lambda_2} = l[c_a(\varepsilon_a^{\lambda_1} - \varepsilon_a^{\lambda_2}) + c_b(\varepsilon_b^{\lambda_1} - \varepsilon_b^{\lambda_2})] \\ \Delta A_b = A^{\lambda_3} - A^{\lambda_4} = l[c_a(\varepsilon_a^{\lambda_3} - \varepsilon_a^{\lambda_4}) + c_b(\varepsilon_b^{\lambda_3} - \varepsilon_b^{\lambda_4})] \end{cases}$$
$$(3)$$

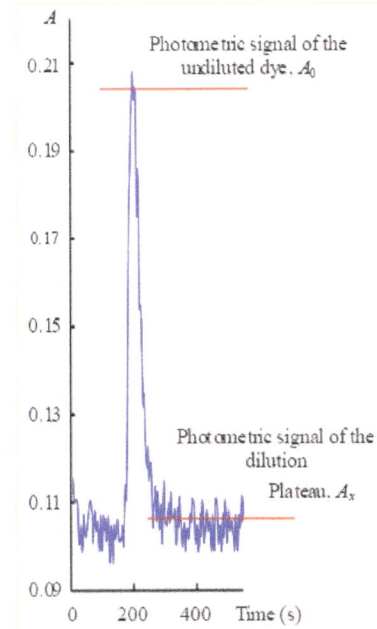

Fig. 3 Dilution curve of a dye (methylene blue) in CBV measurements.

Here, A is absorbance acquired from spectrophotometric measurements. Maxima $\varepsilon_a^{\lambda_1} / \varepsilon_b^{\lambda_1} \times \sqrt{\varepsilon_a^{\lambda_1} \varepsilon_b^{\lambda_1}} = f(\lambda)$ and

$\varepsilon_b^{\lambda_1}/\varepsilon_a^{\lambda_1} \times \sqrt{\varepsilon_a^{\lambda_1}\varepsilon_b^{\lambda_1}} = f(\lambda)$ were used as the wavelengths for Vierordt's method. For the overdetermined Vierordt's system, (3), λ_1 and λ_3 are at the maxima, and λ_2 and λ_4 are at the minima of the absorption spectra of a and b components. The calculations of dye concentration were made taking into account the condition c_a/c_b=const, which is correct for preliminarily prepared two-dye mixtures injected into a flow [48].

2.5 Procedures

2.5.1 Tests of the flow manifold

For each dye, the solutions with a concentration of 0.01% wt − 0.1% wt. were prepared, and the solution volume varied as 250.0 mL, 500.0 mL, 1000.0 mL, 2000.0 mL, 4000.0 mL, or 6000.0 mL. The solutions were placed in a vessel connected to the flow manifold and the photometric signals were measured at a flow rate of 35 mL/min (velocity, 2 cm/s). Next, aliquots (20 mL) were probed and measured at batch conditions. Standard deviation and RSD for the same concentration and the same dye were measured for batch and flow conditions.

2.5.2 Spectrophotometric determination of label dyes in the flow

The 250-mL vessel of the flow manifold was filled with distilled water, which started to circulate continuously through the manifold and the flow cell. After zeroing the absorbance, an aliquot (0.2 mL) of the working solution of the dye was injected and the absorbance was continuously measured, at the working wavelength (663 nm, 624 nm, and 590 nm for MB, BG, and CV, respectively) until the constant value was attained. Next, another aliquot was injected into the same solution. The manifold was washed with 0.5 L of distilled water after measurements.

2.5.3 Differential photometric determination of label dyes in the flow

The 250-mL vessel of the flow manifold was

filled with a solution of the dye with absorbance of A=1.0 − 2.0 (2.00 mL − 3.50 mL of the stock solution was diluted to 250 mL) at the working wavelength (663 nm, 624 nm, and 590 nm for MB, BG, and CV, respectively). Next, the actions were similar to Procedure 2.5.2.

2.5.4 Absorption spectra of label dyes in blood

Dye solutions in blood were prepared in photometric cells: 0.02 mL of the stock solution of the dye and 0.40 mL of blood were placed in a cell and mixed with a micro-stirrer followed by the determination of absorption spectra.

2.5.5 *In vitro* determination of label dyes under batch conditions

Stock solutions of selected dyes in blood (0.40 mL of the stock solution of the corresponding dye and 0.60 mL of blood) were used. This freshly prepared stock solution was diluted in the photometric cells alike in Procedure 2.5.4 by adding a 0.01 mL − 0.05 mL of this solution to 0.40 mL of blood. The calibrations were made at wavelengths 615 nm, 630 nm, 663 nm, and 690 nm, and the limits of detection and other performance parameters were calculated.

2.5.6 *In vitro* assessment of circulating blood volume

The main glass vessel of the manifold was filled with the precisely measured blood volume, and the blood started circulating through the manifold. When the regular flow through the cell was established, the zero absorbance was calibrated. Next, 0.4 mL of an aliquot of stock solutions of MB and CV or their mixture was injected, and the absorbances at 615 nm, 630 nm, 663 nm, and 690 nm were recorded until constant absorbance values were reached. The concentrations of both labels were determined from (3).

2.5.7 Photometric assessment of circulating blood volume in a model system

The main glass vessel of the manifold (Fig. 1)

was filled with the precisely measured blood volume, and the blood started circulating through the manifold. When the regular flow through the cell was established, the zero absorbance was calibrated. Next, 0.4 mL of a mixture of stock solutions of MB and CV was injected, and the absorbances at 615 nm, 630 nm, 663 nm, and 690 nm were recorded until constant absorbance values were reached. The concentrations of both labels were determined from (3).

3. Results and discussion

3.1 Initial label dye tests

Multiple absorbing dyes including MB, BG, CV, congo red, indigo carmine, Evans blue, and bromsulphalein (most previously tested on humans) [25, 77–82] were tested in the presence and absence of blood to select the optimal clinically relevant dye and the wavelength combination with minimal overlapping spectral effects, low concentration (i.e. minimum toxicity), and required accuracy.

As the aim was to develop a system for rapid analysis with trace labels, we excluded indigo carmine, bromsulphalein, and Evans blue as the sensitivity of their spectrophotometric determination was low and CBV assessment would require their high concentrations. Also we excluded ICG and trypan blue, which were previously successfully tested with PAFC to expand the usable label selection [48].

The absorption spectra of the remaining labels show rather significant absorbance over blood background (Fig. 4) and do not change in the pH range 6.0–8.0 characteristic to blood.

Spectrophotometric determination of MB, CV, and BG in aqueous solutions (Procedure 2.5.2) results in limits of detection of 1×10^{-7} M. Differential determination of these label dyes against backgrounds of 0.5–2.0 absorbance units showed a decrease in the sensitivity by an order, which can be considered satisfactory (Table 1).

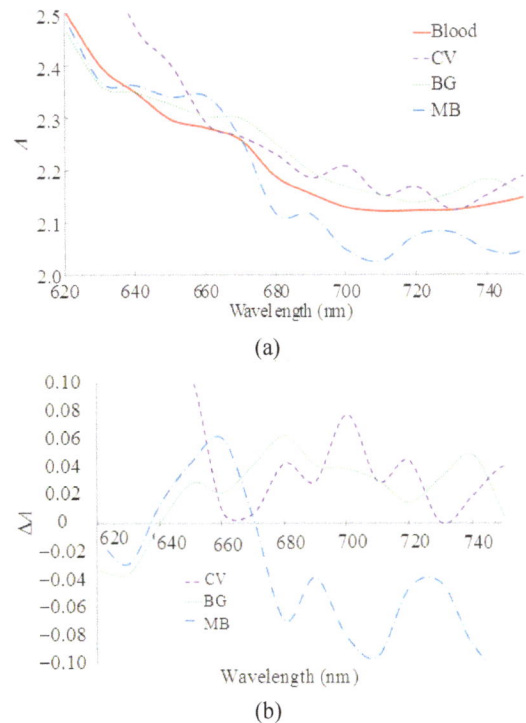

(a)

(b)

Fig. 4 Absorption spectra of blood and selected dyes in blood (a) absolute and (b) differential; crystal violet (CV, 1.2×10^{-4} M), Brilliant Green (BG, 9.8×10^{-5} M), and methylene blue (MB, 1.3×10^{-4} M); $l = 1$ cm.

Table 1 Differential spectrophotometric determination of contrast dyes, pH 7.5, $n = 5$, and $P = 0.95$.

Dye (λ_{max})	A_{bkgr}	c_{min} ($\times 10^7$ M)	Lin. calib. range ($\times 10^6$ M)	r
MB (663 nm)	0.78	1	1–19	0.9961
	0.90	1	4–25	0.9933
	1.25	1	5–36	0.9721
	1.44	2	7–60	0.9687
	1.91	2	20–150	0.9383
	2.17	20	50–500	0.9561
BG (624 nm)	0.54	1	1–25	0.9965
	0.73	1	1–28	0.9886
	1.04	1	6–42	0.9819
	1.33	2	5–50	0.9696
	1.71	10	30–200	0.9042
	1.88	20	40–500	0.8826
CV (590 nm)	0.75	1	4–28	0.9716
	1.07	1	7–38	0.9667
	1.43	2	7–77	0.9511
	1.66	3	10–120	0.8922
	1.91	10	20–160	0.9677
	1.98	10	20–230	0.9637

As absorbance spectra of any dual mixtures of selected dyes overlap, Vierordt's method should be used for data treatment. The comparison of variants of Vierordt's methods showed that the best results were obtained when the ratio of molar concentrations for both labels was constant during all the tests.

According to our previous findings, a two-wavelength Vierordt's system provided the error of 20% for micromolar concentrations of both dyes in a two-dye mixture; however, the use of a four-wavelength overdetermined system decreases the error to 7% [48].

The results for the selected dyes (Tables 2, 3, and 4) confirmed this: a two-wavelength system resulted in 6% − 15% of errors in the determination of the concentration and, as a result, CBV provided a significant increase in the precision (replicability). In this case, the limits of detection of all the three dyes differ insignificantly from single dye solutions. The flow spectrophotometric determination of individual labels in the transfusion manifold (Procedure 2.5.3) showed negligible difference from the batch conditions.

Table 2 Comparison of results of determination of components of two-dye mixtures of MB and BG at pH 7.5, n=3, and P=0.95. [c_{add} is added concentration, c_{calc} is a concentration found from (3) (concentrations in μM)].

c_{add}	c_{calc}	
	663+690+624+550 (nm)	663 + 624 (nm)
	MB	
4.3	4.4 ± 0.4 (3.6%)	4.4 ± 0.3 (3.8%)
6.4	6.3 ± 0.5 (5.5%)	6.2 ± 0.4 (2.7%)
8.6	8.8 ± 0.4 (3.0%)	8.8 ± 0.3 (2.5%)
11	10 ± 1 (5.8%)	10 ± 1 (6.1%)
13	12 ± 1 (5.0%)	12 ± 1 (5.0%)
	BG	
3.3	3.5 ± 0.3 (5.0%)	3.4 ± 0.4 (3.3%)
5.0	5.2 ± 0.2 (3.6%)	5.2 ± 0.3 (3.6%)
6.6	6.6 ± 0.1 (0.1%)	6.7 ± 0.2 (0.6%)
8.3	8.3 ± 0.3 (2.6%)	8.1 ± 0.3 (2.2%)
10	9.3 ± 0.4 (6.2%)	9.5 ± 0.5 (4.8%)

Table 3 Comparison of results of determination of components of two-dye mixtures of MB and CV at pH 7.5, n=3, and P=0.95. [c_{add} is added concentration, c_{calc} is a concentration found from (3) (concentrations in μM)].

c_{add}	c_{calc}	
	663+690+630+590 (nm)	663+590 (nm)
	MB	
4.3	4.6 ± 0.3 (6.3%)	4.1 ± 0.3 (4.1%)
6.4	6.3 ± 0.4 (2.1%)	4.2 ± 0.2 (2.7%)
8.6	8.9 ± 0.4 (3.5%)	8.1 ± 0.4 (6.0%)
11	10 ± 1 (2.0%)	10 ± 1 (9.1%)
13	12 ± 1 (4.1%)	12 ± 1 (6.4%)
	CV	
3.9	4.2 ± 0.3 (7.0%)	4.5 ± 0.3 (15.2%)
5.9	6.2 ± 0.3 (6.3%)	6.3 ± 0.3 (6.8%)
7.8	6.6 ± 0.3 (4.3%)	8.8 ± 0.7 (12.6%)
10	9.8 ± 0.4 (0.3%)	10 ± 1 (6.8%)
12	12 ± 1 (1.5%)	12 ± 1 (5.0%)

For two-dye mixtures, the relative standard deviation of the determination is below 10% and is 2% − 5% for micromolar dye concentrations while using four-wavelength determination with an overdetermined Vierordt's system of equations.

The model flow manifold emulated a transfusion system: a 0.3 cm i.d, and the flow rate of 35 mL/min; the flow cell had the same diameter. The changes in the circulating blood volume within the range 250 mL − 6000 mL (Procedure 2.5.1) showed that the absorbance levels, the reproducibility, and accuracy of measurements for haemoglobin and all the dyes did not depend on the volume. CV and MB showed negligible absorption on the materials of the transfusion system, while BG was intensively absorbed by the manifold.

Table 4 Comparison of results of determination of components of two-dye mixtures of CV and BG at pH 7.5, n=3, and P=0.95. [c_{add} is added concentration, c_{calc} is a concentration found from (3) (concentrations in μM)].

c_{add}	c_{calc}	
	630+590+624+550 (nm)	624+590 (nm)
	BG	
3.3	3.4 ± 0.3 (2.4%)	4.0 ± 0.6 (19.9%)
5.0	5.0 ± 0.1 (0.1%)	5.6 ± 0.5 (12.8%)
6.6	6.6 ± 0.1 (0.04%)	7.8 ± 0.4 (16.9%)
8.3	8.2 ± 0.2 (1.7%)	8.8 ± 0.3 (6.8%)
10	10.0 ± 0.3 (0.02%)	10 ± 1 (5.2%)
	CV	
3.9	4.1 ± 0.3 (4.7%)	3.4 ± 0.3 (12.8%)
5.9	5.8 ± 0.4 (2.2%)	5.1 ± 0.6 (12.7%)
7.8	8.0 ± 0.2 (2.3%)	6.6 ± 0.5 (15.9%)
10	10 ± 1 (2.2%)	8.2 ± 0.4 (15.9%)
12	11 ± 1 (4.7%)	10 ± 1 (13.6%)

3.2 Assessment of dyes in blood

Brilliant green shows a very indistinct spectrum in the target wavelength range (600 nm − 650 nm) in blood, while MB and CV show very good differential spectra in 630 nm − 680 nm and 610 nm − 690 nm, respectively (Procedure 2.5.4). No correlation with their concentrations over 690 nm is due to haemoglobin absorption. The limits of detection of dyes in blood (Procedure 2.5.5) are 3×10^{-6} M − 8×10^{-6} M (Table 5); the error of determination is low, providing assessing both dyes in blood with relative standard deviations below 5% at 10^{-5} M.

Table 5 Performance parameters for the assessment of contrast dyes in blood, n=5 and P=0.95.

Dye	λ (nm)	c_{min} ($\times 10^6$ M)	Lin.cal. range ($\times 10^5$ M)	r
	690	20		0.7830
	680	10		0.9381
	663	10		0.9694
	630	10		0.9822
MB	630 + 690	10	4–20	0.9896
	630 + 680	10		0.9786
	630 + 663	10		0.9268
	663 + 690	20		0.7435
	663 + 680	40		0.7648
	630	5		0.9934
	615	3		0.9569
CV	615 + 690	2	3–20	0.9766
	615 + 663	2		0.9726
	630 + 690	3		0.9897
	630 + 663	3		0.9857

Varying the wavelengths of Vierordt's system, we selected the optimum wavelengths for the overdetermined system, 630 nm and 690 nm for MB and 615 nm and 663 nm for CV (Procedure 2.5.5). For blood volumes 0.3 L – 6 L, the error of assessment was below 4%.

The optimum wavelengths fit the wavelengths of serial diode lasers, which can be used in compact detection schematics for photometric CBV assessment. We tested such a system with commercial diode lasers with 610 nm, 635 nm, 660 nm, and 690 nm. The error of measurements increased to 10%, which can be considered a satisfactory result.

3.3 Laser-based measurements of dyes in blood

The above values were confirmed using the developed laser-based flow spectrometer tested *in vitro* in batch and flow conditions using the available flow module as a simple model for a blood vessel and an absorption spectroscopy unit to real-time measure of absorption spectra of dye alone and in mixtures.

The setup (Fig. 2) was built using the previously optimized schematics taking into account the precision of measurements and the linear range of the thermal lens signal. The main idea is to combine photometric and PA/PT measurements in a single setup. This idea was supported by our previous data on the use of the thermal-lens spectrometer as a single-wavelength photometer providing enough measurement precision [73]. The probe wavelength of 808 nm (marked grey in Fig. 2) was selected as it provided the minimum light absorption of the blood components and selected dyes and low scattering. The excitation beam was made of four beams of diode lasers with the same beam parameters and the wavelengths of 610 nm, 635 nm, 660 nm, and 690 nm. In a more advanced mode, the diode laser with the same beam parameters and the wavelength of 532 nm was added. The lasers were chopped in sequence synchronized with the main light chopper at a frequency of ψ; thus, every excitation on-off cycle of the thermal lens corresponded to the certain wavelength, and the frequency of the excitation with the same wavelength was $\psi/4$ (or $\psi/5$ for a five-wavelength mode).

In order to get the maximum information for a flowing sample (at a flow rate of 35 mL/min), high chopper frequencies are preferable, while the sensitivity of PT measurements decrease with the chopper frequency [83]. Thus, a compromise frequency of 10 Hz was used, which corresponded to 2.5 Hz frequency for a single wavelength. This corresponded to the measurement of ca. 0.15 mL of the flowing sample with a certain wavelength and 0.45 mL with other wavelengths, in cycles.

To implement dual-mode measurements, we used a dichroic mirror to separate the probe beam from the excitation. The whole excitation beam after the cell penetrated the dichroic mirror and was gathered with a focusing lens by the primary photometric detector, which was compared with pre-calibrated power meter used as the absorbance ground signal. The selection of the optimum parameters of PT measurements were discussed previously [73, 84].

The results obtained using laser-based photometry at the selected wavelengths were in good agreement with the performance parameters in

Table 5. The estimations of CBV in the test manifold similarly to Procedure 2.5.1 showed the error of assessment is below 5% for blood volumes 0.3 L – 5 L and above concentrations of dyes diminished compared with photometric measurements. The error of measurements agrees well with our previous data on PAFC measurements of CBV [48].

4. Conclusions

As an additional verification of our proof-of-concept of the PAFC [48, 68, 85−90] for CBV assessment, we performed optical photometric CBV measurements *in vitro* with a specially designed multi-wavelength photothermal-lens spectrometer and a procedure for photometric determination of two dyes (methylene blue and crystal violet) in a blood-transfusion system. The precision of measurements is somewhat worse than in PDD [3, 24, 25], but better than photometric measurements [13−15, 20] and does not require any a priori data on the test system and is simple and inexpensive. Tests of single dyes and their mixtures in a flow system simulating a blood transfusion system showed micromolar limits of detection in blood with low errors. These data complements our previous findings for *in vivo* PAFC [48].

Moreover, one can see a decrease in the error of measurements by a factor of 3 − 4 compared with existing CBV techniques due to the independent determination of two dyes instead of one at two different wavelengths. The measurement provides a decrease in the contrast agent concentrations by a factor of at least 30 due to the high absorption sensitivity of PA/PT spectroscopy and low influence of scattering effects [91]. Moreover, we expect a decrease in the error of measurements by a factor of 3 − 4 compared with existing CBV techniques due to the independent determination of two dyes instead of one at two different wavelengths [48]. The determination of Hb [72] simultaneously with dye dilution will provide lower measurement time (2 − 4

fold) compared with photometric CBV techniques and PDD [73]. If the determination of selected dyes against high blood backgrounds ($1 \, cm^{-1} - 10 \, cm^{-1}$) [55] shows a decrease in the sensitivity by more than an order of magnitude compared with the buffer solution, the differential PT schematics can be used in the future, which previously showed a significant accuracy (3% vs. 5% under the same conditions) and sensitivity (2 orders compared with non-differential schematics) [92, 93]. When necessary, we can increase the number of dyes to 3 or 4 in the mixture to additionally improve the accuracy of CBV measurements.

It is possible to compare the sensitivities of photometric and PT measurements for the same detector and the same laser used for measuring absorption and PT excitation. The minimum detectable linear absorption coefficients for PT and photometric (PM) measurements were calculated according to the equations previously deduced from the theory of these two methods for the conditions of shot noise determining the measurement precision [94]

$$\alpha_{min}^{PT} = \sqrt{\frac{2h\nu_p}{\eta P_p} \Delta f} \frac{B_\infty \omega_0^2 \psi}{4 P_e E_0 D_T l} \qquad (4)$$

$$\alpha_{min}^{PM} = \sqrt{\frac{h\nu_e}{\eta P_e} \Delta f} \frac{1}{l}. \qquad (5)$$

where h is the Planck constant, ν_e and ν_p are frequencies of the excitation and probe beams; η is the detector quantum yield; Δf is the detection channel bandwidth; ω_0 is the excitation beam radius; ψ is the chopper frequency in PT measurements; P_e and P_p are the excitation and probe powers, D_T is the thermal diffusivity, ω_{0e} is the excitation beam waist radius, and B_∞ is the steady-state geometry constant of the setup optical scheme. The factor E_0 is the enhancement factor of PT measurements for unit excitation laser power

$$E_0 = \left(-dn/dT\right)/\lambda_p k \qquad (6)$$

where k is the thermal conductivity. The comparison of sensitivities for the same detector type [η and Δf parameters in (4) and (5)] and the same source of absorption/PT excitation is given by the equation

$$\frac{\alpha_{min}^{PT}}{\alpha_{min}^{PM}} = \sqrt{\frac{2\lambda_e}{\lambda_p P_p}} \frac{B_\infty \omega_0^2 \psi}{4E_0 D_T} \frac{1}{\sqrt{P_e}}. \qquad (7)$$

This shows that an increase in the sensitivity depends on the geometry parameters of the setup and the parameters of the medium, in which the thermal lens is bloomed. For the experimental conditions discussed above the calculation shows that the minimum detectable linear absorption coefficient is about 250-fold lower that for photometry. This means that the significant diminishing of dye doses to get reliable PT signals compared with current clinical doses (for PDD, 2.5 mg/mL − 5 mg/mL) can be made when shifting from optical photometric to PT measurements of CBV.

We anticipate that this non-invasive rapid real-time CBV assessment is valuable for monitoring surgical patients in the operating room, especially those undergoing surgery with predictable substantial blood loss. The high capacity of a multi-wavelength PTFC technique would provide CBV assessment with expected advantages compared with existing assays. Successful completion of these specific aims will provide a novel method for CBV measurements *in vivo* and will shift clinical paradigms by achieving high sensitivity (50-fold − 100-fold higher than that using existing techniques), rapid turnaround (a few minutes vs. hours due to exclusion of any a priori information on the patient), and high accuracy (4% − 7% vs. 15% − 30%). The proposed platform can be further applied to PA or PAFC measurements of multiple blood parameters including a total Hb amount, Ht, oxygenation, abnormal blood cells (e.g., sickle), and Hb composition e.g., meta-, carboxy-, nitroso- or HbS (e.g., in sickle cells) during various

diseases like anaemia as well as on use of encapsulated dyes (e.g. ICG). This system can be also used along with optical-fiber sensors [95]. After proof-of-concept on a small animal model, we will plan a preclinical validation with a lager animal model (e.g., sheep), eventually clinical validation, and feasibility testing (e.g., evaluation of measurement sites including finger, nose, lip, ear, and hand veins).

Acknowledgment

The work is supported by the Russian Science Foundation, grant no. 14-23-00012. We are grateful to Agilent Technologies − Russia and its CEO, Dr. Konstantin Evdokimov, for Agilent equipment used in this study.

References

[1] A. Lozhkin, T. Makedonskaya, G. Pakhomova, and O. Loran, "Estimation of the trauma severity degree in injured with associated injuries depending on the blood loss," *Vox Sanguinis*, 2010, 99(Suppl 1): 435−436.

[2] D. M. Jr. Takanishi, E. N. Biuk-Aghai, M. Yu, F. Lurie, H. Yamauchi, H. C. Ho, *et al.*, "The availability of circulating blood volume values alters fluid management in critically ill surgical patients," *American Journal of Surgery*, 2009, 197(2): 232−237.

[3] W. Baulig, E. O. Bernhard, D. Bettex, D. Schmidlin, and E. R. Schmid, "Cardiac output measurement by pulse dye densitometry in cardiac surgery," *Anaesthesia*, 2005, 60(10): 968−973.

[4] M. Kroon, A. B. Groeneveld, and Y. M. Smulders, "Cardiac output measurement by pulse dye densitometry: comparison with pulmonary artery thermodilution in post-cardiac surgery patients," *Journal of Clinical Monitoring and Computing*, 2005, 19(6): 395−399.

[5] S. Henschen, M. W. Busse, S. Zisowsky, and B. Panning, "Determination of plasma volume and total blood volume using indocyanine green: a short review," *Journal of Medicine*, 1993, 24(1): 10−27.

[6] L. Donner and V. Maly, "Total blood volume in some blood diseases. II. results in pernicious anemia, anemia following hemorrhage, polycythemia vera, secondary polyglobulia and leukemia," *Sbornik Lekarsky*, 1955, 57(5): 125−136.

[7] L. K. Vricella, J. M. Louis, E. Chien, and B. M. Mercer, "Blood volume determination in obese and normal-weight gravidas: the hydroxyethyl starch method," *American Journal of Obstetrics and Gynecology*, 2015, 213(3): 408.e1−408.e6.

[8] A. Joffe, N. Khandelwal, M. Hallman, and M. Treggiari, "Assessment of circulating blood volume with fluid administration targeting euvolemia or hypervolemia," *Neurocritical Care*, 2015, 22(1): 82−88.

[9] D. Konno, R. Nishino, Y. Ejima, E. Ohnishi, K. Sato, and S. Kurosawa, "Assessment of the perioperative factors contributing to the hemodynamic changes during surgery in ten patients with pheochromocytoma," *Masui. Japanese Journal of Anesthesiology*, 2013, 62(4): 421−425.

[10] K. Tanaka, T. Sato, C. Kondo, I. Yada, H. Yuasa, M. Kusagawa, *et al.*, "Hematological problems during the use of cardiac assist devices: clinical experiences in Japan," *Artificial Organs*, 1992, 16(2): 182−188.

[11] N. M. Keith, L. G. Rowntree, and J. T. Geraghty, "A method for the determination of plasma and blood volume," *Archives of Internal Medicine*, 1915, 16(4): 547−557.

[12] W. Jegier, J. Maclaurin, W. Blankenship, and J. Lind, "Comparative study of blood volume estimation in the newborn infant using I-131 labeled human serum albumin (Ihsa) and T-1824," *Scandinavian Journal of Clinical and Laboratory Investigation*, 1964, 16: 125−132.

[13] N. N. Uglova, A. I. Volozhin, and V. E. Potkin, "Method for determination of the circulating blood volume with Evans blue T-1824," *Patologicheskaia Fiziologiia i Eksperimentalnaia Terapiia*, 1972, 16(2): 80−82.

[14] S. A. Glants and V. V. Shevchuk, "A micromethod for the determination of blood volume in laboratory animals," *Laboratornoe Delo*, 1963, 16: 49.

[15] O. A. Kovalev and V. N. Grishanov, "Determination of the volume of circulating blood by using Evans blue dye," *Laboratornoe Delo*, 1976, 11: 664−667.

[16] C. M. Gillen, A. Takamata, G. W. Mack, and E. R. Nadel, "Measurement of plasma volume in rats with use of fluorescent-labeled albumin molecules," *Journal of Applied Physiology*, 1994, 76(1): 485−489.

[17] G. R. Tudhope and G. M. Wilson, "A comparison of 86Rb, 32P and 51Cr as labels for red blood cells," *Journal of Physiology*, 1955, 128(3): 61−62.

[18] T. P. Sivachenko, V. K. Kalina, V. P. Ishchenko, A. K. Belous, and V. I. Kapustnik, "Repeated semi-automatic determination of circulating blood volume," *Vrachebnoe Delo*, 1977, (7): 25−28.

[19] S. J. Gray and K. Sterling, "Determination of circulating red cell volume by radioactive chromium," *Science*, 1950, 112(2902): 179−180.

[20] B. M. Datsenko, N. I. Pilipenko, V. I. Gubskii, and R. A. Sherlanov, "The determination of the volume of circulating plasma using the indicator T-1824," *Laboratornoe Delo*, 1990, 11: 32−34.

[21] J. G. Gibson, A. M. Seligman, W. C. Peacock, J. C. Aub, J. Fine, and R. D. Evans, "The distribution of red cells and plasma in large and minute vessels of the normal dog, determined by radioactive isotopes of iron and iodine," *Journal of Clinical Investigation*, 1946, 25(6): 848−857.

[22] V. K. Modestov and A. T. Tsygankov, "Thyroid function tests using triiodothyronine labeled with I-131," *Meditsinskaia Radiologiia*, 1965, 10: 11−13.

[23] I. A. Frid, V. I. Stoliarov, A. I. Evtiukhin, and M. I. Bernshtein, "Hemodynamic indices and the volume of circulating blood in the surgical treatment of cancer of the esophagus and cardial portion of the stomach," *Vestnik Khirurgii Imeni I. I. Grekova*, 1976, 117(10): 92−96.

[24] M. Haruna, K. Kumon, N. Yahagi, Y. Watanabe, Y. Ishida, N. Kobayashi, and T. Aoyagi, "Blood volume measurement at the bedside using ICG pulse spectrophotometry," *Anesthesiology*, 1998, 89(6): 1322−1328.

[25] T. Imai, K. Takahashi, H. Fukura, and Y. Morishita, "Measurement of cardiac output by pulse dye densitometry using indocyanine green: a comparison with the thermodilution method," *Anesthesiology*, 1997, 87(4): 816−822.

[26] J. A. Tichy, M. Loucka, Z. M. Trefny, M. Hojerova, J. Svacinka, J. Muller, *et al.*, "New clearance evaluation method for hepatological diagnostics," *Physiological Research*, 2009, 58(2): 287−292.

[27] C. K. Hofer, M. T. Ganter, and A. Zollinger, "What technique should I use to measure cardiac output?," *Current Opinion in Critical Care*, 2007, 13(3): 308−317.

[28] R. W. Goy, J. W. Chiu, and C. C. Loo, "Pulse dye densitometry: a novel bedside monitor of circulating

blood volume," *Annals of the Academy of Medicine, Singapore*, 2001, 30(2): 192−198.

[29] H. Sugimoto, O. Okochi, M. Hirota, N. Kanazumi, S. Nomoto, S. Inoue, *et al.*, "Early detection of liver failure after hepatectomy by indocyanine green elimination rate measured by pulse dye-densitometry," *Journal of Hepato-Biliary-Pancreatic Surgery*, 2006, 13(6): 543−548.

[30] R. G. Hoff, G. W. van Dijk, A. Algra, C. J. Kalkman, and G. J. Rinkel, "Fluid balance and blood volume measurement after aneurysmal subarachnoid hemorrhage," *Neurocritical Care*, 2008, 8(3): 391−397.

[31] K. Sha, M. Shimokawa, M. Morii, K. Kikumoto, S. Inoue, K. Kishi, *et al.*, "Optimal dose of indocyanine-green injected from the peripheral vein in cardiac output measurement by pulse dye-densitometry," *Masui. Japanese Journal of Anesthesiology*, 2000, 49(2): 172−176.

[32] N. Taguchi, S. Nakagawa, K. Miyasaka, M. Fuse, and T. Aoyagi, "Cardiac output measurement by pulse dye densitometry using three wavelengths," *Pediatric Critical Care Medicine*, 2004, 5(4): 343−350.

[33] Y. Fujita, T. Yamamoto, M. Fuse, N. Kobayashi, S. Takeda, and T. Aoyagi, "Pulse dye densitometry using indigo carmine is useful for cardiac output measurement, but not for circulating blood volume measurement," *European Journal of Anaesthesiology*, 2004, 21(8): 632−637.

[34] T. Imai, C. Mitaka, T. Nosaka, A. Koike, S. Ohki, Y. Isa, *et al.*, "Accuracy and repeatability of blood volume measurement by pulse dye densitometry compared to the conventional method using 51Cr-labeled red blood cells," *Intensive Care Medicine*, 2000, 26(9): 1343−1349.

[35] T. Kunihara, Y. Wakamatsu, A. Adachi, M. Koyama, N. Shiiya, S. Sasaki, *et al.*, "Clinical evaluation of hepatic blood flow and oxygen metabolism during thoracoabdominal aortic surgery using pulse dye-densitometry combined with hepatic venous oxygen saturation," *Kyobu Geka. Japanese Journal of Thoracic Surgery*, 2000, 53(7): 551−557.

[36] T. Hori, S. Yagi, T. Iida, K. Taniguchi, K. Yamagiwa, C. Yamamoto, *et al.*, "Stability of cirrhotic systemic hemodynamics ensures sufficient splanchnic blood flow after living-donor liver transplantation in adult recipients with liver cirrhosis," *World Journal of Gastroenterology*, 2007, 13(44): 5918−5925.

[37] H. Akita, Y. Sasaki, T. Yamada, K. Gotoh, H. Ohigashi, H. Eguchi, *et al.*, "Real-time intraoperative assessment of residual liver functional

reserve using pulse dye densitometry," *World Journal of Surgery*, 2008, 32(12): 2668−2674.

[38] R. E. Stauber, D. Wagner, V. Stadlbauer, S. Palma, G. Gurakuqi, D. Kniepeiss, *et al.*, "Evaluation of indocyanine green clearance and model for end-stage liver disease for estimation of short-term prognosis in decompensated cirrhosis," *Liver International: Official Journal of the International Association for the Study of the Liver*, 2009, 29(10): 1516−1520.

[39] T. Takazawa, K. Nishikawa, I. Watanabe, and F. Goto, "Preoperative evaluation of hemodynamics using indocyanine green clearance meter in patients with peritonitis from gastrointestinal perforation," *Masui the Japanese Journal of Anesthesiology*, 2005, 54(3): 260−264.

[40] R. Hoff, G. Rinkel, B. Verweij, A. Algra, and C. Kalkman, "Blood volume measurement to guide fluid therapy after aneurysmal subarachnoid hemorrhage: a prospective controlled study," *Stroke*, 2009, 40(7): 2575−2577.

[41] O. Okochi, T. Kaneko, H. Sugimoto, S. Inoue, S. Takeda, and A. Nakao, "ICG pulse spectrophotometry for perioperative liver function in hepatectomy," *Journal of Surgical Research*, 2002, 103(1): 109−113.

[42] E. C. Bradley and J. W. Barr, "Determination of blood volume using indocyanine green (cardio-green) dye," *Life Sciences*, 1968, 7(17): 1001−1007.

[43] H. Fukuda, M. Kawamoto, and O. Yuge, "A comparison of finger and nose probes in pulse dye-densitometry measurements of cardiac output, blood volume and mean transit time," *Masui the Japanese Journal of Anesthesiology*, 2001, 50(12): 1351−1356.

[44] F. Bremer, A. Schiele, and K. Tschaikowsky, "Cardiac output measurement by pulse dye densitometry: a comparison with the Fick's principle and thermodilution method," *Intensive Care Medicine*, 2002, 28(4): 399−405.

[45] D. A. Nedosekin, E. I. Galanzha, E. Dervishi, A. S. Biris, and V. P. Zharov, "Super-resolution nonlinear photothermal microscopy," *Small*, 2014, 10(1): 135−142.

[46] D. A. Nedosekin, M. Sarimollaoglu, E. I. Galanzha, R. Sawant, V. P. Torchilin, V. V. Verkhusha, *et al.*, "Synergy of photoacoustic and fluorescence flow cytometry of circulating cells with negative and positive contrasts," *Journal of Biophotonics*, 2013, 6(5): 425−434.

[47] Y. A. Menyaev, D. A. Nedosekin, M. Sarimollaoglu, M. A. Juratli, E. I. Galanzha, V. V. Tuchin, *et al.*,

"Optical clearing in photoacoustic flow cytometry," *Biomedical Optics Express*, 2013, 4(12): 3030‒3041.

[48] M. A. Proskurnin, T. V. Zhidkova, D. S. Volkov, M. Sarimollaoglu, E. I. Galanzha, D. Mock, *et al.*, "*In vivo* multispectral photoacoustic and photothermal flow cytometry with multicolor dyes: A potential for real-time assessment of circulation, dye-cell interaction, and blood volume," *Cytometry Part A the Journal of the International Society for Analytical Cytology*, 2011, 79A(10): 834‒847.

[49] S. E. Bialkowski, *Photothermal spectroscopy methods for chemical analysis*. New York: Wiley-Interscience, 1996.

[50] L. V. Wang, "Multiscale photoacoustic microscopy and computed tomography," *Nature Photonics*, 2009, 3(9): 503‒509.

[51] S. Mallidi, T. Larson, J. Tam, P. P. Joshi, A. Karpiouk, K. Sokolov, *et al.*, "Multiwavelength photoacoustic imaging and plasmon resonance coupling of gold nanoparticles for selective detection of cancer," *Nano Letters*, 2009, 9(8): 2825‒2831.

[52] V. P. Zharov and V. S. Letokhov, *Laser optoacoustic spectroscopy*. Berlin-Heidelberg: Springer-Verlag, 1986.

[53] V. P. Zharov, "Laser optoacoustic spectroscopy in chromatography," in *Laser analytical spectrochemistry*, Boston, MA: Bristol, pp. 229‒271, 1986.

[54] M. Harada, M. Shibata, T. Kitamori, and T. Sawada, "Application of coaxial beam photothermal microscopy to the analysis of a single biological cell in water," *Analytica Chimica Acta*, 1995, 299(3): 343‒347.

[55] L. V. Wang, *Photoacoustic imaging and spectroscopy*. New York: Taylor & Francis/CRC Press, 2009.

[56] M. A. Proskurnin, "Photothermal spectroscopy, " in *Laser spectroscopy for sensing: fundamentals, techniques and applications*. Cambridge: Woodhead Publ Ltd, 2014, pp. 313‒361.

[57] V. P. Zharov and D. O. Lapotko, "Photothermal imaging of nanoparticles and cells," *IEEE Journal of Selected Topics in Quantum Electronics*, 2005, 11(4): 733‒751.

[58] I. Y. Petrova, R. O. Esenaliev, Y. Y. Petrov, H. P. Brecht, C. H. Svensen, J. Olsson, *et al.*, "Optoacoustic monitoring of blood hemoglobin concentration: a pilot clinical study," *Optics Letters*, 2005, 30(13): 1677‒1679.

[59] Y. Y. Petrov, I. Y. Petrova, I. A. Patrikeev, R. O. Esenaliev, and D. S. Prough, "Multiwavelength optoacoustic system for noninvasive monitoring of cerebral venous oxygenation: a pilot clinical test in the internal jugular vein," *Optics Letters*, 2006, 31(12): 1827‒1829.

[60] R. G. Kolkman, W. Steenbergen, and T. G. van Leeuwen, "*In vivo* photoacoustic imaging of blood vessels with a pulsed laser diode," *Lasers in Medical Science*, 2006, 21(3): 134‒139.

[61] S. Ermilov, A. Stein, A. Conjusteau, R. Gharieb, R. Lacewell, T. Miller, *et al.*, "2007 Detection and noninvasive diagnostics of breast cancer with 2-color laser optoacoustic imaging system, " in *Proc. SPIE*, vol. 6437, pp. 643703‒643711, 2007.

[62] S. E. Vaartjes, J. C. G. van Hespen, J. M. Klaase, *et al.*, "2007 First clinical trials of the Twente photoacoustic mammoscope (PAM)," in *Proc. SPIE*, vol. 6629, pp. 662912‒662917, 2007.

[63] D. Razansky, M. Distel, C. Vinegoni, R. Ma, N. Perrimon, R. W. Koster, *et al.*, "Multispectral opto-acoustic tomography of deep-seated fluorescent proteins *in vivo*," *Nature Photonics*, 2009, 3(7): 412‒417.

[64] V. V. Tuchin, A. Tárnok, and V. P. Zharov, "*In vivo* flow cytometry: a horizon of opportunities," *Cytometry Part A*, 2011, 79A(10): 737‒745.

[65] A. V. Brusnichkin, D. A. Nedosekin, E. I. Galanzha, Y. A. Vladimirov, E. F. Shevtsova, M. A. Proskurnin, *et al.*, "Ultrasensitive label-free photothermal imaging, spectral identification, and quantification of cytochrome c in mitochondria, live cells, and solutions," *Journal of Biophotonics*, 2010, 3(12): 791‒806.

[66] E. I. Galanzha, M. S. Kokoska, E. V. Shashkov, J. W. Kim, V. V. Tuchin, and V. P. Zharov, "*In vivo* fiber-based multicolor photoacoustic detection and photothermal purging of metastasis in sentinel lymph nodes targeted by nanoparticles," *Journal of Biophotonics*, 2009, 2(8‒9): 528‒539.

[67] E. I. Galanzha, J. W. Kim, and V. P. Zharov, "Nanotechnology-based molecular photoacoustic and photothermal flow cytometry platform for *in-vivo* detection and killing of circulating cancer stem cells," *Journal of Biophotonics*, 2009, 2(12): 725‒735.

[68] V. P. Zharov, E. I. Galanzha, E. V. Shashkov, J. W. Kim, N. G. Khlebtsov, and V. V. Tuchin, "Photoacoustic flow cytometry: principle and application for real-time detection of circulating single nanoparticles, pathogens, and contrast dyes *in vivo*," *Journal of Biomedical Optics*, 2007, 12(5): 051503-1‒051503-14.

[69] A. D. Modestov, Y. V. Pleskov, V. P. Varnin, and I. G. Teremetskaya, "Synthetic semiconductor diamond

electrodes: a study of electrochemical activity in a redox system solution," *Russian Journal of Electrochemistry*, 1997, 33(1): 55−60.

[70] E. I. Galanzha and V. P. Zharov, "*In vivo* photoacoustic and photothermal cytometry for monitoring multiple blood rheology parameters," *Cytometry Part A*, 2011, 79A(10): 746−757.

[71] S. A. Lozhkin, "Depth of Boolean functions in a complete basis," *Vestnik Moskovskogo Universiteta Seriya 1 Matematika Mekhanika*, 1996, 51(2): 80−83.

[72] A. Brusnichkin, D. Nedosekin, E. Ryndina, M. Proskurnin, E. Gleb, D. Lapotko, *et al.*, "Determination of various hemoglobin species with thermal-lens spectrometry," *Moscow University Chemistry Bulletin*, 2009, 64(1): 45−54.

[73] M. A. Proskurnin, A. G. Abroskin, and D. Y. Radushkevich, "A dual-beam thermal lens spectrometer for flow analysis," *Journal of Analytical Chemistry*, 1999, 54(1): 91−97, 1999.

[74] A. A. Riley, Y. Arakawa, S. Worley, B. W. Duncan, and K. Fukamachi, "Circulating blood volumes: a review of measurement techniques and a meta-analysis in children," *ASAIO Journal*, 2010, 56(3): 260−264.

[75] J. Karpinska, A. Sokol, and M. Rozko, "Applicability of derivative spectrophotometry, bivariate calibration algorithm, and the vierordt method for simultaneous determination of ranitidine and amoxicillin in their binary mixtures," *Analytical Letters*, 2009, 42(8): 1203−1218.

[76] E. Dinc and F. Onur, "Comparative study of the ratio spectra derivative spectrophotometry, derivative spectrophotometry and Vierordt's method applied to the analysis of oxfendazole and oxyclozanide in a veterinary formulation," *Analusis*, 1997, 25(3): 55−59.

[77] M. A. Yaseen, J. Yu, B. Jung, M. S. Wong, and B. Anvari, "Biodistribution of encapsulated indocyanine green in healthy mice," *Molecular Pharmaceutics*, 2009, 6(5): 1321−1332.

[78] V. Saxena, M. Sadoqi, and J. Shao, "Enhanced photo-stability, thermal-stability and aqueous-stability of indocyanine green in polymeric nanoparticulate systems," *Journal of Photochemistry and Photobiology. B: Biology*, 2004, 74(1): 29−38.

[79] N. M. Shestakov, "Complexity and inadequacy of current methods of determining circulating blood volume and the feasibility of a simpler and faster method of determining it," *Terapevticheskii Arkhiv*, 1977, 49(3): 115−120.

[80] C. Tsopelas and R. Sutton, "Why certain dyes are

useful for localizing the sentinel lymph node," *Journal of Nuclear Medicine*, 2002, 43(10): 1377−1382.

[81] A. B. Dawson, H. M. Evans, and G. H. Whipple, "Blood volume studies: III. behavior of large series of dyes introduced into the circulating blood," *American Journal of Physiology*, 1920, 51(2): 232−256.

[82] K. Shoemaker, J. Rubin, G. L. Zumbro, and R. Tackett, "Evans blue and gentian violet: alternatives to methylene blue as a surgical marker dye," *Journal of Thoracic and Cardiovascular Surgery*, 1996, 112(2): 542−544.

[83] A. Smirnova, M. A. Proskurnin, S. N. Bendrysheva, D. A. Nedosekin, A. Hibara, and T. Kitamori, "Thermooptical detection in microchips: From macro- to micro-scale with enhanced analytical parameters," *Electrophoresis*, 2008, 29(13): 2741−2753.

[84] M. A. Proskurnin and A. G. Abroskin, "Optimization of optical system parameters in dual-beam thermal lens spectrometry," *Journal of Analytical Chemistry*, 1999, 54(5): 401−408.

[85] D. A. Nedosekin, M. Sarimollaoglu, E. V. Shashkov, E. I. Galanzha, and V. P. Zharov, "Ultra-fast photoacoustic flow cytometry with a 0.5 MHz pulse repetition rate nanosecond laser," *Optics Express*, 2010, 18(8): 8605−8620.

[86] D. A. Nedosekin, E. V. Shashkov, E. I. Galanzha, L. Hennings, and V. P. Zharov, "Photothermal multispectral image cytometry for quantitative histology of nanoparticles and micrometastasis in intact, stained and selectively burned tissues," *Cytometry A*, 2010, 77(11): 1049−1058.

[87] E. V. Shashkov, M. Everts, E. I. Galanzha, and V. P. Zharov, "Quantum dots as multimodal photoacoustic and photothermal contrast agents," *Nano Letters*, 2008, 8(11): 3953−3958.

[88] V. P. Zharov, E. I. Galanzha, Y. Menyaev, and V. V. Tuchin, "*In vivo* high-speed imaging of individual cells in fast blood flow," *Journal of Biomedical Optics*, 2006, 11(5): 054034.

[89] V. P. Zharov, E. I. Galanzha, E. V. Shashkov, N. G. Khlebtsov, and V. V. Tuchin, "*In vivo* photoacoustic flow cytometry for monitoring of circulating single cancer cells and contrast agents," *Optics Letters*, 2006, 31(24): 3623−3625.

[90] V. P. Zharov, E. I. Galanzha, and V. V. Tuchin, "Photothermal flow cytometry *in vitro* for detection and imaging of individual moving cells," *Cytometry A*, 2007, 71(4): 191−206.

[91] S. E. Bialkowski, *Photothermal spectroscopy*

methods for chemical analysis. New York: A Wiley-Interscience publication, 1996.

[92] M. A. Proskurnin and M. E. Volkov, "Mode-mismatched dual-beam differential thermal lensing with optical scheme design optimized using expert estimation for analytical measurements," *Applied Spectroscopy*, 2008, 62(4): 439−449.

[93] S. E. Bialkowski, X. Gu, P. E. Poston, and L. S. Powers, "Pulsed-laser excited differential photothermal deflection spectroscopy," *Applied*

Spectroscopy, 1992, 46(9): 1335−1345.

[94] A. Y. Luk'yanov, G. B. Vladykin, M. A. Novikov, and Y. I. Yashin, "Comparison of the capability limits of some optical detectors for liquid chromatography," *Journal of Analytical Chemistry*, 1999, 54(7): 633−638.

[95] P. Roriz, A. Ramos, J. Santos, and J. Simões, "Fiber optic intensity-modulated sensors: a review in biomechanics," *Photonic Sensors*, 2012, 2(4): 315−330.

Shift Endpoint Trace Selection Algorithm and Wavelet Analysis to Detect the Endpoint Using Optical Emission Spectroscopy

Sihem BEN ZAKOUR[1*] and Hassen TALEB[2]

[1]*Higher Institute of Management Tunis, University of Tunis, Tunisia*

[2]*Higher institute of Business and Accounting Bizerte, University of Carthage, Tunisia*

*Corresponding author: Sihem BEN ZAKOUR   E-mail: Sihembenzakour@yahoo.com

Abstract: Endpoint detection (EPD) is very important undertaking on the side of getting a good understanding and figuring out if a plasma etching process is done on the right way. It is truly a crucial part of supplying repeatable effects in every single wafer. When the film to be etched has been completely erased, the endpoint is reached. In order to ensure the desired device performance on the produced integrated circuit, many sensors are used to detect the endpoint, such as the optical, electrical, acoustical/vibrational, thermal, and frictional. But, except the optical sensor, the other ones show their weaknesses due to the environmental conditions which affect the exactness of reaching endpoint. Unfortunately, some exposed area to the film to be etched is very low (<0.5%), reflecting low signal and showing the incapacity of the traditional endpoint detection method to determine the wind-up of the etch process. This work has provided a means to improve the endpoint detection sensitivity by collecting a huge numbers of full spectral data containing 1201 spectra for each run, then a new unsophisticated algorithm is proposed to select the important endpoint traces named shift endpoint trace selection (SETS). Then, a sensitivity analysis of linear methods named principal component analysis (PCA) and factor analysis (FA), and the nonlinear method called wavelet analysis (WA) for both approximation and details will be studied to compare performances of the methods mentioned above. The signal to noise ratio (SNR) is not only computed based on the main etch (ME) period but also the over etch (OE) period. Moreover, a new unused statistic for EPD, coefficient of variation (CV), is proposed to reach the endpoint in plasma etches process.

Keywords: Dimension reduction; OES; plasma etching process; wavelet analysis; CV; SNR

1. Introduction

Plasma is partially ionized gas [1]. Therefore, it contains electron energy which excites the atoms and molecules then de-energizes in emitting photons. Under those circumstances, the plasma thus emits light. On the temperature scale, plasma has the three following classical states, solid, liquid, and gas [2]. Plasma is used for the surface treatment through transforming the electrical energy into a chemical energy by separating molecules [3]. Thus, it contains not only radicals and reactive atoms but also ions which can be accelerated by an electric field applied to bombard surfaces. The plasma process is used in many industrial fields such as biomedical, food, textile, automotive, and micro-electronics. In the biomedical sector, plasma is used to sterilize instruments or modify surface properties to make them bio-compatible, thus limiting the risk rejection by the human body. Plasma is also used for the deposition of protective layers on the biomedical tool surfaces. During the etch process, when the desired layer material is clear, the gas of plasma should be stopped to avoid the over etch of the underlying layer. At this moment, a signal will appear indicating that the required clearing is done [4]. The most popular method for detecting the endpoint is to monitor the trace of the reactive species emission or volatile products emission through optical emission spectrometer (OES) [5 – 8]. At the start of the endpoint phenomenon, the augmented intensity in a particular channel signal corresponds to a growth in the concentration of reactant in the plasma etch process, considering that the reactant species is less used in the surface reaction of the integrated circuit. In contrast, any decrease in the intensity of wavelength channel is assigned by a slack in product concentration, because the under product species is contrived in the integrate circuit (IC) surface reaction [9]. As the etched surface becomes more and more small, the collection of huge number of spectra is unavoidable in the aim of improving the detection of endpoint. The implementation of EPD system allows having multi-OES and then a precise stop procedure in a specific layer, which increases throughput and yield [10, 11]. In this paper, a new algorithm is proposed to select the important fifty endpoint traces named shift endpoint trace selection (SETS) from the full spectra in the first section. Then the linear and nonlinear dimension reduction techniques are applied named principal component analysis (PCA), factor analysis (FA), and wavelet analysis (WA), in Section 3, respectively. The results and the sensitivity analysis is done based on mean and coefficient of variation (CV) statistics through the use of signal to noise ratio (SNR) in Section 4. Finally, the concluding remarks are given in Section 5. Table 1 shows the list of abbreviations used in this work.

Table 1 Abbreviation lists.

Abbreviation	Definition
IC	Integrated circuit
SETS	Shift endpoint trace selection
PCA	Principal component analysis
FA	Factor analysis
WA	Wavelet analysis
CV	Coefficient of variation
SNR	Signal to noise ratio
EPD	Endpoint detection
ME	Main etch
OE	Over etch
M	Mean
SD	Standard deviation

2. Shift endpoint trace selection (SETS) algorithm

2.1 Endpoint states and traces

Endpoint detection is employed to identify when the etched film has been cleared to the underlying film. At this moment, the process can be stopped or modified to a more selective etch. To detect the endpoint, when the film will be removed, without falling on over etch state in other words without damaging or removing the underlying film, and being sure about avoiding also the under etch state, that is the film being etched has not been completely removed, as shown in Fig. 1.

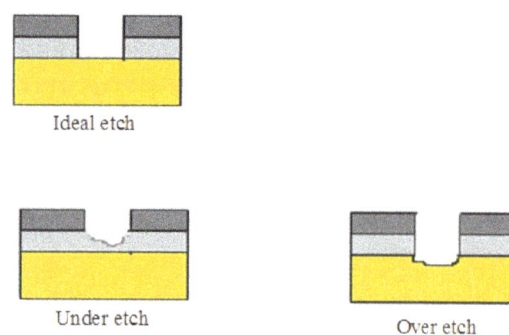

Fig. 1 Devices have been ideally etched, of which one has been over etched and the other under etched.

The ideal endpoint trace for an etch process has plotted intensity like a step change [12] as depicted in Fig. 2. This ideal case in the plasma etch process has no noise, no drift, and with uniform clearing of features across the wafer. In reality, the etch process is affected by some variations, and those variations in the etch rate will produce non-uniform clearing. Hence, the endpoint trace will contain error and drift as shown in the aforementioned figure. In general, the endpoint detection does not occur at a specific time but refers to the range of times over which the film is cleared. The starting of endpoint is named the start of clear, and the finishing of endpoint is named the end of clear. During any chemical process, there is typically a transient state which starts at the beginning of any plasma process, which refers to the initial transient. Then, the signal generally obtains a steady state before detecting the endpoint, named the main etch [4].

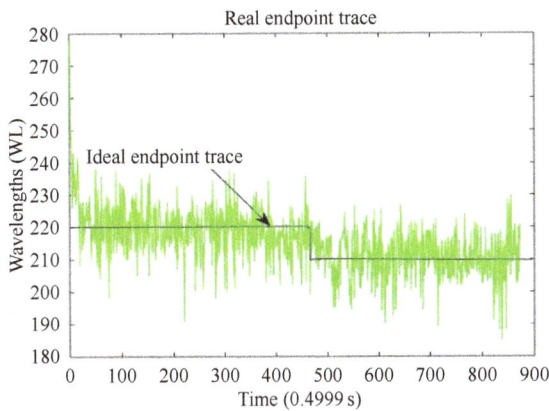

Fig. 2 Real endpoint trace can have various sources of noise and drift, with a change that occurs not in a single step, but gradually over some time span, due to non-uniform clearing of features and the ideal endpoint trace has step-wise change at endpoint and no noise in the signal.

2.2. Proposed algorithm named shift endpoint trace selection (SETS) algorithm

The growth on the collected data leads to a very large databases, high complexity, and long time execution [4, 6, 12]. The size reduction is one of the main tasks on the multivariate analysis [12]. It abates a large observed set of dimensions into a smaller features set. The major and the significant purposes of dimensionality reduction techniques are to visualize, compress, de-noise, and reduce the size of the data. As the importance of plasma etch process on the production of integrated circuit (IC) and on the side to understand and detect endpoint in the plasma etching process, collecting a huge number data (about of $1201 \times 872 \times 5 = 4695910$ intensities) is unavoidable. All spectra intensities are presented in time resolution and spectral resolution.

Despite the benefits of having a lot of information about all process details and progresses, this collection could handle the exactness of monitoring the endpoint. For this reason, the selection of the most important OES light is a decisive and essential task. A new proposed algorithm, named shift endpoint trace selection (SETS) to select the nearly meaningful time traces, is given as follows:

> *For all run Plasma etch step*
>   *For time endpoint trace*
>   *Compute |difference| between endpoint range*
>   *Rank Difference with an increasing order*
>   *Selecting the first fifty differences*
>   *ENDFOR*
> *ENDFOR.*

3. Dimension reduction techniques

The use of multivariate methods for endpoint detection is unavoidable to monitor multi-wavelength channels. In this section, the multivariate tools are investigated. The matrices notations are given as a basic fact to master the multivariate analysis. And an introduction to matrix (linear) algebra is essential in order to better understand the next coming multivariate algorithms. The endpoint optical data are arranged in two-dimensional array (matrix) and given by the matrix below:

$$\mathbf{X} = \begin{bmatrix} x_{11} & x_{12} & \cdots\cdots & x_{1n} \\ x_{21} & x_{22} & \cdots\cdots & x_{2n} \\ \cdot & & & \\ \cdot & \cdot & \cdots\cdots & \cdot \\ \cdot & & & \\ x_{m1} & x_{m2} & \cdots\cdots & x_{mn} \end{bmatrix} \quad (1)$$

where $\mathbf{X}$ is the intensity matrix having m time samples and n wavelength channels. Each sample x represents the spectra intensity for the ith time sample and the jth wavelength channel. It is often commodious to divide the matrix into row and column vectors. The column of the matrix $\mathbf{X}$ refers to a particular wavelength trace, noted as x_j. Hence, the endpoint traces can expressed with $\mathbf{X}$ by

$$\mathbf{X} = [x_1 \quad x_2 \quad \cdots \quad x_j \quad x_{n-1} \, x_n]. \qquad (2)$$

The row vector of the matrix, $x_{i,}$, refers to spectrum at a specific time sample i. The matrix $\mathbf{X}$ could be expressed by using row vectors as follows:

$$\mathbf{X} = \begin{bmatrix} x_1 \\ x_2 \\ . \\ . \\ x_m \end{bmatrix}. \qquad (3)$$

As it was mentioned previously, the endpoint occurs seldom instantaneously, and in the most cases it occurs during a small time interval not on a specific point. On all occasions, the endpoint represents a mean shift from the main etch mean to the over-etch mean [4]. If this shift is much larger than some boundary which is computed from the etch data state, the endpoint is detectable. The matrix formulation of endpoint problem is given by the matrix $\mathbf{X}$ as a matrix containing two partitions, the main etch data and the endpoint data.

$$\mathbf{X} = \begin{bmatrix} X_{(ME)} \\ X_{(EP)} \end{bmatrix} \qquad (4)$$

where $X_{(ME)}$ contains the main etch data and $X_{(EP)}$ contains the endpoint data. The starting idea of principal component analysis (PCA) is to fractionate correlated data into a new set of uncorrelated measurements. The principal component analysis (PCA) is the most used method to reduce data [13–15]. References [16, 17] employed PCA to analyze in-situ spectroscopy data, and PCA is also used as a feature selection by [18, 19] in order to have information about processes and detect faults when there is no sufficient historical data. While the major aims of factor analysis (FA) is to identify the most significant data set to explain correlations among factors. There are several references that treat the factor analyses [20]. Reference [21] employed FA to evaluate of semiconductor ray spectra. Hence, the

factor analysis serves to identify the correlation between the process variables and the common factors (latent variables). The main difference between PCA and FA is that the first relates variables into a small number of PCs and studies all variance while the second produces the factors and analyzes only the shared variance. The employment of PCA and FA which transform data on linear combinations of variables to analyze OES data represents a constraint themselves of linearity. A common form of multivariate non-linear analysis is the wavelet analysis. A wavelet is a waveform, with limited duration and having an average value of zero, and with irregular and asymmetric properties. As a result, there are different types of wavelets such as the Haar, Daubechies, Coiflets, Symlet sand, and biorthogonal wavelets [22]. For each aforementioned wavelet, they have their wavelet filters (low pass and high pass) while the Haar is the most simplest and its filter has only two coefficients in both low pass and high pass. The others such as Daubechies and Coiflet, have more vanishing moments not symmetric and also more coefficients both in low pass and high pass side. The Haar wavelet is a perfect choice in studying the time domain (compactly supported, small support, only 2 taps) but not in the frequency domain. In addition, the Haar wavelet has an efficient memory exactly reversible (easy reconstruction) and it is computationally the cheapest one. Wavelet theory, discovered by [23], has been employed in different scientific fields, such as physics, engineering and mathematic, data compression, and speech analysis. The wavelet analysis decomposes a function into frequency components that represent different degrees of function smoothness, with high frequency components capturing the least smooth function behavior while low frequency components capture the most smooth function behaviors, which makes it easy to extract the information exclusively in the time-frequency domain, as shown in Fig. 3.

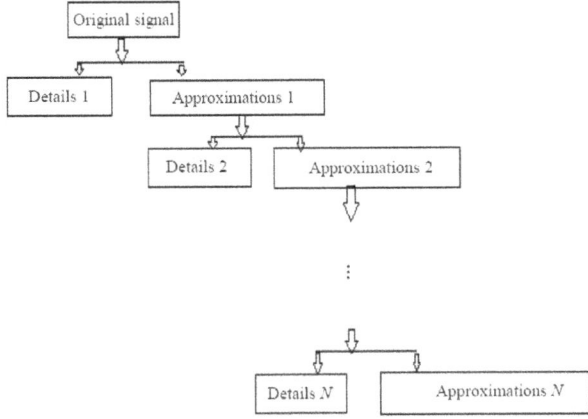

Fig. 3 Multi-resolution analysis showing the decomposed signal into approximation and details at Level N.

The wavelet analysis employs a linear combination of basis functions (wavelets), including time and frequency resolutions. For any function in L^2, the wavelet could be presented as follows [23]:

$$f(t) = \sum_{k \in Z} C_k \Phi_k(t) + \sum_{j=L}^{\infty} \sum_{k \in Z} d_{j,k} \Psi_{j,k}(2^j t - k) \quad (5)$$

where j and k are dilation and translation indices, respectively, and C_k and $d_{j,k}$ refer to the approximation and detail coefficients, respectively. $\Phi_{L,k}(t)$ is the father wavelets representing the low frequency and smooth parts of a signal, however, $\Psi_{j,k}(t)$ refers to the mother wavelet having high frequency and detail part of a signal. Their expressions of father and mother wavelet functions are given, respectively, as follows:

$$\Phi_{L,k}(t) = 2^{L/2} \Phi(2^L t - k) \quad (6)$$

$$\Psi_{j,k}(t) = 2^{j/2} \Psi(2^L t - k) \quad (7)$$

where $j, k, L \in Z$ and $2^{L/2}$ (and $2^{j/2}$) are needed to normalize the function $f(t)$, where, L (and j) corresponds to the level of time resolution (i.e, the width of the time interval) and k corresponds to the shift in the time location. The wavelet coefficients $c_{L,k}$ and $d_{j,k}$ are defined as inner products of $f(t)$ and the corresponding wavelet functions (mother and father) are called the discrete wavelet transformation of the signal $f(t)$, respectively. They are expressed as follows:

$$C_{L,k} = \langle f(t), \Phi_{L,k} \rangle = \int f(t) \Phi_{L,k} dt \quad (8)$$

$$d_{j,k} = \langle f(t), \Psi_{j,k} \rangle = \int f(t) \Psi_{j,k} dt. \quad (9)$$

The highest level of decomposition corresponds to the level after which there is a significant drop in the energy content, and the expression of energy content is given as follows:

$$EN_j = \int f_j^2(t) = \sum_{k=1}^{n} \left[d_{j,k} \Psi(2^j t - k) \right]^2. \quad (10)$$

The reconstructed signal is accurate only if the criterion of threshold selection is optimized. The threshold value using the Visushrink method (or [24–27] universal threshold rule) is given as follows:

$$t_j = \sigma_j \sqrt{2 \lg(n)} \quad (11)$$

where n is the signal length and σ_j is the standard deviation of the noise at scale j. Only the significant wavelet coefficient situated outside of the threshold limits are extracted by applying soft or hard thresholding. In hard thresholding, the wavelet coefficient (at each level) above threshold will be unchanged (keep the same value for the coefficients that exceed the threshold), and the values which are lower than the threshold are made zero, which can cause large variance in the reconstructed signal and sometimes artifacts with an roughness appearance of the signal after reconstruction. However, it can better represent peaks and discontinuities. While the soft thresholding is an extension of hard thresholding, of which the thresholded coefficients are set to zero when the absolute values of wavelet coefficients are lower than the threshold (t_j) and adjusted by the following expression $sign(d_{j,k})(|d_{j,k}| - t_j)$ if coefficients are upper than t_j. This method of thresholding gives better visual filtering quality. Indeed, it affects the detail threshold coefficients in a smooth way without making a radical change in its value. And the final step in the wavelet analysis is the reconstruction. Through inverse wavelet transforms, the signal $f(t)$ is reconstructed from the threshold wavelet coefficients. After determination of the threshold details and approximation at Level j, they will be used as inputs, to calculate the coefficients at Level $(j-1)$ until getting the signal with the noise eliminated.

The summary of the main three steps in wavelet analysis:

Decompose: Choose a wavelet. Choose the Level *J*. Calculate the wavelet decomposition of the signals at the Level *J*.

Threshold: For each level from 1 to *J*, select a threshold and apply soft thresholding to the detail coefficients.

Reconstruct: Through the approximation coefficients of Level *J* and the thresholded detail coefficients the wavelet reconstruction is done.

4. Experimental results and discussion

4.1 Results

In this paper, the optical emission spectrometer (OES) is employed. And physically, the root of the optical emission is the light emitted through a chemical element, when the high energy state decreases to the lower one. In the plasma etch process, many chemical species have several emission spectra. The observed optical emission spectra display the chemical species and their variations. An optical emission spectroscopy should be able to resolve three components of plasma gas: (1) spectral resolution, (2) temporal resolution, and (3) spatial resolution. Hence, the study of the full spectral range OES is a challenging task. In this work, the sensor collects an array of measurements having 1201 channels of data, with over 827 units of time, since there are about approximately million data points in a single processing step. In other words, an optical emission spectroscopy is implemented in order to scan 1201 wavelengths (200 nm – 800 nm) from 0.4999 s to 435999 s. Given the extra data size, it is logical to ameliorate the sensitivity of the endpoint detection. And it is recommended to compress the data into a smaller subset that contains the most valuable information about the process, and at the same time minimizing the space on the hard drives by using dimension reduction techniques. The collected channels are gathered and analyzed in order to reach the real EP.

The first fifty rows (from 0.499 s to 24.999 s) referring to the initial state of plasma etch (Fig. 4) will be suppressed in order to avoid bias results (Fig. 5). Based on the new proposed algorithm named shift endpoint trace selection (SETS), only the first fifty endpoint traces having the highest intensity difference are selected. As the experimental OES data are coming from 5 etch steps, the total retained endpoint traces are equal to one hundred (50×5). Then reduction dimension techniques noted before will be applied to the retained traces to improve the picked-out endpoint traces. Moreover, the spectra are pre-processed to remove noise and reduce dimensionality.

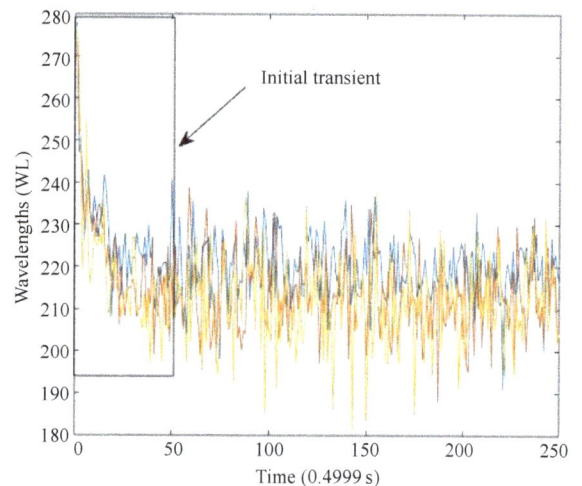

Fig. 4 Plot of an endpoint traces showing the initial transition state.

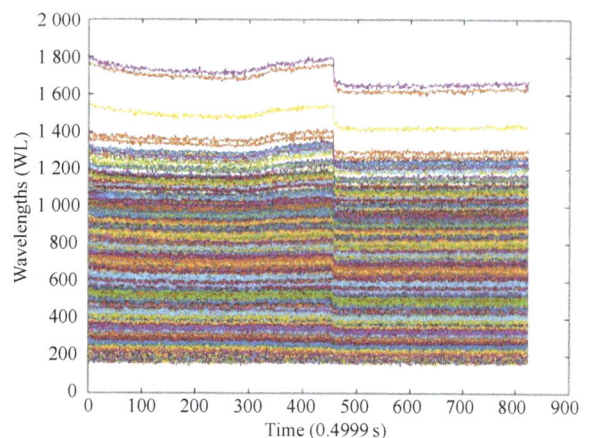

Fig. 5 Plasma etch endpoint traces from Step 3 after suppressing intensities from 0.499 s to 24.999 s.

The PCA is commonly used in the data analysis. The first fifty principal components are studied which catch most of the original data variation, even for large numbers of wavelengths (>1000). After applying PCA, the five retained endpoint traces from the fourth etch run notice that the endpoint is detected in 250.999 s to 252.499 s. The same procedure done on PCA is done on FA, hence the new proposed algorithm is preceded then FA is applied (Fig. 6).

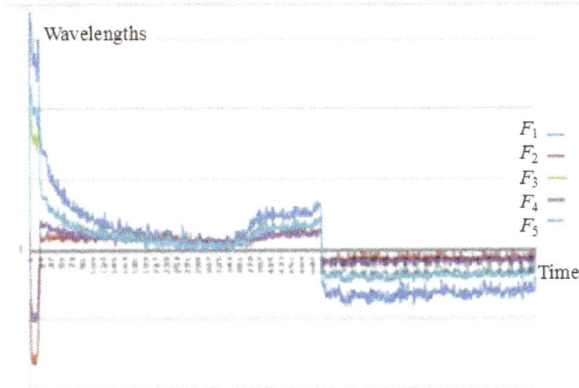

Fig. 6 Plotting the five significant retained factors.

The SETS algorithm is applied to the optical endpoint traces then the denoising procedure is applied by using the wavelet analysis. The chosen wavelet here is Haar wavelet. As mentioned previously, it is the most appropriate to describe the step change. Here, the obtained endpoint traces from the shift endpoint trace selection algorithm are then denoised and decomposed by using the wavelet analysis. The mean and CV of each endpoint trace of ME and OE are computed separately. From the obtained mean column presenting mean of all kept spectra, the mean will be decomposed at Level 3. This level is chosen based on the energy function drop. It should be noted that based on the gathered data, if the level of decomposition increases signal at a higher level, the signal will be smoother and may lose a lot of information about the right moment of endpoint detection and the species (gas) of the plasma etch process. Also, to plot endpoint traces,

the reconstructed approximation coefficients will be used for those reasons noted below. (1) It is the denoised reconstruct original signal. (2) The endpoint detection is done based on the mean shift. (3) The detail coefficients represent high variance

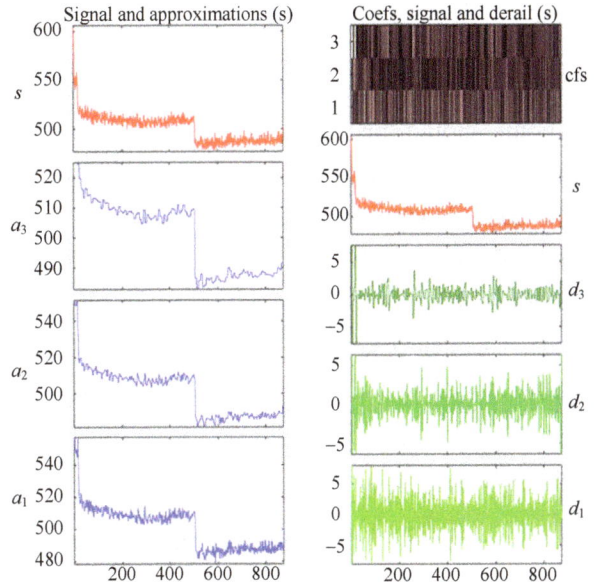

Fig. 7 Mean approximation signal at Level 3.

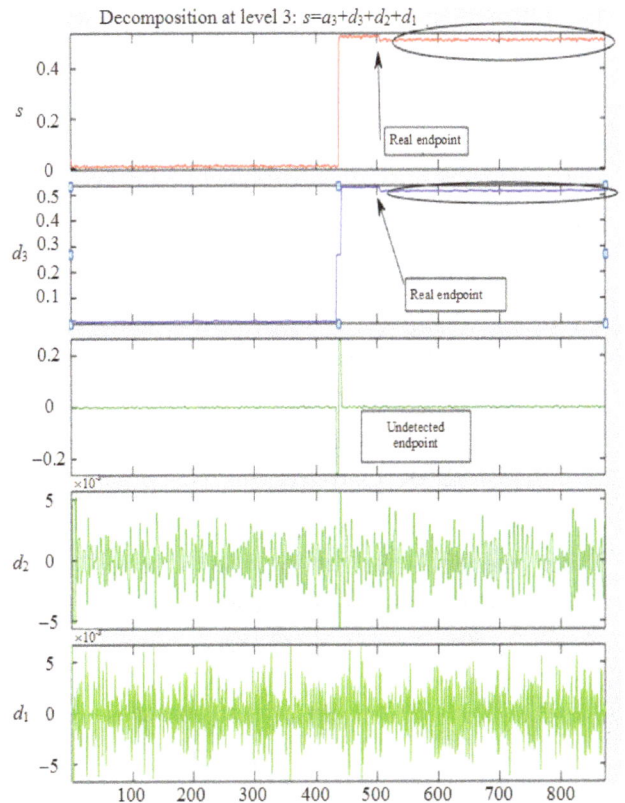

Fig. 8 CV approximation and details signal at Level 3.

and noise. The mean and the CV of each endpoint trace of ME and OE are computed separately, both of which will be decomposed at Level 3. Figure 7 shows the approximated mean wavelength at Level 3 that the endpoint is reached at the interval 250.999 s to 252.499 s. Based on Fig. 8, the WA-CV-approx at Level 3 records a meaningful shift before the real endpoint (under etched device). Therefore, the endpoint should be monitored based on the OE interval. The latter is more stable, and the first significant shift is detected at the real endpoint (250.999 s to 252.499 s). The WA-CV-details do not allow the detection of endpoint while the WA-CV-approx can detect endpoint if it is computed based on the coefficients of approximation in the OE interval.

4.2 Comparing result

As mentioned before, after initial transient and during the main etch step, a stable signal exists during the ME for each of the whole channels, but the intensity of the signal changes (decreases or increases) after the onset of endpoint. Any increase in the intensity of the signal refers to an increase in reactant in the plasma chamber, while the decrease in intensity of spectral channel refers to product. The SNR for the main etch period is the amount of signal compared with the noise on the main etch, which is used to compare the performance of the aforementioned preprocessing methods and is expressed as follows:

$$\left| SNR_{ME} \right| = \left| \frac{M_{OE} - M_{ME}}{\sigma_{ME}} \right|. \qquad (12)$$

When the SNR for the over etch period is the amount of signal compared with the noise on the over etched device, it is expressed as follows:

$$\left| SNR_{OE} \right| = \left| \frac{M_{OE} - M_{ME}}{\sigma_{OE}} \right|. \qquad (13)$$

A comparative result is summarized in Tables 2

and 3. Table 2 computes M, SD, and CV based on the main and over Etch intervals. Table 3 compares PCA, FA, WA-mean, WA-CV-approx, and WA-CV-details based on SNR. The SNR is computed during the main etch period and the over etch. CV coefficients are computed based on the approximated signal (approximation) and details.

Table 2 Mean, SD, and CV results based on Main and over etch intervals.

	M_{ME}	M_{OE}	SD_{ME}	SD_{OE}	CV_{ME}	CV_{OE}
PCA	40.48	41.017	53.451	6.7949	1.320	0.165
FA	5042.627	6917.3	6566.9	455.80	1.302	0.0659
WA-mean	511.1609	487.17	9.651	3.6795	18.88^{-3}	7.552^{-3}
WA-CV-approx	1.945	1.953	59^{-3}	14.1^{-3}	30.74^{-3}	7.2^{-3}
WA-CV-details	0.5309	0.5143	272.24	252.965	512.722	491.873

Table 3 SNR results based on Main etch interval and the Over etch interval.

	SNR_{ME}	SNR_{OE}
PCA	0.01034	0.079
FA	0.285	4.113
WA-mean	2.4857	6.52
WA-CV-approx	0.1099	0.496
WA-CV-details	6.095^{-5}	6.05^{-5}

4.3. Discussion of results

Based on the obtained results and with the most compelling evidence, the wavelet analysis method for the mean outperforms all other methods. And this is due to the characteristics of the WA-mean method, which is based on the flexible signal with no restriction about linearity, stationarity and symmetry of studied traces. In contrast, the wavelet coefficient of variation does not give us a better understanding about the endpoint detection, especially if it is computed based on the detail coefficients. As the CV is computed in terms of SD which tends to be non-stationary statistic having an increasing then decreasing trend, then the endpoint detection is not possible with statistic CV. The high variance hinders the endpoint detection. FA preserves the shape of most data, therefore, all the retained five factors detect endpoint clearly. While the data are mean centered in PCA, the FA detects mean shift better than PCA based on SNR. The obtained result could be explained according to the differences between PCA and FA given below: (1) PCA works in the variable space while FA surpasses the variable space; (2) PCA resolves non-segmented variance while FA resolves common variance only; (3) PCA is an empirical summarizing technique keeping m components while FA is a theoretical modeling method suiting fixed number m factors to the data; (4) PCA is a dimension reduction technique only when FA is also a clustering technique which tries to find coherent variables; (5) FA is a more statistical technique used to translate an observed dataset into new axes, similarly to PCA. While PCA refines to combine variables into tiny PCs, FA examines the structure underlying the original variables. The SNR is a relative measure of the magnitude of a data set to the standard deviation. If the SNR is larger, the magnitude of the signal is relatively larger than the amount of noise which is quantified by the standard deviation. Then in this case, the studied signal is deemed to be significant signal. There is a negative relation between CV and SNR, such as the WA-CV-details for ME is 512.722 and its SNR is 6.095 e^{-6}. Hence, an inverse correlation is detected between them. The small peaks with SNR give a large CV while the largest SNR gives small details and approximations. Using details coefficients there is a high variance compared with the mean which is very small, therefore the SNR is very low. Also, there is a significant improvement of SNR for all used methods if this ratio is computed based on the variance of the OE period.

5. Conclusions and future perspective

Based on Fig. 9, the worst result is given by WA-CV-details because the details are generally used to monitor variance, and the monitored variance is very small compared with mean shift. Based on [28], for CV<0.5, the influence function response will have negative values. There is a negative correlation between CV and SNR. The WA-CV-approx surpasses PCA because the former is computed based on the mean and the variance of the approximate signal which are proportionally significant. In addition to that, the WA-CV-approx has no constraints such as linearity and mean centering data, which are the main postulates of PCA. Furthermore, WA-CV-approx has less performance than FA, because both methods do not need to mean central data. While for the linearity assumption, both are also appropriate but FA is the most appropriate because it is already designed for linear transformation. The ratio CV has a small amount compared with mean, therefore FA gives us better SNR results than WA-CV-approx. To detect EP, it is advantageous to use directly the approximation coefficients which identify quickly the mean shift (EP). Those results remain the same in both intervals (ME and OE) but it should be noted that there is an improvement of SNR for OE range because the variance during the aforementioned period is more stable and smaller compared with the ME period. In relation to our current results, one can investigate more OE periods to detect the EP and hence consider the plasma etch process CV for moving from unstable to stable one.

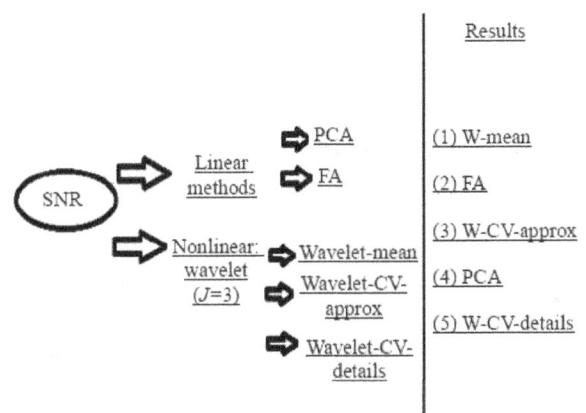

Fig. 9 Schematic presenting the obtained results.

Acknowledgment

The authors would like to thank reviewers and the editors for their valuable remarks and their time and every person helped us to do this work.

References

[1] M. Mitchner and C. H. Kruger, *Partially ionized gases*. New York: Wiley, 1973.

[2] ChemWiki: The dynamic chemistry hypertext. http://chemwiki.ucdavis.edu/Analytical_Chemistry/Qualitative_Analysis/Classification_of_Matter.

[3] W. Taylor, "Technical synopsis of plasma surface treatments," Dissertation for the Degree of University of Florida, Gainesville, FL, December, 2009.

[4] B. Goodlin, "Multivariate endpoint detection of plasma etching processes," Ph.D. dissertation, Dept. Massachusetts Institute of Technology, U. S. A., 2002.

[5] R. Yang and R. Chen, "Real-time plasma process condition sensing and abnormal process detection," *Sensors*, 2010, 10(6): 5703–5723.

[6] J. Yang, C. McArdle, and S. Daniels, "Dimension reduction of multivariable optical emission spectrometer datasets for industrial plasma processes," *Sensors*, 2014, 14(1): 52–67.

[7] J. Yang, "Multivariable OES data analysis for plasma semiconductor etching process," Ph.D. dissertation, Dept. Dublin City University, Republic of Ireland, 2014.

[8] D. Mercier, M. Bouttemy, J. Vigneron, P. Chapon, and A. Etcheberry, "GD-OES and XPS coupling: a new way for the chemical profiling of photovoltaic absorbers," Applied Surface Science, 2015, 347 (2015): 799–807.

[9] H. K. Chiu, "Method of controlling plasma etch process," US patent 6703250 B2, 2004 March 9.

[10] P. M. Cederstav. "Increase vacuum processing throughput and yield using advanced downstream pressure control methods," in *The 44th Annual Technical Conference Proceedings*, Philadelphia, pp. 501–502, 2001.

[11] Z. Fekete, "Development and characterisation of silicon microfluidic components and systems." Ph.D. dissertation, Dept. Institute for Technical Physics and Materials Science Research Centre for Natural Sciences, Hungarian Academy of Sciences, Budapest, 2012.

[12] C. J. Pugh, "End point detection in reactive ion etching," Ph.D. dissertation, Dept. University College London, London, 2013.

[13] B. M. Wise, N. B. Gallagher, S. W. Butler, D. D. White, and G. G. Barna, "A comparison of principal component analysis, multiway principal component anal ysis, trilinear decomposition and parallel factor analysis for fault detection in a semiconductor etch process," *Journal of Chemometrics*, 1999, 13(13): 379–396.

[14] B. M. Wise and N. B. Gallagher, *PLS Toolbox Version 2.0*, Manson, WA: Eigenvector Research, Inc., 1998.

[15] I. T. Joliffe, *Principal component analysis*, New York: Springer-Verlag, 2002.

[16] B. Kim and M. Kwon, "Optimization of PCA-applied in-situ spectroscopy data using neural network and genetic algorithm," *Applied Spectroscopy*, 2008, 62(1): 73–77.

[17] H. S. Kim, Y. J. Sung, D. W. Kim, T. Kim, M. D. Dawson, and G. Y. Yeom, "Etch end-point detection of GaN-based device using optical emission spectroscopy," *Materials Science and Engineering: B*, 2001, 82(1–3): 159–162.

[18] J. Chen and J. Liu, "Derivation of function space analysis based PCA control charts for batch process monitoring," *Chemical Engineering Science*, 2001, 56(10): 328–3304.

[19] S. Yoon and J. F. MacGregor, "Statistical and causal model-based approaches to fault detection and isolation," *Aiche Journal*, 2000, 46(9): 1813–1824.

[20] R. L. Gorsuch, *Factor analysis (2nd ed.)*. Hillsdale, NJ: Lawrence Erlbaum Associates, 1983.

[21] S. Krnac and P. P. Povinec, "Factor analysis of semiconductor γray spectra," *Applied Radiation and Isotope*, 1996, 47(9–10): 905–910.

[22] E. Brannock, M. Weeks, and R. Harrison, "The effect of wavelet families on watermarking," *Journal of Computers*, 2009, 4(6): 554–566.

[23] I. Daubechies, "Orthonormal bases of compactly supported wavelets," *Communications on Pure and Applied Mathematics*, 1988, 41(7): 909–996.

[24] D. L. Donoho, I. M. Johnstone, G. Kerkyacharian, and D. Picard, "Wavelet shrinkage: asymptopia?" *Journal of the Royal Statistical Society-Series B*, 1995, 57(2): 301–337.

[25] D. L. Donoho and I. M. Johnstone, "Ideal spatial adaptation by wavelet shrinkage," *Biometrica*, 1994, 81(3): 425–455.

[26] D. L. Donoho and I. M. Johnstone, "Adapting to unknown smoothness via wavelet shrinkage," *Journal of the American Statistical Association*, 1995, 90(432): 1200–1224.

[27] D. L. Donoho and I. M. Johnstone, "Minimax estimation via wavelet shrinkage," *The Annals of Statistics*, 1998, 26(3): 879–921.

3

Study of a Distributed Feedback Diode Laser Based Hygrometer Combined Herriot-Gas Cell and Waterless Optical Components

Yubin WEI[1,2], Jun CHANG[1*], Jie LIAN[1], Qiang WANG[1], and Wei WEI[1]

[1]School of Information Science and Engineering and Shandong Provincial Key Laboratory of Laser Technology and Application, Shandong University, Jinan, 250100, China

[2]Laser Research Institute of Shandong Sciences Academy, Jinan, 250014, China

[*]Corresponding author: Jun CHANG   E-mail: changjun@sdu.edu.cn

Abstract: A distributed feedback diode laser (DFB-DL) based hygrometer combined with a long-path-length Herriot gas cell and waterless optical components was proposed and investigated. The main function of this sensor was to simultaneously improve the measurement reliability and resolution. A comparison test between a 10-cm normal transmission-type gas cell and a 3-m Herriot gas cell was carried out to demonstrate the improvement. Reliability improvement was achieved by influence suppression of water vapor inside optical components (WVOC) through combined action of the Herriot gas cell and waterless optical components. The influence of WVOC was suppressed from 726 ppmv to 25 ppmv using the Herriot gas cell. Moreover, combined with waterless optical components, the influence of WVOC was further suppressed to no more than 4 ppmv. Resolution improvement from 11.7 ppmv to 0.32 ppmv was achieved mainly due to the application of the long-path-length Herriot gas cell. The results show that the proposed sensor has a good performance and considerable potential application in gas sensing, especially when probed gas possibly permeates into optical components.

Keywords: Herriot gas cell; hygrometer; wavelength modulation spectroscopy; waterless optical components; water vapor inside optical components

1. Introduction

Reliable and high-resolution measurement of gases of interest becomes more and more important in the field of scientific research and facility manufacture. Particularly, trace water vapor qualification analysis is indispensable in a wide of applications, ranging from polymer fabrication [1], environment analysis [2] to chemical production [3]. Because of remote detection capability, safety in hazardous environments, and immunity to electromagnetic radiation, the distributed feedback diode laser (DFB-DL) based hygrometer attracted the widest applications. In the recent years, many techniques were developed to improve the measurement reliability and resolution. For example, the utilization of differential value of two adjacent absorption peaks solved the difficulty of choosing reference point [4]. The utilization of combination of wavelength scanning and intensity modulation significantly improved the detection resolution [5]. The application of wavelength modulation spectroscopy (WMS) could achieve high resolution detection of trace gases at atmospheric pressure [6]. The utilization of the improved WMS technique achieved rapid measurement of water vapor at high temperature [7].

Almost all of DFB-DL based hygrometers consist of several optical components, such as light source (DFB-DL), fiber collimators, and photoelectric detector. However, it was found in [8] that water vapor in these optical components seriously influenced the measurement result not only in resolution but also in reliability. To realize reliable and high-resolution measurement of water vapor, an approach, namely the photo detector (PD) matching, had been reported to achieve preliminary suppression for the influence of water vapor inside optical components (WVOC). However, it is found in practical application the suppression effectiveness of the above approach depends on the match level between the two PDs, which is tedious and not guaranteed.

In this paper, an alternative approach consisting of Herriot cell and waterless optical components is investigated. It can observably improve the reliability and resolution simultaneously. The influence of WVOC can be diluted by introducing a Herriot cell with an effective length of 3 meters. Here we describe the experimental system and the mathematical justification of this approach and demonstrate the validity through exhibiting the improved measurements made by this approach.

2. Experimental details

Figure 1 shows the experimental system. A DFB-DL (with a linewidth of 3 MHz) is used as the light source. The inside temperature is controlled by a chip LTC 1923. The driving current consists of a 20-Hz sawtooth waveform and a 2-kHz sinusoidal waveform, and it is produced by a home-built signal generator using a chip LPC 1758. The sawtooth waveform ranges from 23 mA to 78 mA. As a result, a sweep range of about 280 pm is given to scan the absorption line of water vapor at $7306.75\,\mathrm{cm}^{-1}$. At the same time, the sinusoidal current is used to accomplish wavelength modulation of the DFB-DL. The output of the DFB-DL propagates through a gas cell to interact with water vapor molecules. To introduce a long optical absorption path, a multiple-pass Herriot cell is used. The 34 traversals inside the Herriot cell give a path length of 3 meters. The long optical path length not only improves the effective absorbance during the interaction, which is beneficial to improve the measurement resolution but also dilute the influence of WVOC, which is very important to improve the measurement reliability. After the interaction, the laser beam is detected by an InGaAs PD whose output is fed to RS850 digital lock-in amplifier (LIA). To implement the WMS technique, the second harmonic detection is performed with the LIA. To avoid variation of the phase difference between the WMS derivative signal and reference signal, the reference signal is synchronously generated by the same home-built signal generator. The final output is observed and captured by a computer.

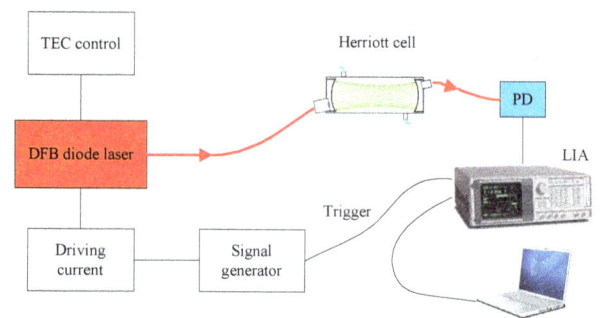

Fig. 1 Experimental setup.

3. Analysis of reliability and resolution improvement

The description of laser beam attenuation due to wavelength dependent absorption for a gas absorption line is given by Beer-Lambert's law [9], $I_t(v)=I_0(v)\exp[-\alpha(v)CL]$, where $I_0(v)$ and $I_t(v)$ are the incident and transmitted laser beam intensities, respectively, $\alpha(v)$ is the wavelength independent absorption coefficient, C is the mole fraction concentration, and L is the optical path length. For WMS, the simultaneous frequency and amplitude of the modulated laser beam can be represented by

$$v(t) = v_c + \Delta v \cos(\omega t) \quad (1)$$

$$I_0(t) = I_{0c} + i_0 \cos(\omega t + \psi) \quad (2)$$

where $v(t)$ is the instantaneous optical frequency, $I_0(t)$ is the laser emission intensity, and ψ is the phase separation [10] between the intensity and optical frequency modulation. v_c and I_{0c} donate the slowly-varying central frequency and intensity, as the laser scans across the absorption features. ω is the sinusoidal modulation frequency, Δv and i_0 are the maximum small-amplitude excursions of the $v(t)$ and $I_0(t)$ around v_c and I_{0c}.

In the implementation of the WMS technique, the signals at the output can be found using Fourier cosine series expansion of the absorbance line shape function, $\alpha(v)CL$ [11]. The result for the primary second harmonic term is shown as

$$I_{2\omega}(v_c) = \frac{i_0}{2} H_3(v_c, \Delta v) \cos(2\psi + \varphi) +$$
$$I_{0c} H_2(v_c, \Delta v) \cos(\psi + \varphi) + \frac{i_0}{2} H_1(v_c, \Delta v) \cos\varphi \quad (3)$$

where H_1, H_2, and H_3 denote three Fourier components obtained for the absorbance line shape function. φ designates the phase difference between the detection signal and the laser emission intensity. Although the components of H_3 and H_1 on the measurement axis of H_2 can distort the detected profile of the detected second harmonic term [12], the amplitude of the detected second harmonic signal at the line center is immune to the components of H_3 and H_1. This is because the amplitudes of H_3 and H_1 are zero at the absorption peak [11]. Thus the amplitude of the detected second harmonic (i.e. the value of the detected second harmonic signal at the absorption peak) can be expressed as

$$I_{2\omega}(v_p) = I_{0p} H_2(v_p, \Delta v) \cos(\psi + \varphi) \quad (4)$$

where v_p and I_{0p} donate the frequency and intensity at the absorption peak. To simplify the expression of the relationship between the detected second harmonic amplitude and absorbance, (4) can be rewritten as

$$I_{2\omega}(v_p) = \kappa \alpha(v_p) CL. \quad (5)$$

The above expression is applicable only when the absorbance is not too high. For the detection of water vapor at 7306.75 cm^{-1} in this research, the above expression is applicable when C is no more than 2000 ppmv. κ in (5) designates the conversation coefficient, which is relative to the beam intensity at the absorption peak, values of ψ and φ, and Fourier component H_2.

According to the analysis in [8], the water vapor inside the optical components can influence the measurement result. Combined with (5) in the WMS technique, the influenced expression of the water vapor concentration should be

$$C = \frac{I_{2\omega}(v_p)}{\kappa \alpha(v_p) L} + \frac{L_{eq}}{L} \times C_{eq} \quad (6)$$

where L_{eq} and C_{eq} donate the equivalent absorption length and concentration of the water vapor inside the optical components. An extended optical absorption length can simultaneously improve the measurement reliability and resolution. Firstly, during the practical application, the detected harmonic signal is directly used to invert the wanted water vapor concentration. However, system noises (optical noise and electrical noise) are accumulated and added in the detected harmonic signal. Generally speaking, the system noises have nothing to do with the absorption length. According to (6), for the same optical gas sensor, the measurement resolution can be significantly improved using a longer optical absorption length. Secondly, the influence of the water vapor inside the optical components is relatively certain and invariable. Similarly, the measurement reliability can be improved by weakening the influence of the water vapor inside the optical components.

4. Experimental results

4.1 Existence of WVOC

To demonstrate the severity of existence of the water vapor inside the optical components, a simple test was organized. The simple schematic of the test is shown in Fig. 2(a). Laser light from DFB-LD directly couples on the PD. The most distinctive characteristic is there is no gas cell in the test. The process of the detected signal is the same with the introduction above.

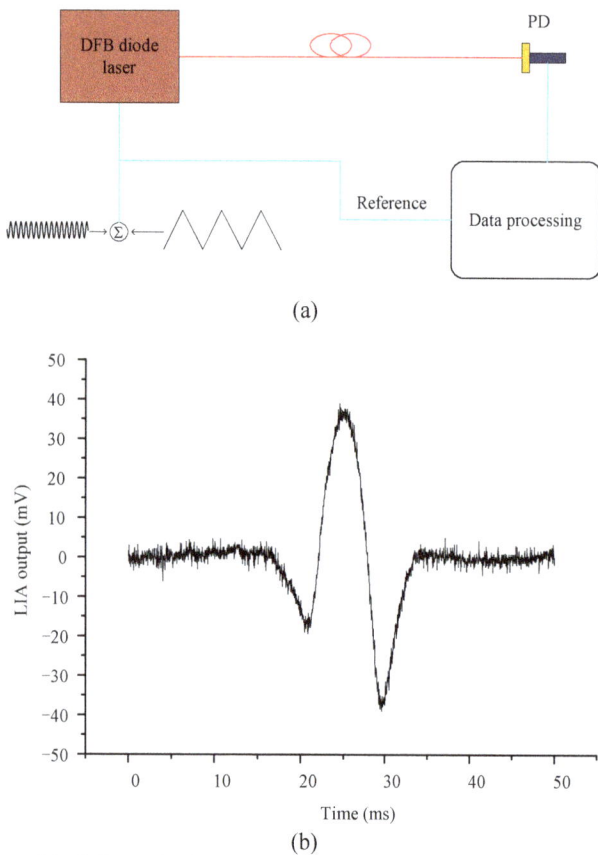

(a)

(b)

Fig. 2 Test of existence of the water vapor inside the optical components: (a) test setup and (b) measured second harmonic signal without the gas cell.

Based on the traditional understanding, there should be no absorption during this setup. However, the detected result shown in Fig. 2 (b) subverts our preliminary cognition. Not only the existence of WVOC badly impacts the measurement result during the practical application of an optical fiber

hygrometer, but also variation of the inside water vapor can further impact the measurement.

4.2 Reliability and resolution improvement

To fully exhibit the influence of the water vapor inside the optical components especially the improvement of reliability and resolution, a comparison research was demonstrated experimentally between the normal transmission-type gas cell with a 10-cm effective path length [8] and Herriot gas cell with a 3-m effective path length. Figure 3(a) shows the structure of the normal gas cell comprising two commercial fiber collimators used in communication. One collimator is used to transfer laser beam into parallel light inside the gas cell, and the parallel light interacts with the probed gas molecules and possesses information of characteristics associated with the probed gas molecules. The other one is used to receive the signal light and couple it into the fiber. The two fiber collimators are fixed on a stainless steel tube to guarantee reliable alignment and achieve low coupling loss (less than 0.5 dB). Holes of the stainless steel tube are used the realize gas exchange.

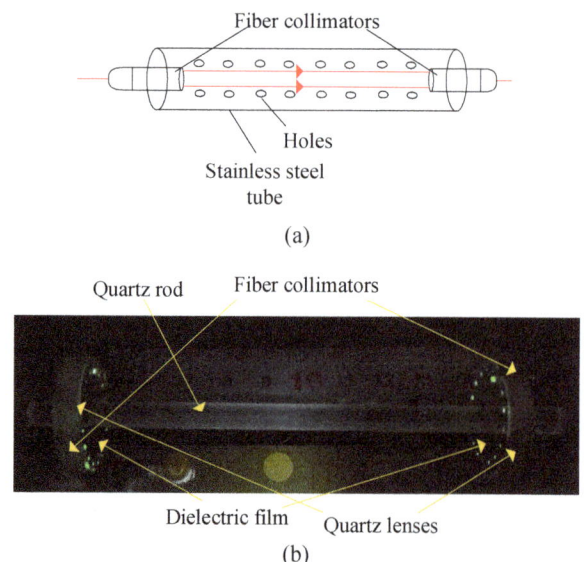

(a)

(b)

Fig. 3 Schematic of gas cell structure: (a) normal transmission-type gas cell with an effective path length of 10 cm and (b) Herriot gas cell with an effective path length of 3 m, the two quartz lenses are coated with the dielectric film.

Figure 3(b) shows the structure of the Herriot cell. Two quartz lenses with a diameter of 25.4 mm are fixed with a straight column quartz rod. The two quartz lenses coated with the dielectric film have the same curvature radius of 100 mm and are separated with a distance of about 90 mm. A fiber collimator with a diameter of 1.3 mm is used to introduce light into the Herriot gas cell. After adjusting the incidence angle, 17 reflections exist for each lens. Another identical fiber collimator is used to collect the signal light and couple it into fiber. As a result, the effective optical path length has been extended to about 3 meters, and the insertion loss is lower than 1 dB.

During the comparison research, the measurement reliability and resolution were studied. The reliability improvement reflected in influence suppression of WVOC. A linearity test was arranged. A water vapor generator (DG-70, Shanghai Hema, China) was used to generate sample gases with different water vapor concentrations ranging from 20 ppmv to 430 ppmv. To change the inside water vapor concentration, sample gases were introduced into the gas cell at a flow rate of 100 mL/min at 1 atm and (296 ± 0.5) K. The overflowing sample gas from gas cell was used to monitor the water vapor concentration by a moisture meter (S8000, MICHELL, UK). Precision of S8000 is within 1 ppmv in the range of 1 ppmv – 2000 ppmv.

The measurement result is shown in Fig. 4. Amplitude (in mV) is proportional to the water vapor concentration. The solid line named "normal transmission-type gas cell" is the linearity test result using the 10-cm normal transmission-type gas cell, and the dot line named "Herriot gas cell" is the linearity test result using the 3-m Herriot gas cell. It can be seen from Fig. 4 that both the lines do not pass through the origin of coordinates (0, 0), and this is caused by the influence of WVOC. That is to say influence of WVOC can be reflected by the intercepts of the measured linearity lines. After the

linear fitting process slope of solid line is calculated to be 0.11 mV/ppmv, thus the intercept of solid line is 79.9 mV corresponding to a water vapor concentration of 726 ppmv in a 10-cm path length. While the slope of dot line is calculated to be 3.38 mV/ppmv, and the intercept of solid line is 85.1 mV corresponding to a water vapor concentration of 25 ppmv in a 3-m path length. Apparently, the influence of WVOC on the measured probed water vapor concentration has been significantly suppressed.

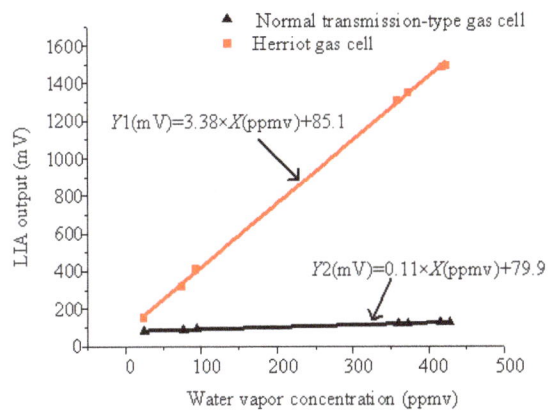

Fig. 4 Reliability improvement test through WVOC suppression (the lower line is for a 10-cm transmission-type gas cell, and the upper line is for a 3-m Herriot gas cell).

As for the resolution improvement test, the resolution [with signal to noise ratio (SNR) =1] is obtained through deducing the SNR. Given the existence of WVOC, the deducing process is not based on a certain water vapor concentration. On the contrary, the signal amplitude in SNR calculation is derived from two different amplitudes (in Fig. 4) corresponding to two water vapor concentrations, which are measured by S8000. The noise amplitude is derived from the WMS profile corresponding to one of the two water vapor concentrations shown in Fig. 5.

From the above test and process, the resolution of measurement using the normal transmission-type gas cell is calculated to be 11.7 ppmv, and that of measurement using the Herriot gas cell is 0.32 ppmv. Thus reliability can obtain significant improvement using a long path length gas cell.

(a)

(b)

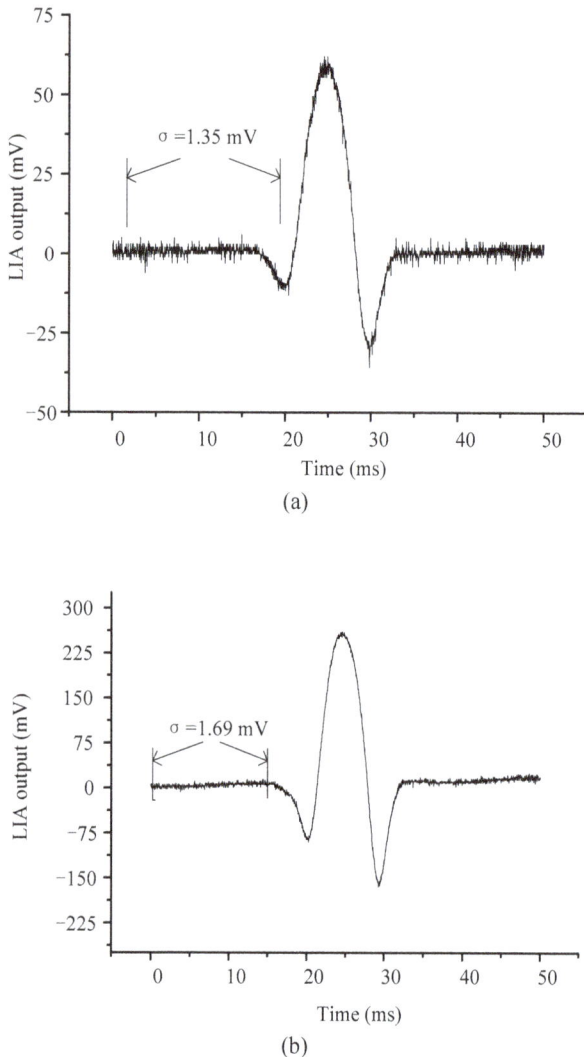

Fig. 5 Detected WMS profiles: (a) corresponds to the normal transmission-type gas cell with a water concentration of 94 ppmv and (b) corresponds to the Herriot gas cell with a water concentration of 92 ppmv.

4.3 Further improvement using waterless optical components

Although a long optical path length gas cell can apparently suppress the influence of WVOC, there still exists an apparent influence, which is shown in Fig. 4. Thus waterless optical components were designed and adopted to further suppress influence of WVOC. A 14-pin butterfly packaged DFB-DL and a PD with the same package were used. The big difference between the two waterless optical components and normal ones is that the inside space is filled of high purity nitrogen. As a result, inside water vapor content could be rather low.

To obtain the efficiency of water vapor suppression using these waterless optical components, a test was carried out, and the test result is shown in Fig. 6 (a). It is apparent that the intercept of the linearity test has been suppressed to 10.9 mV corresponding to no more than 4 ppmv. Figure 6(b) shows the detected LIA output without the gas cell and a result of fitting procedure based on the Fourier analysis of the Lorentzian profile.

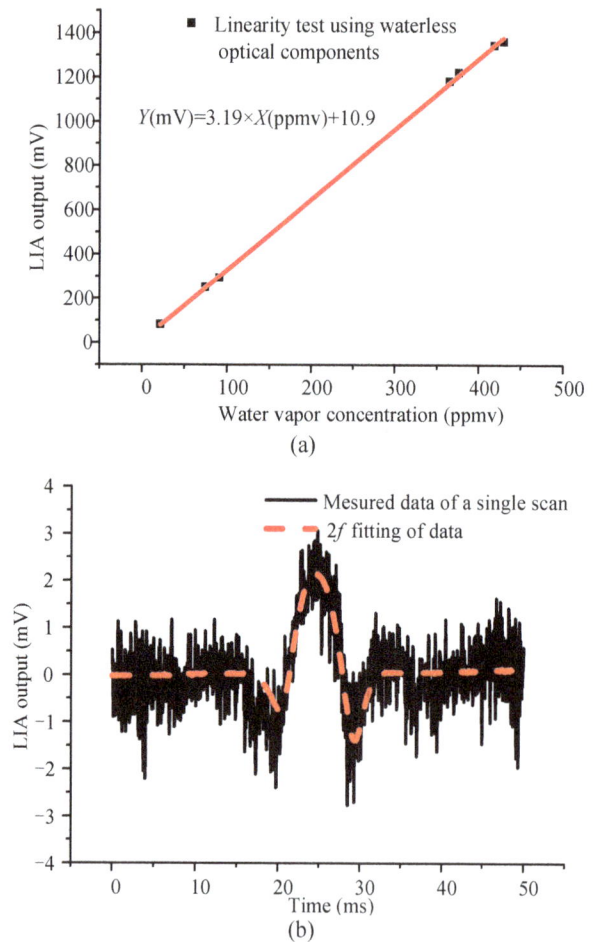

(a)

(b)

Fig. 6 Further suppression results for influence of WVOC using waterless optical components: (a) linearity test and (b) situation of no gas cell.

5. Conclusions

We proposed and demonstrated a technique comprising Herriot Cell waterless optical components. Reliability and resolution of the DFB-DL based hygrometer can be simultaneously improved by this technique. A comparison test between a 10-cm normal transmission-type gas cell and this 3-m Herriot gas cell was carried out. As a result, the influence of WVOC had been suppressed from 726 ppmv to 25 ppmv. Moreover, waterless optical components were used to further suppress the influence of WVOC, and the influence had been further suppressed to be no more than 4 ppmv. In addition, the measurement resolution had been improved from 11.7 ppmv to 0.32 ppmv.

In summary, the reliability improvement had been achieved through suppressing the influence of WVOC using a long path length Herriot gas cell and waterless optical components. The resolution improvement had been achieved through enhancing absorbance using a long path length Herriot gas cell. The proposed sensor has potential applications in the design of the reliable and high-resolution gas sensor, especially when permeation of probed gas into optical components exists.

Acknowledgment

This work was supported by the National Natural Science Foundation of China (60977058 & 61475085), the Science and Technology Development Project of Shandong Province (2014GGX101007), and the Fundamental Research Funds of Shandong University (2014YQ011).

References

[1] W. Zhang, D. J. Webb, and G. D. Peng, "Investigation into time response of polymer fiber Bragg grating based humidity sensors," *Journal of Lightwave Technology*, 2012, 30(8): 1090−1096.

[2] J. Goldmeer and D. J. Kane, "Real-time 2-D imaging of water vapor in diffusion flames," in *Laser Applications to Chemical and Environmental Analysis*, Santa Fe, New Mexico United States, 2000.

[3] S. Wu, F. Deng, Y. Hao, M. Shima, X. Wang, C. Zheng, *et al.*, "Chemical constituents of fine particulate air pollution and pulmonary function in healthy adults: the healthy volunteer natural relocation study," *Journal Hazardous Materials*, 2013, 260C(6): 183−191.

[4] Q. Wang, J. Chang, C. G. Zhu, C. Li, F. J. Song, Y. N. Liu, *et al.*, "Detection of water vapor concentration based on differential value of two adjacent absorption peaks," *Laser Physics Letters*, 2012, 9(6): 421−425.

[5] Q. Wang, J. Chang, Z. Wang, C. Tian, S. Jiang, and G. Lv, "Study of an optical fiber water vapor sensor based on a DFB diode laser: combined wavelength scanning and intensity modulation," *Journal of Modern Optics*, 2014, 61(18): 1538−1544.

[6] A. Upadhyay and A. L. Chakraborty, "Residual amplitude modulation method implemented at the phase quadrature frequency of a 1650-nm laser diode for line shape recovery of methane," *IEEE Sensors Journal*, 2015, 15(2): 1153−1160.

[7] H. Li, A. Farooq, J. B. Jeffries, and R. K. Hanson, "Near-infrared diode laser absorption sensor for rapid measurements of temperature and water vapor in a shock tube," *Applied Physics B*, 2007, 89(2): 407−416.

[8] Q. Wang, J. Chang, F. J. Song, F. P. Wang, C. G. Zhu, Z. Liu, *et al.*, "Measurement and analysis of water vapor inside optical components for optical fiber H_2O sensing system," *Applied Optics*, 2013, 52(26): 6445−6451.

[9] X. Liu, J. B. Jeffries, and R. K. Hanson, "Measurements of spectral parameters of water-vapour transitions near 1388 and 1345 nm for accurate simulation of high-pressure absorption spectra," *Measurement Science and Technology*, 2007, 18(5): 1185−1194.

[10] X. Zhu and D. T. Cassidy, "Modulation spectroscopy with a semiconductor diode laser by injection-current modulation," *Journal of the Optical Society of America B*, 1997, 14(14): 1945−1950.

[11] L. C. Philippe and R. K. Hanson, "Laser diode wavelength-modulation spectroscopy for simultaneous measurement of temperature, pressure, and velocity in shock-heated oxygen flows," *Applied Optics*, 1993, 32(30): 6090−6103.

[12] K. Duffin, A. J. McGettrick, W. Johnstone, G. Stewart, and D. G. Moodie, "Tunable diode laser spectroscopy with wavelength modulation: a calibration-free approach to the recovery of absolute gas absorption line shapes," *Journal of Lightwave Technology*, 2007, 25(10): 3114−3125.

4

Research on Pavement Roughness Based on the Laser Triangulation

Wenxue CHEN[*], Zhibin NI, Xinhan HU, and Xiaofeng LU

Equipment Academy of the Rocket Force, Beijing, 100094, China

[*]Corresponding author: Wenxue CHEN   E-mail: 200231030072@whu.edu.cn

Abstract: Pavement roughness is one of the most important factors for appraising highway construction. In this paper, we choose the laser triangulation to measure pavement roughness. The principle and configuration of laser triangulation are introduced. Based on this technology, the pavement roughness of a road surface is measured. The measurement results are given in this paper. The measurement range of this system is 50 μm. The measurement error of this technology is analyzed. This technology has an important significance to appraise the quality of highway after completion of the workload.

Keywords: Pavement roughness; laser triangulation; measurement

1. Introduction

In recent years, with the development of highway construction, the testing and maintenance on highway construction after completion has became an important work. Thus, the requirements to the road detection level become higher. Among many indicators which affect the car comfort and safety, the pavement roughness is an important factor which represents the quality of highway driving. According to the full knowledge of the international standard [1, 2], the pavement roughness is chosen to appraise the quality of highway in this paper.

The laser triangulation has been widely used in industrial field with the characteristics of high speed, small size, automation, large scale, non-contact measurement, and high precision [3-10]. In this paper, the laser triangulation is chosen to measure the pavement roughness.

In this paper, the principle and configuration of the laser triangulation are introduced. The pavement roughness is measured based on the laser triangulation. The measurement error of the laser triangulation is analyzed.

2. Setup

The detecting system of pavement roughness is shown in Fig. 1. The laser triangulation is installed on the chassis to reduce the influence induced by the jounce. The laser triangulation emits the laser to measure the distance between the jounce and road surface. The variation of measuring distance represents the pavement roughness. In order to reduce the jounce, the car runs at a low speed.

The configuration of the laser triangulation is shown in Fig. 2. Here, we use the laser diode (LD) as the light source. Firstly, the LD emits the laser. The transmitted light passes through the diaphragm. After that, the laser beam reaches the measured object. One part of the laser, which is called the scattered light, is reflected to go through the light filter. The scattered light contains the displacement information of the measured object. It goes through the lens and reaches the complementary metal oxide semiconductor (CMOS). When the position of the measured object varies, the position of the scattered light on the CMOS will change. Thus, the displacement of the measured object can be detected by measuring the position of the scattered light. This is the measurement principle of the laser triangulation.

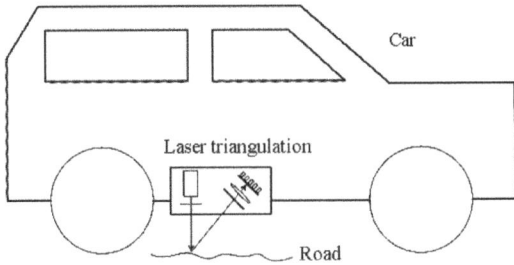

Fig. 1 Measurement setup of pavement roughness.

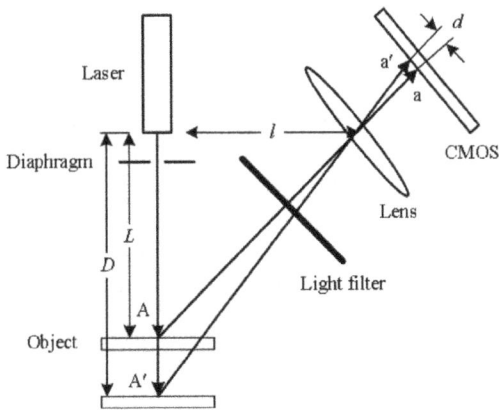

Fig. 2 Configuration of the laser triangulation.

The configuration of the laser triangulation can be divided into four parts: a light source, a filter, a signal detecting, and a signal acquisition system.

A laser diode with the wavelength of 655 nm is used as the light source. The intensity of laser is 0.56 mW. The radiation angle and diameter of the laser are 0.4 mrad and 1.6 mm, respectively. The working voltage of the LD is 3.3 V. The electric current is 35 mA. The working temperature of the LD is 0 ℃. The LD which is chosen as the optical source in our system has the advantages of small size, tiny weight, and small radiation angle. That ensures the narrow ray width of LD laser to improve the measurement precision.

The filter is made up of the diaphragm and optical filter. The diaphragm is used to reduce the diameter of ray. The central wavelength of the optical filter is 655 nm, and the transmittance ratio of the optical filter is larger than 90% at the wavelength of 655 nm. The transmittance ratio is almost zero at the other wavelength.

The signal detecting part is formed by the lens and CMOS.

A signal acquisition system is used to acquire the output of CMOS and deal with the data. The acquire rate of the signal is 3 kHz. The output voltage of the system is between 0 and 5 V. The relationship between the displacement of the measured object and the output voltage of the system is linear. Thus, the output voltage V represents the displacement D of the measured object. The formula is given as (1):

$$D = 0.16V + 0.2. \tag{1}$$

Based on the relationship between the object and image, the distance between the measured object and the LD laser is expressed as

$$D = \frac{d \times (f \times \sqrt{L^2 + l^2} - l^2) - f \times l \times L}{d \times L - f \times l}. \tag{2}$$

The sign meanings in (2) are shown in Fig. 2.

3. Results

We measure the pavement roughness based on the laser triangulation, and the result obtained is displayed in Fig. 3. In the measurement, 2000 seconds data are acquired to evaluate the road quality. In order to reduce the error induced by the jounce, the car runs at a speed about 30 km/h.

As shown in Fig. 3, the maximum value of the roughness is 18 mm.

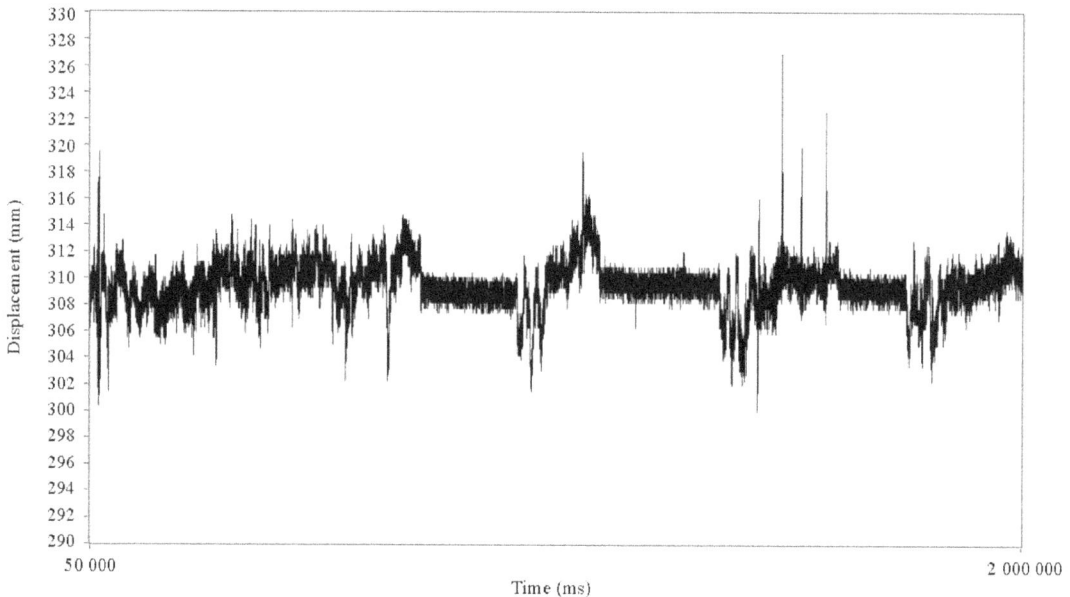

Fig. 3 Measurement result of the pavement roughness.

4. Error analysis

The error is mainly induced by the instability of the light source, installation error, and so on.

The wavelength of LD laser is influenced by the temperature. The variation of the wavelength will change the ray diameter. This will introduce the error during the image processing.

The installation processing will introduce the error into the measurement system. For example, the Scheimpflug condition of the position relationship among the CMOS, the ray axis of lens, and laser ray is not satisfied.

5. Summary

In this paper, the laser triangulation is applied to measure the pavement roughness. The principle and configuration of the laser triangulation are introduced. The pavement roughness is measured, and the maximum roughness is 18 mm. The measurement error is analyzed. The work in this paper has a contribution to the measurement of pavement roughness.

Acknowledgment

This work is supported by the National Natural Science Foundation of China (NSFC) (Grant No. 61505255).

Reference

[1] L. Shen, D. Li, and F. Luo, "Study on laser speckle correlation method applied in triangulation displacement measurement," *Journal of Optical Technology*, 2013, 80(10): 604–610.

[2] D. Wu, Q. Lu, X. Chen, and B. Ge, "Analysis on measurement range and precision of the oil film thickness measurement sensor based on differential laser trigonometry," *Journal of Tianjin University (Science and Technology)*, 2013, 46(11): 999–1002.

[3] T. Cui, X. Shen, J. Hu, D. Li, and J. Li, "Tightness testing of gas meter valve based on laser triangulation," *Opto-Electronic Engineering*, 2013, 40(7): 71–76.

[4] J. Chu, D. Li, F. Luo, and M. Liu, "Fuzzy adaptive control of light in laser triangulation displacement measurement," *Medicina Clínica*, 2013, 124(3): 106–107.

[5] M. Sayers, T. Gillespie, and C. Queiroz, *The international road roughness experiment: establishing*

*correlation   and   calibration   standard   for
measurements*. Washington: World Bank, 1986.

[6] M. Sayers, *Two quarter-car model for defining road
roughness: IRI   and   HRI*.   Washington,   DC:
Transportation Research Board, pp. 165–172, 1989.

[7] R. G. Dorsch, G. H. Usler, and J. M. Herrmann, "Laser
triangulation: fundamental uncertainly in distance
measurement,"   *Applied   Optics*,   1994,   33(7):
1306–1314.

[8] Z. Ji and M. C. Leu, "Design of optical triangulation
deveices," *Optics and Laser Technology*, 1989, 21(5):
335–338.

[9] K. Goh, N. Phillips, and R. Bell, "The applicability of
a   laser   triangulation   probe   to   non-contacting
inspection," *International   Journal   of   Production
Research*, 1986, 24(6): 1331–1348.

[10] T. Clarke, "The use of optical triangulation for high
speed acquisition of cross section or profiles of
structures," *Photogrammetric Record*, 1990, 13(7):
523–532.

5

A Harmful-Intrusion Detection Method Based on Background Reconstruction and Two-Dimensional K-S Test in an Optical Fiber Pre-Warning System

Fukun BI[1], Tong ZHENG[1*], Hongquan QU[1], and Liping PANG[2]

[1]*College of Information Engineering, North China University of Technology, Beijing, 100144, China*

[2]*School of Aeronautic Science and Engineering, Beijing University of Aeronautics and Astronautics, Beijing, 100191, China*

*Corresponding author: Tong ZHENG   E-mail: dolly115dolly@163.com

Abstract: The key technology and main difficulty for optical fiber intrusion pre-warning systems (OFIPS) is the extraction of harmful-intrusion signals. After being processed by a phase-sensitive optical time-domain reflectometer (Φ-OTDR), vibration signals can be preliminarily extracted. Generally, these include noises and intrusions. Here, intrusions can be divided into harmful and harmless intrusions. With respect to the close study of signal characteristics, an effective extraction method of harmful intrusion is proposed in the paper. Firstly, in the part of the background reconstruction, all intrusion signals are first detected by a constant false alarm rate (CFAR). We then reconstruct the backgrounds by extracting two-part information of alarm points, time and amplitude. This ensures that the detection background consists of intrusion signals. Secondly, in the part of the two-dimensional Kolmogorov-Smirnov (K-S) test, in order to extract harmful ones from all extracted intrusions, we design a separation method. It is based on the signal characteristics of harmful intrusion, which are shorter time interval and higher amplitude. In the actual OFIPS, the detection method is used in some typical scenes, which includes a lot of harmless intrusions, for example construction sites and busy roads. Results show that we can effectively extract harmful intrusions.

Keywords: Optical fiber intrusion pre-warning; extraction of harmful-intrusion signals; two-dimensional K-S test

1. Introduction

Intrusion prediction for pipelines (IPP) is used to monitor the transport process of petroleum and natural gas. It is well known that leakage pre-warning can prevent many fatal losses. Optical fibers buried along the pipeline are generally used to transmit data. Moreover, the vibration measurement of these optical fibers is an effective and convenient way to detect intrusion. In other words, we can locate harmful intrusions and predict damage with the proposed method [1−5].

At present, there are two types of regime in the field of optic fiber intrusion, namely reflection and interference. The Mach-Zehnder (MZ) method is an interference method and applied to the detection of synthesis signals across a whole optical fiber. However, the detection performance will be rapidly distorted if there are concurrent vibration signals. Unfortunately, the optical fiber intrusion pre-warning systems (OFIPS) have many concurrent intrusion signals. On the other hand, the phase-sensitive optical time-domain reflectometer (Φ-OTDR), which is a reflection method, can be used to detect concurrent intrusions, even at very small resolutions [6]. However, the Φ-OTDR system also has a limitation in its sensitivity [7–12], thus interference and harmful intrusions are difficult to identify. After a close study, we find that the signal can be divided into three parts: noise, and harmless and harmful intrusions. Here, noise and harmless intrusion are interference. Hence, we need to hierarchically extract different kinds of signals and locate harmful intrusion. The support vector method (SVM) is used to identify different interference in Tian Jin University of China [10].

The background distribution of OFIPS can be divided into two orthogonal Gaussian noises, and the envelope of the synthesis signal obeys a Rayleigh distribution. This character of background distribution is similar to radar signals. For this reason, the constant false alarm rate (CFAR) method in radar detection is introduced here to detect the intrusion signal of optical fibers. In the spatial domain detection, CFAR is divided into several categories, including a mean level CFAR (ML-CFAR), ordered statistics CFAR (OS-CFAR), and adaptive CFAR (A-CFAR). On the one hand, optimal statistics of OS-CFAR depend on prior information [13]. In actual conditions, the prior information is difficult to obtain. On the other hand, A-CFAR can adapt to a variable background, but the running time for this is quite long due to a complicated algorithm [14]. However, CA-CFAR, categorized within ML-CFAR, has a simple algorithm and outstanding performance in a homogeneous background. In view of these facts, the CA-CFAR is used to detect vibration signals [15–17]. In our system, all intrusion signals, including harmless and harmful intrusions, can be

detected by the CFAR method. It is obvious that the extraction of harmful intrusion is a significant part of this paper.

The proposed method includes two parts, namely background reconstruction and a two-dimensional K-S test. Harmful intrusion signals can be properly extracted. As harmful intrusion exists in all intrusion signals, we plan to reconstruct the background. The disposal of the background reconstruct ensures that later detected backgrounds consist entirely of intrusion signals. Moreover, the harmful intrusion signals are continuous within a period of time, so we need to detect the system in time dimensions. It is well known that one CFAR detected method in time dimensions is a clutter map. However, this has the shortcoming of self-shielding [17–18] and cannot adapt to a variable background. According to studies, the background of OFIPS is unstable. In view of this, in the proposed method, we firstly employ the Kolmogorov-Smirnov (K-S) test, which belongs to a non-parametric test, to extract harmful-intrusion signals. The statistics of K-S quantify the distance between the empirical distribution function (EDF) of the sample and the cumulative distribution function (CDF) of the reference distribution, or between the EDFs of two samples. Under the K-S test method, statistical information, such as mean, variance, and distribution type, do not need to be calculated before the test. Hence, it has wide applications [19–23]. Compared with the harmless intrusion, a harmful one has a shorter time interval and higher amplitude. According to the character of the harmful intrusion, we design a detected method, a K-S test in two dimensions. The intrusion type, whether harmful or harmless, can be confirmed by this method.

The remainder of the paper is organized as follows. Section 2 gives an introduction of the proposed method. Section 3 is devoted to the background construction and reconstruction. The method of the K-S test in two dimensions is proposed in Section 4. In Section 5, experiments on real data demonstrate the effectiveness of the proposed method. The discussion and conclusion are provided in Section 6.

2. Summary of the overall process

The sensing signal includes noise, and harmless and harmful intrusions. Here, noise is caused by the self-vibration of the ground, and it is inevitable. Harmless intrusion, which is generated by the construction on the site or vehicles, has no ability to damage the underground pipeline or optical fiber. Harmful intrusion becomes the target of extraction, such as excavation at the ground. The overall detection process of harmful intrusion is shown in Fig. 1.

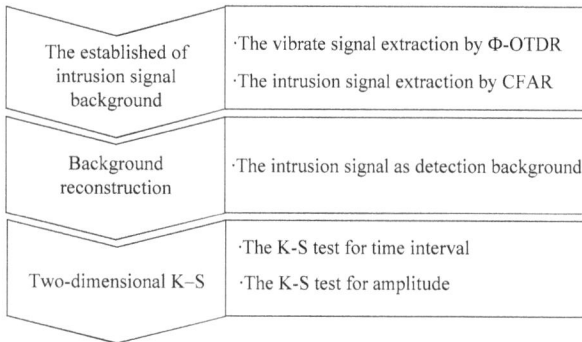

Fig. 1 Overall process of the system.

In the part of background establishment, the sensing signals are preliminarily extracted by Φ-OTDR. It consists of noise, and harmless and harmful intrusions. An intrusion background is established. In addition, all intrusion signals can be obtained, and all interference is eliminated by the CA-CFAR detector.

In view of this, the key point of this paper is the extraction of the harmful intrusion. Firstly, the detection background needs to be reconstructed. In other words, all intrusion signals become new background for later detection. In actual conditions, harmful-intrusion signals have a shorter time interval and higher amplitude than harmless ones. In the view of these characteristics, we apply the K-S test method in two dimensions to detect harmful-intrusion signals.

We provide a two-dimensional K-S test method

to detect the system. K-S tests for the time interval and amplitude are used to extract the harmful intrusion. We do not recognize the intrusion signal as harmful until outcomes of two-dimensional tests both have satisfied corresponding requirements.

3. Construction and reconstruction of an intrusion signal background

3.1 Extraction of vibration signal based on Φ-OTDR

The principal of OFIPS is shown in Fig. 2. If there are any above-ground intrusions, optical fiber buried along the pipeline will perceive the vibration. The vibration can be collected and treated by the hardware system. Finally, the result of intrusion detection needs to be displayed in software. In view of the elastic-optical effect, the refractive index of the optical fiber changes suddenly. This may cause variations in the intensity of back-scattered light. Differences in the intensity of background light measured at different time can be used to locate the vibration [6]. Hence, the Rayleigh backscattered light power is calculated as follows:

$$P_{RB} = 2 \cdot P_i \cdot RC \left\{ 1 + \cos\left[\Phi + \Delta\Phi(t) \right] \right\} \qquad (1)$$

where P_{RB} is the power of the backscattered light, P_i is the power of source light transmitted from the first end, RC is the backscattered Rayleigh coefficient, Φ is the phase between the rising and trailing edges of light pulse, and $\Delta\Phi(t)$ is the change in phase following time. According to (1), the light power of optical fibers on the intrusion position can change after an aboveground intrusion.

3.2 Extraction of an intrusion signal based on CA-CFAR

The CFAR methods consist of mean-level CFAR, mean level CFAR [14], ordered statistics CFAR [15], adaptive CFAR [16], and so on. Considering the appliance difficulty, the CA-CFAR which belongs to mean-level CFAR is used in our system. The principle of CA-CFAR is shown in Fig. 3. The estimated background noise level is the mean value of the reference cells. The adaptive threshold is $S=TZ$, where T is the threshold multiplier, and Z represents the noise level. In the detection, the test cell is compared to an estimated threshold. The sample is judged as an intrusion when it is greater than the threshold. However, the sample is a noise that does not need to be detected. Clearly, the extraction of the intrusion signal by CA-CFAR is effective and efficient.

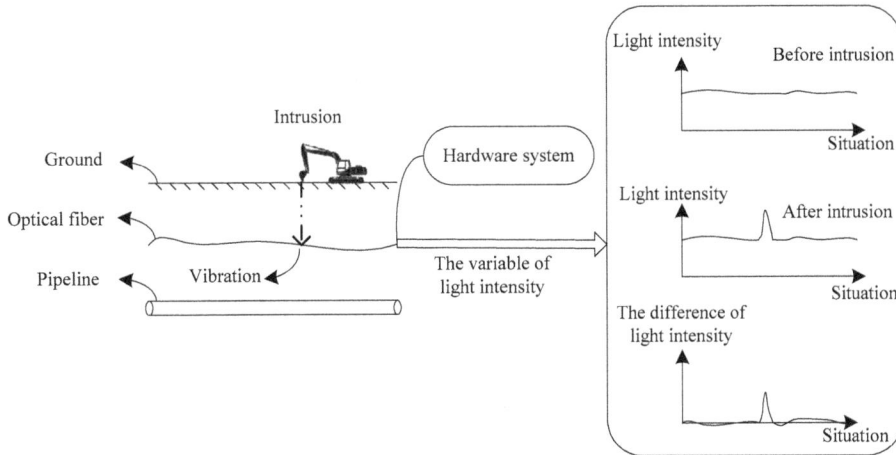

Fig. 2 Principle of OFIPS.

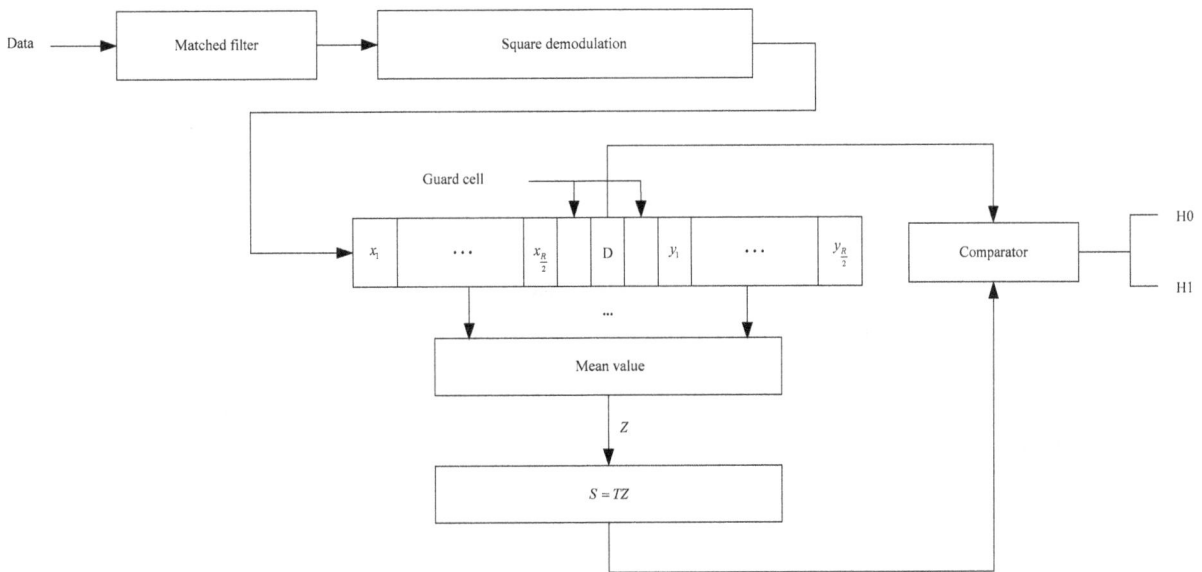

Fig. 3 Principle of CA-CFAR.

3.3 Reconstruction of detection background

The sensing signals can be obtained by Φ-OTDR, and the detection background is established in this way. After the CA-CFAR detection, all intrusion signals are extracted. Intrusions consist of harmless- and harmful-intrusion signals. We need to reconstruct the background and guarantee that the new one consists of all intrusion signals. The reconstruction method is shown in Fig. 4.

The plane is the detection background before reconstruction in the left side of Fig. 4. The abscissa represents the situation, and the ordinate is time. Intrusions detected by CA-CFAR are marked as "o". If we assume that every sample set includes four samples, the method of background reconstruction will collect every four samples extracted by the same situation point and adjacent time point as a sample set. The above process is shown in the right side of Fig. 4. The sample numbers in each sample set, N, can be adjusted according to the concrete condition. Finally, we can obtain many sample sets, which consist of intrusion samples. For the latter detection, the information of time interval and amplitude needs to be recorded.

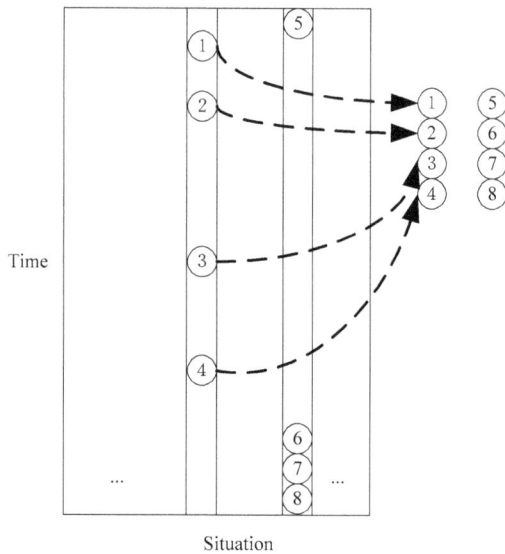

Fig. 4 Schematic diagram of background reconstruction.

4. Harmful-intrusion signal detection based on the K-S test in two dimensions

4.1 Principle of the K-S test

In statistics, the K-S test is a non-parametric test of the equality of distributions. It can be used to compare a sample with a reference probability distribution (a one-sample K-S test) or to compare two samples (a two-sample K-S test). The K-S statistic quantifies the distance between the energy distribution function (EDF) of sample and the cumulative density function (CDF) of the reference distribution, or between the EDFs of two samples. When applying a two-sample K-S test, the distance D between two EDFs $F_{N1}(x)$ and $R_{N2}(x)$ is calculated as [19–23]

$$D = \sup \left| F_{N1}(x) - R_{N2}(x) \right| \qquad (2)$$

where N_1 and N_2 are the numbers of data points in the first and second samples, respectively. Clearly, the distance approaches zero following the approach of EDFs. The distance D is used to judge whether the two distributions are homologous.

4.2 Method of the K-S test in two dimensions

Compared with the harmless intrusion, harmful-intrusion signals have the shorter time interval and higher amplitude. According to these characteristics, we design a two-dimensional K-S test method. Firstly, the time-interval K-S test needs to be implemented. If the outcome of the K-S test is that two distributions are homologous, the test sample set is a harmless one. Otherwise, it may be a harmful one, and the amplitude K-S test needs to be executed. Only under the condition that two amplitude distributions are different, the test sample can be recognized as a harmful-intrusion set.

4.2.1 Harmful-intrusion K-S test based on the characteristic of time interval

Considering the characteristics of the harmful intrusion within a shorter time interval, a time-interval detection method has been designed in this paper. The overall process is shown in Fig. 5. Firstly, we perform time-interval K-S tests between every sample set and its adjacent reference harmless set. Secondly, we can judge whether every sample set is harmless or suspected harmful. Finally, the K-S test

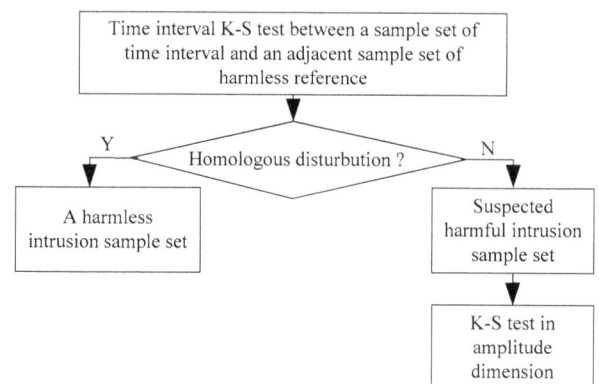

Fig. 5 K-S test in the time interval.

in the amplitude dimensional needs to be employed for every suspected harmful-intrusion sample set.

4.2.2 Harmful-intrusion K-S test based on the amplitude characteristic

In view of the characteristics of the harmful intrusion with higher amplitude, the amplitude K-S test has been designed in the paper. The overall process is shown in Fig. 6. Firstly, we perform amplitude K-S tests between every sample set and its adjacent reference harmless set. Secondly, we can judge whether every sample set is harmless or harmful. If the test outcome is that of a homologous distribution, the testing set is a harmless one. Otherwise, it is a harmful-intrusion sample set. Harmful intrusion can be extracted by this method.

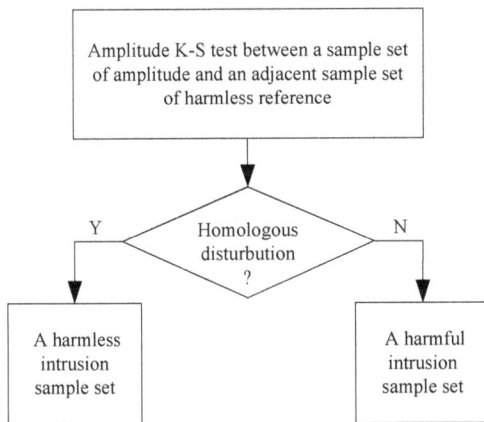

Fig. 6 K-S test in amplitude.

4.2.3 Fusion of the K-S test in time interval and amplitude

According to above two descriptions, the harmful intrusion has the shorter time interval and higher amplitude than harmless one. In other words, if the two outcomes of the K-S test both display with different distributions, the sample set can be recognized as a harmful one. If the sample set is one with a short time interval and low amplitude, the damage to the underground pipeline is negligible. Hence, the condition is defined as a harmless one. On the other hand, a sample set with the long time interval and high amplitude is harmless too. Because

the intrusion is not continuous, damage is generally deemed ignorable. Hence, in the condition that two requirements are satisfied, the testing sample set can be recognized as a harmful-intrusion sample set.

5. Experiments and analysis

The real data for OFIPS were collected from the construction site and busy road of the Da Gang oilfield. The vibration data for roads were obtained on Jin Qi road, and the construction site data were acquired in Yi He village, on the west side of Jin Qi road. The collection hardware system is shown in Fig. 7. This includes pulse light launch, photoelectric conversion, and intrusion-signal acquisition. The actual conditions of fiber burial and tests in the field are shown in Figs. 8 and 9, respectively.

5.1 Confirmation of key parameters

It is clear that the main algorithms of OFIPS are background reconstruction and the two-dimensional K-S test. Here, we confirm the key parameters by simulation and a real data test.

In the part of background reconstruction, we choose 40 as the sample number for each sample set. In other words, the two-dimensional K-S test cannot be implemented until the requirement that each sample set has 40 intrusion points has been first guaranteed. There are two reasons to confirm the parameter. On the one hand, if there are few intrusion points in every sample set, the characteristic of distribution has not been accurately reflected. On the other hand, the time interval of the adjacent vibration point is 1 ms in our system. The running time will be long, and difficulties in intrusion display will emerge, if the intrusion number of each sample set becomes excessive. After exhaustive experimentation, we set the parameter as 40. This parameter can effectively balance two potential threats. Complete distribution can be revealed, and running time is acceptable with this condition.

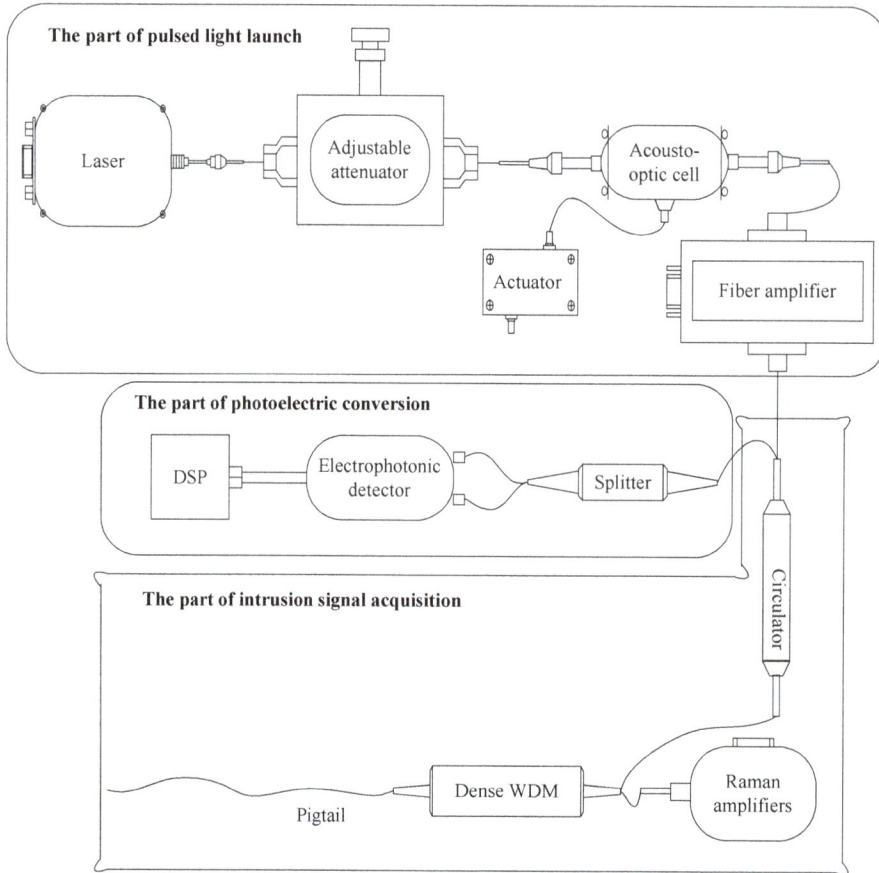

Fig. 7 Hardware system.

Fig. 8 Fiber buried in the field.

Fig. 9 Field tests.

5.2 Detection test on real data

The testing personnel dug at the construction site using a jackhammer. The test was implemented over a position of 200 m in Yi He village. The length of the excavation time was 10 s. After preliminary detection by CFAR, the outcome is shown in Fig. 10(a). Then we extract the harmful intrusion by background reconstruction and K-S test in two dimensions. The final outcome is displayed in Fig. 10(b).

It is clear that there are many intrusion signals after the CA-CFAR detection. The revealed outcome consists of harmless- and harmful-intrusion signals. According to Fig. 10(b), the outcome treated by the proposed method just includes harmful ones. We can intuitively obtain situations and time of harmful intrusions.

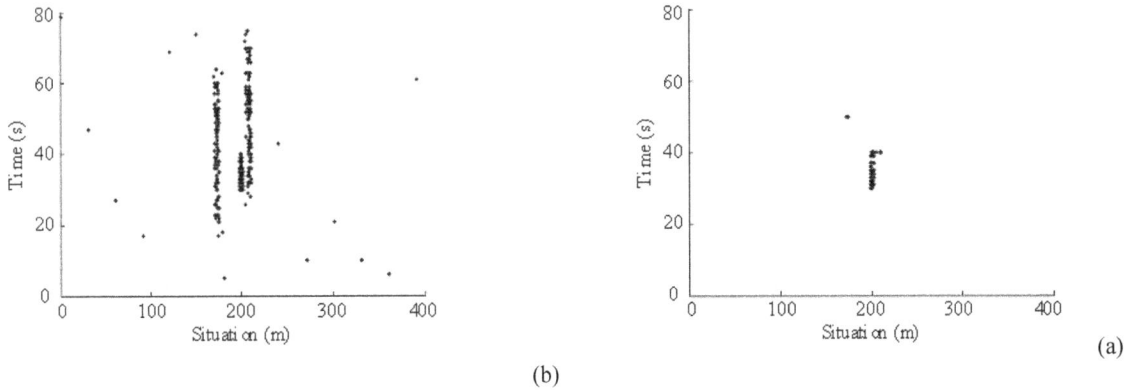

Fig. 10 Detection outcome at the construction site: (a) outcome after the CA-CFAR detecting and (b) outcome after treatment with the proposed method.

On Jin Qi road, testing personnel also dug using a jackhammer. The concrete position was 150 m, and the excavation lasted for 10 s. The intrusion outcomes after the CFAR detection and the proposed method are displayed in Figs. 11(a) and 11(b), respectively.

It is clear that there are many diagonal intrusions

after the CA-CFAR detection, caused by driving. Diagonal intrusions are harmless intrusions. Hence, we need to remove harmless intrusions. In general, the outcome treated by the proposed method includes only harmful ones. We can directly obtain the situations and time of harmful intrusions.

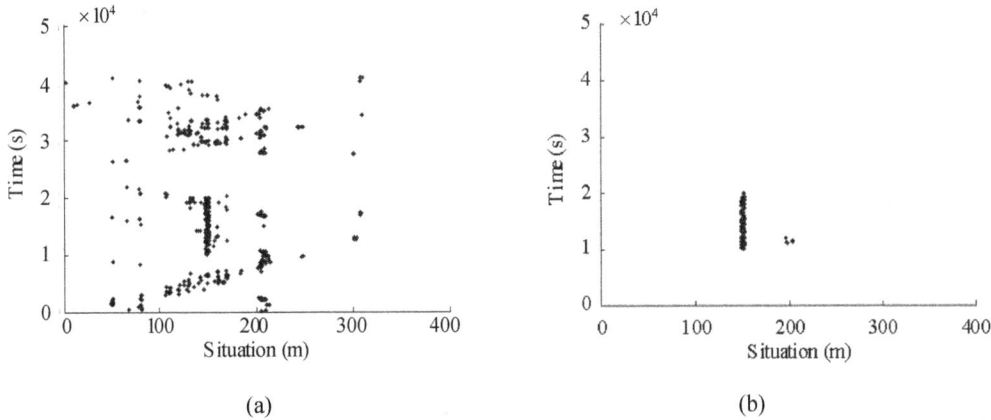

Fig. 11 Detection outcome on the road: (a) outcome after the CA-CFAR detection and (b) outcome after treatment by the proposed method.

6. Discussion and conclusions

Because signals of OFIPS include noise, harmless and harmful intrusions, the method of extracting harmful intrusions from all signals that is provided in the paper is important. We preliminarily extract vibration signals by Φ-OTDR, and the detection background can be established. The CA-CFAR detector is used to extract intrusion signals from all vibration data. The alarm signals consist of both harmless and harmful intrusions.

Hence, the background reconstruction is proposed. Finally, we use the method of the K-S test in two dimensions to detect the harmful-intrusion sample set.

Here, differences in the intensity of background light measured at different time can be used to locate the vibration in Φ-OTDR even at very small resolutions. We apply the CA-CFAR to detect intrusion signals. All alarm points, which include both harmless and harmful intrusions, are detected.

The background reconstruction classifies every adjacent one of 40 intrusion points as a sample set. The time interval and amplitude of each intrusion set should be recorded, and they are used in the later K-S test.

The two-dimensional K-S test includes time interval and amplitude. Firstly, we perform time-interval K-S tests between every sample set and its adjacent set of harmless reference. If the distribution is different, the amplitude K-S test needs to be used. Otherwise, the testing set is a harmless sample set. The amplitude K-S test is similar with a time interval. If amplitude distribution is homologous, the testing set is a harmless one. If not, a harmful-intrusion sample set is extracted.

In sum, we can effectively detect harmful intrusions by the proposed method. The method can be used in many scenes, some of which possess massive harmless intrusions, such as construction sites and busy roads. According to experiments, we find the detection effects to be high quality.

Acknowledgment

The authors are grateful to the anonymous reviewers for their critical and constructive review of the manuscript. This work was funded by the Scientific Pre-research Foundation, the National Natural Science Foundation of China (No. 61571014), and the Scientific Pre-research Foundation of NCUT.

References

[1] Z. Qu, H. Feng, Z. Zeng, J. Zhuge, and S. Jin, "A SVM-based pipeline leakage detection and pre-warning system," *Measurement*, 2010, 43(4): 513–519.

[2] J. Kang and Z. Zou, "Time prediction model for pipeline leakage based on grey relational analysis," *Physics Procedia*, 1989, 25(2): 2019–2024.

[3] W. Liang, L. Lu, and L. Zhang, "Coupling relations and early-warning for 'equipment chain' in long-distance pipeling," *Mechanical Systems and Signal Processing*, 2013, 41(1–2): 335–347.

[4] W. Liang, L. Zhang, Q. Xu, and C. Yan, "Gas pipeline leakage detection based on acoustic technology," *Engineering Failure Analysis*, 2013, 31(6): 1–7.

[5] T. Zhang, Y. Tan, H. Yang, J. Zhao, and X. Zhang, "Locating gas pipeline leakage based on stimulus-response method," *Energy Procedia*, 2014, 61: 207–210.

[6] Q. Lv, L. Li, H. Wang, Q. Li, and X. Zhong, "Influences of laser on fiber-optical distributed disturbance sensor based on Φ-OTDR," *Infrared and Laser Engineering*, 2014, 43(12): 3918–3923.

[7] H. F. Martins, S. Martin-Lopez, P. Corredera, M. L. Filograno, O. Frazão, and M. Gonzáez, "Coherent noise reduction in high visibility phase-sensitive optical time domain reflectometer for distributed sensing of ultrasonic waves," *Journal of Lightwave Technology*, 2013, 31(23): 3631–3637.

[8] Q. Li, C. Zhang, L. Li, and X. Zhong, "Localization mechanisms and location methods of the disturbance sensor based on phase-sensitive OTDR," *Optik – International Journal for Light and Electron Optics*, 2014, 125(9): 2099–2103.

[9] Q. Lin, C. Zhang, and C. Li, "Fiber-optic distributed sensor based on phase-sensitive OTDR and wavelet packet transform for multiple disturbances location," *Optik – International Journal for Light and Electron Optics*, 2014, 125(24): 7235–7238.

[10] Z. Qu, H. Feng, Z. Zeng, J. Zhuge, and S. Jin, "A SVM-based pipeline leakage detection and pre-warning system," *Measurement*, 2010, 43(4): 513–519.

[11] A. R. Bahrampour and F. Maaoumi, "Resolution enhancement in long pulse OTDR for application in structural health monitoring," *Optical Fiber Technology*, 2010, 16(4): 240–249.

[12] L. Lu, Y. Song, X. Zhang, and F. Zhu, "Frequency division multiplexing OTDR with fast signal processing," *Optics & Laser Technology*, 2012, 44(7): 2206–2209.

[13] H. Rohling, "Radar CFAR thresholding in clutter and multiple target situations," *IEEE Transactions on Aerospace and Electronic Systems*, 1983, 19(4): 608–621.

[14] S. D. Himonas and M. Barkat, "Automatic censored CFAR detection for nonhomogeneous environments," *IEEE Transactions on Aerospace and Electronic Systems*, 1992, 28(1): 286–304.

[15] R. Zhang, W. Sheng, and X. Ma, "Improved switching CFAR detector for non-homogeneous environments," *Signal Processing*, 2013, 93(1): 35–48.

[16] G. V. Weinberg, "Management of interference in Pareto CFAR processes using adaptive test cell analysis," *Signal Processing*, 2014, 104(104): 264–273.

[17] B. Shi, C. Hao, C. Hou, X. Ma, and C. Peng, "Parametric Rao test for multichannel adaptive detection of range-spread target in partially homogeneous environments," *Signal Processing*, 2015, 108: 421−429.

[18] R. Nitzberg, "Clutter map CFAR analysis," *IEEE Transactions on Aerospace and Electronic Systems*, 1986, 22(4): 419−421.

[19] X. Wang and V. Makis, "Auotoregressive model-based gear shaft fault diagnosis using the Kolmogorov-Smirnov test," *Journal of Sound and Vibration*, 2009, 327(3−5): 413−423.

[20] B. Swiderski, S. Osowski, M. Kruk, and J. Kurek, "Testure characterization based on the Kolmogorov-Smirnov distance," *Expert Systems With Applications*, 2015, 42(1): 503−509.

[21] J. Rajan, A. J. den Dekker, and J. Sijbers, "A new non-local maximum likelihood estimation method for Rician noise reduction in magnetic resonance images using the Kolmogorov-Smirnov test," *Signal Processing*, 2014, 103(10): 16−23.

[22] Z. Drezner and O. Turel, "Normalizing variables with too-frequent values using a Kolmogorov-Smirnov test: a practical approach," *Computer & Industrial Engineering*, 2011, 61(4): 1240−1244.

[23] R. Gong and S. Huang, "A Kolmogorov-Smirnov statistic based segmentation approach to learning from imbalanced datasets: with application in property refinance prediction," *Expert Systems with Applications*, 2012, 39(6): 6192−6200.

6

Fiber Cavity Ring Down and Gain Amplification Effect

Susana SILVA[1], Regina MAGALHÃES[1], Rosa Ana PÉREZ-HERRERA[2],
Manuel LOPEZ-AMO[2], M. B. MARQUES[1], and O. FRAZÃO[1*]

[1]*1INESC TEC, Rua do Campo Alegre 687, 4169-007 Porto, Portugal and Dept. de Física e Astronomia da Faculdade de Ciências da Universidade do Porto, Rua do Campo Alegre 687, 4169-007 Porto, Portugal*

[2]*Public University of Navarra, Dept. of Electric and Electronic Engineering, Campus de Arrosadía, Pamplona, Spain.*

[*]Corresponding author: Orlando FRAZÃO   E-mail: ofrazao@inesctec.pt

Abstract: The effect of an erbium-doped fiber amplifier (EDFA) placed inside the fiber ring of a cavity ring down (CRD) configuration is studied. The limitations and advantages of this configuration are discussed, and the study of the ring-down time as a function of the current applied and gain to the EDFA is also presented. In this case, the power fluctuations in the output signal are strongly dependent on the cavity ring-down time with the EDFA gain.

Keywords: Cavity ring down; EDFA; multimode laser; optical fibers.

1. Introduction

Over the past thirty years, cavity ring down (CRD) spectroscopy has been a strong subject of research. In fact, the CRD scheme was firstly developed for quantifying high-reflectivity mirrors that were difficult to characterize by other means [1, 2]. A few years later, the CRD technique gained popularity due to its ability of measuring absorption in real time, with highly sensitivity, using pulsed light sources [3]. Nowadays, CRD spectroscopy is widely used for chemical and molecular analysis in real time [4]. This technology quickly gave rise to the development of fiber optic-based CRD schemes, where a fiber loop is used as resonant cavity [5]. Although its focus is in optical spectroscopy, the CRD technique has also been used to the measurement of physical parameters, such as strain [6], pressure [7], temperature [8], refractive index

[9], and biochemical sensing [10].

A common drawback of the CRD systems is an almost 100% coupling loss when light is coupled into the cavity, either using reflective layers ($R > 99.9\%$) or optical fiber couplers with high splitting ratios (99:1) in the case of fiber loop configurations. One example to overcome this issue was the use of an erbium-doped fiber amplifier (EDFA) for loss compensation, allowing the use of open-path micro-optic cells, increasing the ring-down time of the system [11]. The study of signal amplification by placing an EDFA inside a fiber ring configuration was also reported [12]. Recently, a fiber-based CRD technique that used a large core multimode fiber-cavity design based on highly reflective gold coatings was demonstrated [13].

In this work, it is demonstrated the effect of using an EDFA for signal amplification inside the fiber ring of a cavity ring-down configuration. The study of the ring-down time as a function of the current applied to the EDFA is presented. The limitations caused by the EDFA placed inside the fiber ring are discussed, and the advantages are also presented.

2. Experimental results

The experimental setup of the proposed CRD system with amplification is presented in Fig. 1. The configuration is composed by a modulated multimode laser source, two standard (2×1) 1:99 optical fiber couplers, a fiber loop with a length of ~1 km (SMF 28), an EDFA, a photodetector (Thorlabs PDA 10CS-EC, 70 dB maximum gain), and an oscilloscope (Tektronix TDS1002C-EDU).

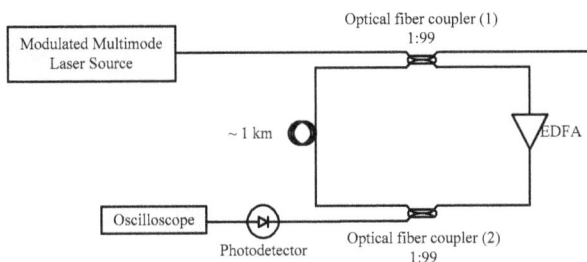

Fig. 1 Experimental setup of the proposed CRD with amplification performed inside the fiber loop by means of an EDFA (the signal is introduced inside the fiber cavity using a modulated multimode laser source and monitored via a photodetector and an oscilloscope).

The modulated multimode laser source is used to send impulses (1 s at 1550 nm) down into the fiber loop—the train of pulses is coupled via 1% arm of the input optical coupler, rings around inside the fiber loop, and is coupled out via 1% arm of the output coupler; the amplitude of the output pulses decays with time due to the total existing losses in the fiber loop (fiber loss, fiber couplers insertion losses), passes through a photodetector (gain of 40 dB), and is monitored in an oscilloscope. The EDFA is made in house, has 2 m of an erbium-doped fiber (losses of 14 dB/m @ 980 nm), and it is inserted in the fiber loop for signal amplification of the CRD configuration.

Figure 2 shows the spectral response of the multimode laser source with and without

amplification, when interrogated by an optical spectrum analyzer. One can observe that the implementation of the EDFA (gain of 10 dB) in the CRD setup increases 10-fold the amplitude of the optical power signal of the multimode laser source. The spectral response of the multimode laser changes because the EDFA gain has a high gain for shorter wavelength in the region presented in Fig. 2.

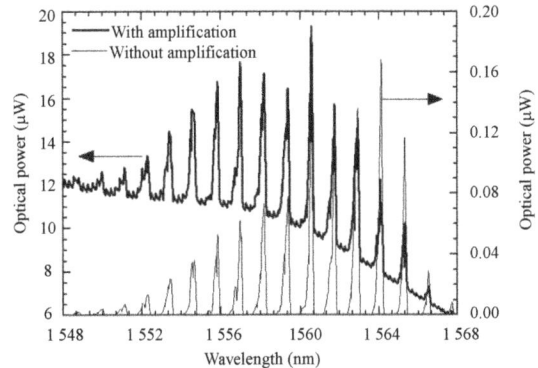

Fig. 2 Optical spectrum of the multimode laser source with (bold line) and without (line) amplification.

Figure 3(a) presents the typical CRD decay waveform, when the maximum current is applied to the EDFA, namely, 56.0 mA. This corresponds to a ring-down time of 82.7 s with the light traveling 60 times inside the fiber loop. The several CRD traces as a function of current applied to the EDFA are depicted in Fig. 3(b). The minimum current used to observe the first decaying signal is 45.1 mA, with a ring-down time of 11.04 s; when applying a current of 52.0 mA, it shows an amplitude signal and ring-down time similar to the one already reported with a decay time response of 31.8 s [14].

It is important to notice that the CRD setup without the EDFA allows reading the output signal with a ring-down time of 32 s when light travels 25 times inside the fiber loop [14]. In this case, the gain of the EDFA compensates its own losses. In the current range of (52.0–56.0) mA, the EDFA gain eliminates losses of the fiber ring itself, such as splices, fiber connectors, and others. For currents above 56 mA, losses are totally eliminated, and lasing is observed. From the results presented in

Fig. (3b), it is possible to obtain the exponential fit for each waveform, and the ring-down time is determined, as shown in Fig. 4.

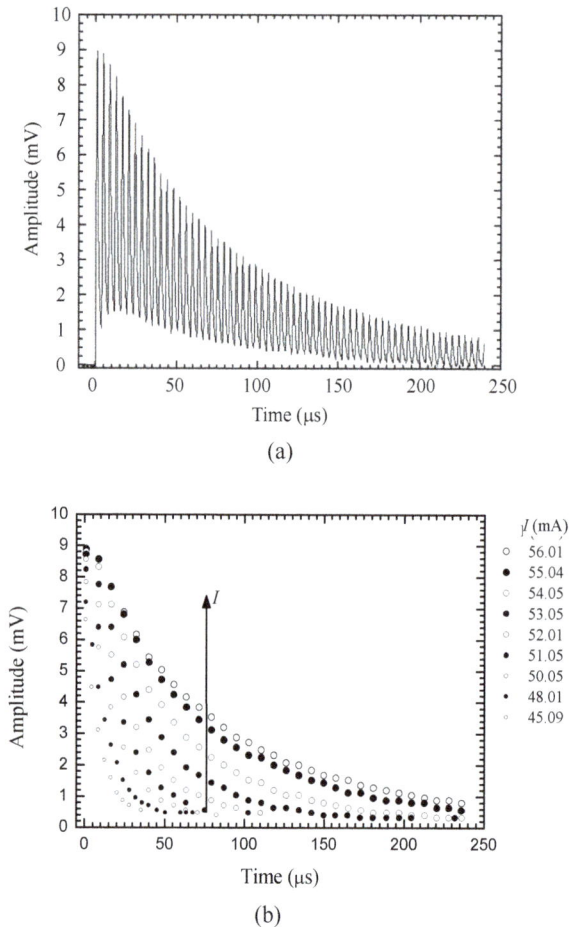

(a)

(b)

Fig. 3 CRD trace obtained for: (a) maximum current applied to the EDFA (56.01 mA) and (b) for different currents applied to the EDFA.

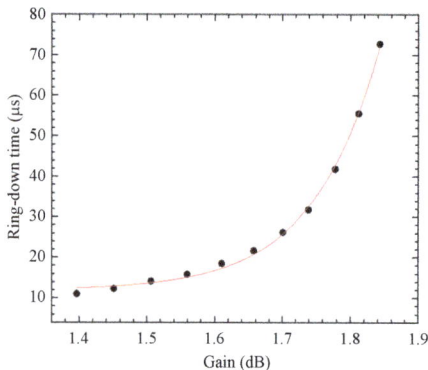

Fig. 4 Ring-down time versus gain of the CRD signal.

Figure 4 presents the relationship between the ring-down time and the gain of the CRD signal. A small variation of the ring-down time is observed for gain signals below 1.65 dB. Therefore, in the case of

any power instability of the EDFA will be negligible in this region. Ranging from 1.65 dB, the ring-down time is actually obtained due to the amplification of the CRD amplitude signal inside the cavity loop, where signal saturation may be observed. In this range, any small variation of the EDFA gain will change significantly the ring-down time. The EDFA may be used in these conditions requiring however high power stability. The ring-down time obtained with the proposed configuration is strongly dependent on the EDFA gain due to its location inside the fiber ring, which in turn influences the output power when read by the photodetector. This is a limitation when placing an EDFA or any other kind of amplification system inside the fiber ring of a CRD configuration. On the other hand, an advantage of the proposed configuration is that the EDFA placed inside the fiber ring allows to work in a larger dynamic range (with power stabilization) when compared with the conventional CRD configuration.

3. Conclusions

This work presents the study of implementing an EDFA inside the fiber ring of a cavity ring-down configuration for signal amplification. The fiber CRD configuration uses a modulated multimode laser source, and signal amplification is performed in the gain range [1.40–1.86] dB. The limitation of using such configuration is that the CRD output signal becomes dependent on the EDFA input power and, consequently, so as the ring-down time achievable in the gain range studied. In this case, the immunity of power fluctuations to the input power is eliminated when it is inserted an EDFA in the ring cavity. An advantage is that a larger dynamic range may be obtained when compared with the conventional CRD configuration.

Acknowledgment

This work was supported by Project "CORAL – Sustainable Ocean Exploitation: Tools and Sensors,

NORTE-01-0145-FEDER-000036, financed by the North Portugal Regional Operational Programme (NORTE 2020), under the PORTUGAL 2020 Partnership Agreement, and through the European Regional Development Fund (ERDF). S.S. received a Pos-Doc fellowship (ref. SFRH/BPD/92418/2013) also funded by FCT – Portuguese national funding agency for science, research and technology.

References

[1] J. M. Herbelin, J. A. McKay, M. A. Kwok, R. H. Uenten, D. S. Urevig, D. J. Spencer, *et al.*, "Sensitive measurement of photon lifetime and true reflectances in an optical cavity by a phase-shift method," *Applied Optics*, 1980, 19(1): 144–147.

[2] D. Z. Anderson, J. C. Frisch, and C. S. Masser, "Mirror reflectometer based on optical cavity decay time," *Applied Optics*, 1984, 23(8): 1238–1245.

[3] A. O'Keefe and D. A. G. Deacon, "Cavity-ring-down optical spectrometer for absorption measurements using pulsed laser sources," *Review of Scientific Instruments*, 1988, 59(12): 2544–2551.

[4] G. Berden, R. Peeters, and G. Meijer, "Cavity ring-down spectroscopy: experimental schemes and applications," *International Reviews in Physical Chemistry*, 2000, 19(4): 565–607.

[5] G. Stewart, K. Atherton, H. Yu, and B. Culshaw, "An investigation of an optical fiber amplifier loop for intra-cavity and ring-down cavity loss measurements," *Measurement Science and Technology*, 2001, 12(12): 843–849.

[6] N. Ni, C. C. Chan, X. Y. Dong, J. Sun, and P. Shum, "Cavity ring-down long period fiber grating strain sensor," *Measurement Science Technology*, 2007, 18(10): 3135–3138.

[7] C. Wang and S. T. Scherrer, "Fiber ring-down pressure sensors," *Optics Letters*, 2004, 29(4): 352–354.

[8] C. Wang, "Fiber ringdown temperature sensor," *Optical Engineering*, 2005, 44(3): 030503.

[9] W. C. Wong, W. Zhou, C. C. Chan, X. Dong, and K. C. Leong, "Cavity ring-down refractive index sensor using photonic crystal fiber interferometer," *Sensors and Actuators B Chemical*, 2013, 161(1): 108–113.

[10] C. M. Rushworth, D. James, J. W. L. Lee, and C. Vallance, "Top notch design for fiber-loop cavity ring-down spectroscopy," *Analytical Chemistry*, 2011, 83(22): 8492–8500.

[11] G. Stewart, K. Atherton, and B. Culshaw, "Cavity-enhanced spectroscopy in fiber cavities," *Optics Letters*, 2004, 29(5): 442–444.

[12] B. Vizoso, C. Vfizquez, R. Civera, M. Lopez-Amo, and M. A. Muriel, "Amplified fiber-optic recirculating delay lines," *Journal of Lightwave Technology*, 1994, 12(2): 294–305.

[13] M. Fabian, E. Lewis, T. Newe, and S. Lochmann, "Optical fiber cavity for ring-down experiments with low coupling losses," *Measurement Science and Technology*, 2010, 21(9): 094034.

[14] D. J. Passos, S. O. Silva, J. R. A. Fernandes, M. B. Marques, and O. Frazão, "Fiber cavity ring-down monitoring with an optical time-domain reflectometer," *Photonic Sensors*, 2014, 4(4): 295–299.

7

Novelty Design in Gain Flattening Filter of ASE Source Based on Fat Ultra-Long Period Fiber Grating

Fereshteh Mohammadi NAFCHI[1], Sharifeh SHAHI[2*],
Mohammad Taha SHAFFAATIFAR[3], Mohammad KANANI[2], and
Hossein NOORMOHAMMADI[2]

[1] Department of Electrical Engineering, Majlesi Branch, Islamic Azad University, Majlesi, Isfahan, Iran

[2] Dental Biophotonics and Laser Research Center (DBLRS), Khorasgan (Isfahan) Branch, Islamic Azad University, Arghavanieh, Isfahan, Iran

[3] Department of Computer Engineering, Najafabad Branch, Islamic Azad University, Najafabad, Isfahan, Iran

*Corresponding author: Sharifeh SHAHI   E-mail: shahilaser@khuisf.ac.ir

Abstract: A new type of gain flattening filter for amplified spontaneous emission (ASE) source based on erbium doped fiber (EDF) is proposed and demonstrated by fabricating and writing two series ultra-long period fiber grating (ULPFG) on single mode fiber (SMF-28). The novelty method in this research is based on writing the two ULPFGs as fat gratings. The LPG is written by a simple and available arc-discharge method. The pump power based on single-pass backward pump configuration is around 100 mW, and the average wavelength is near to 974 nm. The gain flattening profile is obtained by 3.4 (±1.7) dB ripple in the wavelength range between 1524 nm and 1565 nm with 41-nm band width.

Keywords: Amplified spontaneous emission (ASE) source; erbium-doped fiber amplifier (EDFA); gain flattening filter; long-period fiber grating (LPFG)

1. Introduction

Erbium doped fiber amplifier is an essential component for increasing capacity of transmission in long-way communication systems [1]. Although these amplifiers are more useful than others, un-flatness gain spectrum of amplified spontaneous emission (ASE) source around 1532 nm leads to decrement in capacity and bandwidth in wavelength division multiplexing (WDM) and dense wavelength division multiplexing (DWDM) systems. Hence, the gain flattening of erbium-doped fiber amplifier (EDFA) is serious for these systems. The using of several pump configurations, such as forward or backward and single or double pass EDFA, has been proposed during years [2–4]. The gain fluctuation was obtained around 0.9 dB for the wavelength between 1560 nm and 1610 nm with 15-m EDF length for the pumping with single pass forward (SPF) configuration [2]. The gain spectrum is a very non-uniform profile in the range of 1530 nm – 1565 nm for this situation, while these amounts of fluctuations are acceptable results in L-band windows. Therefore, the SPF is not a suitable way for the C-band region. According to the experimental tests carried out by J. Yang *et al.*, the ripple around ±0.4 dB on the gain spectrum was achieved in 1570 nm – 1610 nm with the using of two-stage pump configuration [4]. Also in the study conducted by M. K. Jazi *et al.*, un-flatness gain spectrum around 1532 nm was still observed with double pass bidirectional ASE source [3]. However, the using of two or more pumps makes systems complex in operation that our concentration is flattening the gain profile in 1530 nm – 1565 nm by simple way and components.

In a general classification, there are two ways for flattening gain spectrum. One is to apply changes to the material environment of erbium doped fibers, for example by doping EDF with Al [5] or Zr [6]. A hybrid configuration with zirconia-based erbium doped fiber and semiconductor optical amplifier produced the gain variation around 4 dB in 1530 nm– 1560 nm. The second general way is to use the external filters, such as acousto-optic filter [7, 8], Mach-Zehnder interferometer [9], and fiber loop mirror [9, 10].

The performance of these filters is creating inverse of ASE spectrum to achieve flatness. The combination of ASE and the inverse one removes the un-flatness region, and then the gain spectrum is uniformed.

Long period fiber gratings (LPFG) that have attractive sensing applications [11, 12] (for example in refractive index [13, 14], strain, and temperature sensors [14, 15]) are suitable filters because of their low insertion and return loss. These filters with different fabrication methods [16, 17] and implementation capabilities on several kinds of fibers (such as photonic crystal fibers [18] and the three-layer fibers [19]) are used with various techniques for equalization.

The LPFG promotes the coupling mode between the propagating light of fundamental core mode (LP01) and the co-propagating cladding modes. Because of scattering losses, the light coupled to the cladding modes decays fast in the guided core mode

observed at the output end of LPFG [19, 20]. Therefore, transmission spectrum of this type of filter has a series of attenuation bands which are employed for the ripple compensation in EDF gain.

There are several gain flattening methods that have employed LPGs. At the effect of acoustic wave [7, 8], temperature [21, 22], twisting [23, 24], and bending process on LPGs, the fluctuations are decreased in around 30 nm on bandwidth by LPG length more than 51 mm. However, we observed gain flattening around 1 dB due to new design of LPG on extended bandwidth by employing simple equipments.

The proposed method in this study is following the study on fabrication of ASE source by Kanani *et al.* [3]. Our purposed work is making a gain flattening filter for this type of source. According to the survey about ASE sources in [3], single pass backward (SPB) ASE sources is suitable for the gain flattening in C-band telecommunication region owning to its stable broad-band spectrum and high output power rather than forward sources.

In this research, by design and fabrication of new kind of ULPFG with fat gratings and combination with another LPG on SMF-28, the flatness gain is obtained. This gain flattening filter is fabricated in the research center by point-to-point arc-discharge method which is simple and affordable.

2. Experimental setup and results

The performed steps for realization and fabrication arrays based on LPG as a filter are described in details.

2.1 Designing setup

The simulation setup is illustrated in Fig. 1 by OptiSystem software based on broad band non-coherent ASE source, LPG, optical spectrum analyser, optical isolator, and other components.

As shown in Fig. 1, the 974-nm laser diode is employed to pump the EDF with 100 mW pump power through the WDM as a C-band SPB ASE source. Figure 2 shows the simulated spectrum of this source in C band region.

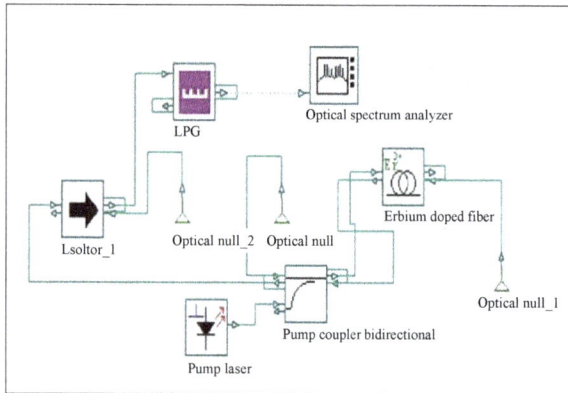

Fig. 1 Configuration of designed set-up in OptiSystem software.

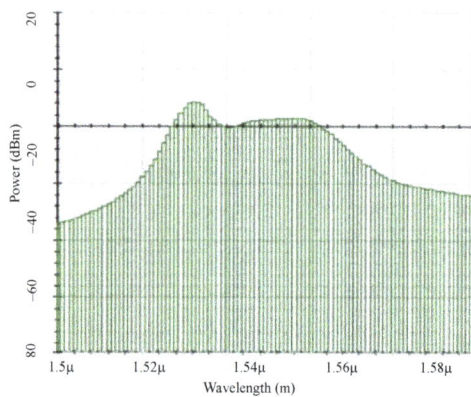

Fig. 2 Simulated spectrum of the proposed ASE source in C-band.

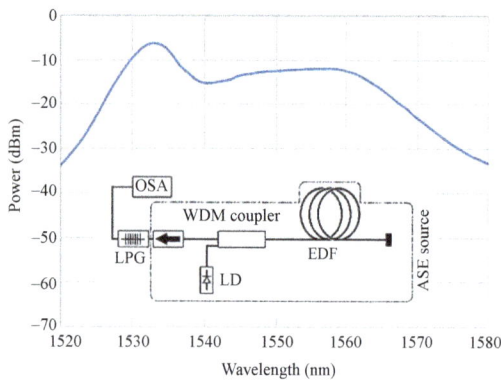

Fig. 3 Experimental spectrum of the proposed ASE source.

A minimum length of EDF (3 m) is used as the active medium in simulation and experimental work where the fiber doping concentration is $1/9 \times 10^{25}$ ions/m^3 with an NA of 0.2. The 3-dB optical coupler or WDM coupler is placed at the end of the laser diode to transfer power to EDF (inset in Fig. 3). As a result, the variation of gain spectrum against pump power due to the experimental ASE

source is shown in Fig. 3. According to this figure, a maximum peak power exists around 1532 nm and the fluctuations of the gain profile in C band interval are around 8.69 dB.

A comparison between simulation and experimental results in Figs. 2 and 3 reveals only a little loss due to experimental conditions.

2.2 Fabrication method of ULPFG

According to the simple available equipment, a point-to-point arc-discharge method for ULPFG fabrication is employed in this study. Since the thickness of induced grating by arc is 959 μm, and distance between each grating is tuned on 630 μm, the period and length of the grating is obtained around 1589 μm and 39 mm, respectively. The operating wavelength is also considered as 1532 nm.

2.3 Performance of fabricated ULPG

After fabricating the mentioned ULPFG in the previous step and entering the light to it, the process that transferring energy from light of fundamental mode (LP-01) into forward propagating cladding mode (LP-02) is done. Designing ULPG leads to create the loss band around 1532 nm and the transmission spectrum of ULPFG experiences the decrement in power in this region. When this filter is set in the following ASE setup, the decrement level of power around the peak takes place and the uniform gain profile is achieved. However, the performance of ULPFG as a gain flattening filter is shown in Fig. 4.

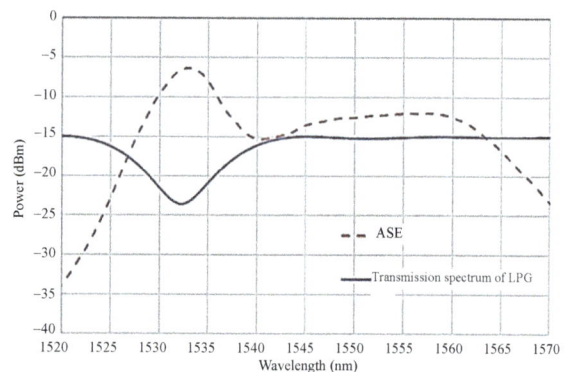

Fig. 4 Performance of ULPG for gain flattening filter.

2.4 Novelty in fabricating LPG

In order to configure LPG for gain equalization, the first stage is writing ULPG with 25 segments and 39-mm length of grating on the SMF-28. The fluctuations of ASE profile decrease to 6.3 dB after passing from this LPG (Fig. 5). The aim of the second stage is to improve this spectrum and decrease the fluctuations greatly. So, a new kind of ULPG is designed and demonstrated in this study.

In the fabrication process of LPFGs, to implement uniform gratings, the fiber is fixed in the device, and the gratings are written on it, so that one end of the fiber is fixed, and another end is strained with a weight. In novelty process, LPFG has made with pushing two ends of the fiber towards each other while writing gratings. So the gratings are made in a fat shape (Fig. 6). However, with minimum 5 fat segments on 8 mm in continuing of previous ULPFG with 25 segments, more flattening gain spectrum is obtained (Fig. 5).

However, two grating series are engraved on the normal fiber. The ASE profiles of these processes are shown in Fig. 5. This figure shows the ripple of ASE spectrum that is greatly reduced by employing of two ULPFG series as a filter.

At the best final condition, by inserting of the 5 fat gratings, the fluctuations around 3.4 dB are obtained.

As mentioned before, Fig. 6 shows the designing model of this new type of ULPFG with fat gratings that have higher refractive index rather than any region of the fiber.

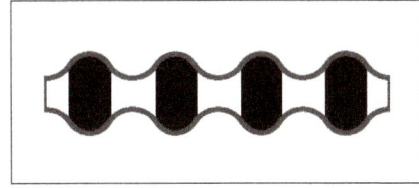

Fig. 6 Novel LPG design for gain flattening with fat grating.

2.5 ASE spectrum with different input powers

Finally, in order to achieve the compact filter, two kinds of ULPG are fabricated on the fiber. The gain profile is compared in Fig. 7, by prepared filter and the source in different pump powers of 100 mW and 300 mW, respectively, the gain profile is compared in Fig. 7. As shown in this figure, the fluctuations of pump power in 100 mW are less than that in 300 mW. The ripple rate for 100 mW is around 3.4 dBm while for 300 mW is around 4.5 dBm. So, for the less power in this study, the gain profile has become more flatten.

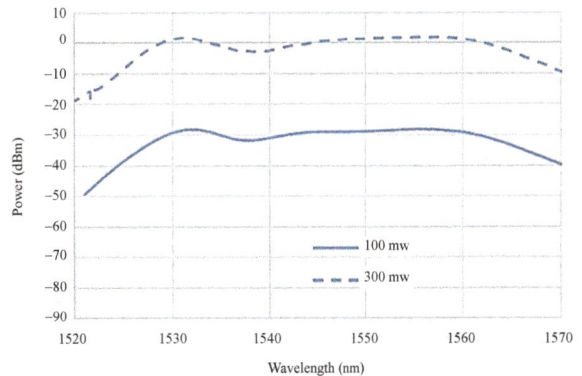

Fig. 7 Comparison of the flattened gain profile in 100 mW and 300 mW pump power output.

3. Comparison of results

Table 1 shows the comparison between results and selection of the best filter for the equalization of ASE profile in this study.

Table 1 Comparing of the fluctuations in gain profile.

Kind of profile	Value of fluctuation (dB)	Optimum state
ASE without filter	8.69	
ASE+ ULPFG (25 segment)	6.3	
ASE+ ULPFG (25 segment)+ ULPFG (5 fat segment)	3.4(±1.7)	✓

Compared with other methods for equalization gain profile, the following points should be considered:

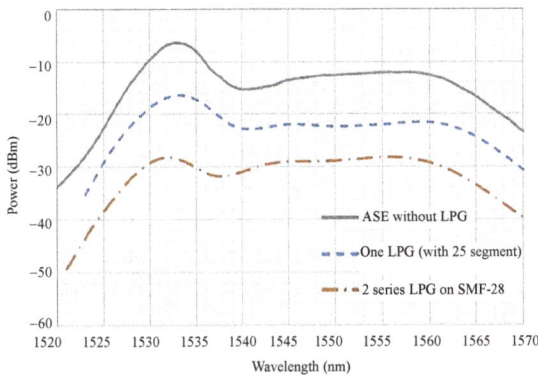

Fig. 5 ASE spectrum in absence of any filter (solid curve), ASE after passing from one ULPG with 25 segment (dash line curve), and ASE after two ULPFG (dash-dot curve).

(1) The minimum length of erbium doped fiber that used (around 3m in compare to other reports >10 m) is so better because of compacting integrated circuits and economic issue.

(2) The length of LPFG that is fabricated is 39 mm while in many studies it is longer.

(3) The important point is a simple fabrication method that is achieved by arc-discharge system and simple equipments. Although, some of investigations have better fluctuations in ASE spectrum, the results with this method are more desirable rather than the complicated fabrication way.

4. Conclusions

In summary, a novel filter was proposed for flattening the gain spectrum of ASE in this research. The SPB ASE source was used and gain flattening profile with employing two stage ULPG was achieved, where the second ULPFG was made of fat gratings. In this way, the input power was 100 mW. With this low level of pump power, the uniform gain profile in C band region with 3.4 (±1.7) dB ripple rate was observed.

Hence, the recombination of two LPGs by using the novel ULPG with fat grating at the comparison with other methods is easy, compact, and affordable for employing in various applications. The ASE flat spectrum is also very applicable for broadband sensors, WDM systems, filters, and so on.

References

[1] K. Toge and F. Ito, "Recent research and development of optical fiber monitoring in communication systems," *Photonic Sensors*, 2013, 3(4): 304–313.

[2] J. Tiana, Y. Yaoa, Y. X. Suna, X. C. Xub, X. H. Zhaoa, and D. Y. Chen, "Flat broadband erbium-doped fiber ASE source based on symmetric nonlinear optical loop mirror," *Laser Physics*, 2010, 20(8): 1760–1766.

[3] M. K. Jazi, S. Shahi, M. J. Hekmat, H. Saghafifar, A. T. Khuzani, H. Khalilian, *et al.*, "The evaluation of various designs upon C&L band super fluorescent sources based erbium doped fiber," *Laser Physics*, 2013, 23(6): 553–559.

[4] J. Yang, X. Meng, C. Liu, and C. Liu, "Gain-flattened two-stage L-band erbium-doped fiber amplifier by weak gain-clamped technique," *Optical Engineering*, 2015, 54(3): 036107-1–036107-5.

[5] H. Lin, W. Fan, and W. Han, "Broad gain of the Er/Al- doped fiber amplifier by pumping with a white light-emitting diode," *Journal of Luminescence*, 2014 146(1): 87–90.

[6] N. A. D. Huri, A. Hamzah, H. Arof, H. Ahmad, and S. W. Harun, "Hybrid flat gain c-band optical amplifier with zr-based erbium-doped fiber and semiconductor optical amplifier," *Laser Physics*, 2011, 21(1): 202–204.

[7] C. A. F. Marques, R. A. Oliveira, A. A. P. Pohl, and R. N. Nogueira, "Adjustable EDFA gain equalization filter based on a single LPG excited by flexural acoustic waves for future DWDM networks," in *International Conference on Fibre Optics and Photonics*, 2012, 180: 3770–3774.

[8] A. A. P. Pohl, R. A. Oliveira, R. E. D. Silva, C. A. F. Marques, P. D. T. Neves Jr, K. Cook, *et al.*, "Advances and new applications using the acousto-optic effect in optical fibers," *Photonic Sensors*, 2013, 3(1): 1–25,.

[9] X. Xue, W. Zhang, L. Yin, S. Wei, Sh. Gao, P. Geng, *et al.*, "All-fiber intermodal Mach-Zehnder interferometer based on a long-period fiber grating combined with a fiber bitaper," *Optics Communications*, 2012, 285: 3935–3938.

[10] N. Kumar and K. Ramachandran, "Mach-Zehnder interferometer concatenated fiber loop mirror based gain equalization filter for an EDFA," *Optics Communications*, 2013, 289(4): 92–96.

[11] A. P. Zhang, S. Gao, G. Yan, and Y. Bai, "Advances in optical fiber Bragg grating sensor technologies," *Photonic Sensors*, 2012, 2(1): 1–13.

[12] K. Kalli, T. Allsop, K. Zhou, G. Smith, M. Komodromos, D. Webb, *et al.*, "Sensing properties of femtosecond laser-inscribed long period gratings in photonic crystal fiber," *Photonic Sensors*, 2011, 1(3): 228–233.

[13] A. Singh, "Long period fiber grating based refractive index sensor with enhanced sensitivity using michelson interferometric arrangement," *Photonic Sensors*, 2015, 5(2): 172–179.

[14] Q. Huang, Y. Yu, Z. Ou, X. Chen, J. Wang, P. Yan, *et al.*, "Refractive index and strain sensitivities of a long period fiber grating," *Photonic Sensors*, 2014, 4(1): 92–96.

[15] R. C. Chaves, A. D. A. P. Pohl, I. Abe, R. Sebem, and A. Paterno, "Strain and temperature characterization of LPGs written by CO2 laser in pure silica LMA photonic crystal fibers," *Photonic*

Sensors, 2015, 5(3): 241–250.

[16] X. Xu, J. Tang, J. Zhao, K. Yang, C. Fu, Q. Wang, *et al.*, "Post-treatment techniques for enhancing mode-coupling in long period fiber gratings induced by CO2 laser," *Photonic Sensors*, 2015, 5(4): 339–344.

[17] M. Melo and P. V. S. Marques, "Fabrication of tailored Bragg gratings by the phase mask dithering/moving technique," *Photonic Sensors*, 2013, 3(1): 81–96.

[18] J. Ju and W. Jin, "Long period gratings in photonic crystal fibers," *Photonic Sensors*, 2012, 2(1): 65–70.

[19] A. Singh, "Study of modeling aspects of long period fiber grating using three-layer fiber geometry," *Photonic Sensors*, 2015, 5(1): 32–42.

[20] Q. Li, F. Yan, P. Liu, W. Peng, G. Yin, and T. Feng, "Analysis of transmission characteristics of tilted long period fiber gratings with full vector complex coupled mode theory," *Photonic Sensors*, 2012, 2(2):

158–165.

[21] J. K. Bae, J. Bae, S. H. Kim, N. Park, and S. B. Lee, "Dynamic EDFA gain-flattening filter using two lpfgs with divided coil heaters," *IEEE Photonics Technology Letters*, 2005, 17(6): 1226–1228.

[22] V. Nascimento, J. D. Oliveira, V. B. Ribeiro, and A. C. Borndonalli, "Dynamic EDFA gain spectrum equalizer using temperature controlled optoceramic filter array," in *IEEE Microwave & Optoelectronics Conference (IMOC)*, Natal, 2011, pp. 273–276.

[23] R. B. Shang, W. G. Zhang, W. B. Zhu, P. C. Geng, J. Ruan, S. C. Gao, *et al.*, "Fabrication of twisted long period fiber gratings with high frequency CO2 laser pulses and its bend sensing," *Journal of Optics*, 2013, 15(7): 75402–75407.

[24] W. Jin and H. Xuan, "Rocking long period gratings in single mode fibers," *IEEE Lightwave Technology*, 2013, 31(18): 3117–3122.

8

Compact Dual-Frequency Fiber Laser Accelerometer With Sub-µg Resolution

Qian CAO, Long JIN[*], Yizhi LIANG, Linghao CHENG, and Bai-Ou GUAN

Guangdong Provincial Key Laboratory of Optical Fiber Sensing and Communications, Institute of Photonics Technology, Jinan University, Guangzhou, 510632, China

[*]Corresponding author: Long JIN   E-mail: iptjinlong@gmail.com

Abstract: We demonstrate a compact and high-resolution dual-polarization fiber laser accelerometer. A spring-mass like scheme is constructed by fixing a 10-gram proof mass on the laser cavity to transduce applied vibration into beat-frequency change. The loading is located at the intensity maximum of intracavity light to maximize the optical response. The detection limit reaches $107\,\text{ng/Hz}^{1/2}$ at $200\,\text{Hz}$. The working bandwidth ranges from $60\,\text{Hz}$ to $600\,\text{Hz}$.

Keywords: Dual-frequency fiber lasers; fiber Bragg grating; photonic accelerometers

1. Introduction

High-resolution accelerometer is an essential device in geophysical applications. Fiber-optic accelerometers have presented inherent advantages including high responsivity, low temperature cross sensitivity, and immunity to electromagnetic interference [1, 2]. To date, photonic accelerometers have been demonstrated in many different schemes to attain higher detection limit. Interferometric accelerometers have been fabricated by using a compliant cylinder or a central/edge-supported flexural disc to elongate optical fibers [3−5]. An accelerometer with a minimum detectable acceleration of $84\,\text{ng/Hz}^{1/2}$ has been achieved with 75-meter long optical fiber [4]. Fiber Bragg grating based accelerometers have attracted great interests due to the much shorter sensing element [6−8]. For example, the fiber Bragg grating (FBG) accelerometer based on a double-diaphragm

structure has exhibited a resolution of $385\,\mu\text{g/Hz}^{1/2}$. Fiber laser accelerometers have higher detection capability as a result of the narrow linewidth of the laser output and the interrogation based on phase detection. For example, a fiber laser accelerometer with $126\,\text{ng/Hz}^{1/2}$ resolution has been achieved by directly stressing the laser with a proof mass [8].

Alternatively, single-longitudinal-mode fiber lasers with the orthogonal polarization output have been exploited as high resolution sensors, by monitoring the beat frequency in the radio frequency (RF) domain [9, 10]. In this paper, we demonstrate a high-resolution dual-polarization fiber laser accelerometer. The accelerometer is fabricated by simply fixing a 10-gram proof mass on the laser cavity at the light intensity maximum, to effectively transduce applied vibration to beat-frequency variation. The accelerometer presents a resolution of $107\,\text{ng/Hz}^{1/2}$ at $200\,\text{Hz}$ and a working bandwidth from $60\,\text{Hz}$ to $600\,\text{Hz}$. This accelerometer presents sub-μg detection limit with a weight of only tens of grams, which is greatly beneficial for geophysical applications.

2. Dual-frequency fiber grating laser

Figure 1 shows the schematic of the sensing element, i.e., a fiber grating laser, fabricated by photoinscribing two wavelength-matched, highly reflective (typically higher than 25 dB) intracore Bragg gratings in an Er-doped fiber. Due to the imperfection in symmetry of the fiber core and the ultraviolet (UV) side illumination, the lasing frequencies v_x and v_y of the two orthogonal polarization modes are somewhat different, yielding a radio-frequency beat signal. The beat frequency is mainly determined by the intracavity birefringence by $v=|v_x-v_y|=cB/n_0\lambda$, where B denotes the birefringence, c is the vacuum light speed, and n_0 represents the average mode index.

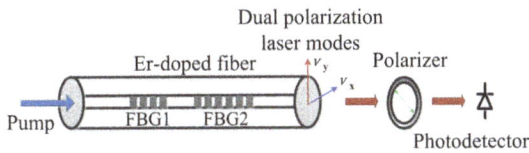

Fig. 1 Schematic of the dual-polarization fiber grating laser.

Figure 2 shows the measured spectrum of the beat signal of a fiber grating laser in an Er-doped fiber (M-12, Fiber Core Ltd.). The gratings are 5.5 mm and 4.5 mm in length and 31 dB and 29.6 dB in coupling strength, respectively. The grating separation is about 2 mm. We intend to reduce the separation as possible to approach the distributed feedback (DFB) structure. The lasing wavelength is 1553 nm, determined by the pitch of the phase mask. The beat signal is measured by use of a vector signal analyzer (MS2692A, Anritsu), which incorporates a frequency demodulator with extremely high frequency resolution, with 1 kHz resolution bandwidth. The beat frequency is 371.65 MHz, corresponding to a birefringence 2.78×10^{-6} of the fiber. The signal-to-noise ratio is higher than 60 dB. The laser can be exploited as a photonic sensor by

transducing the measurands into beat frequency variation.

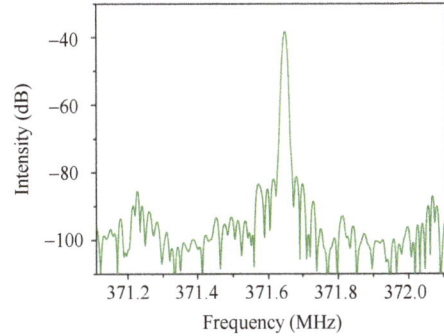

Fig. 2 Measured frequency spectrum of the output beat signal.

3. Sensitivity and noise

The accelerometer has a spring-mass configuration, which involves transversely loading the cavity with a proof mass. The frequency responsivity can be characterized by use of the mechanical susceptibility, expressed by [11]

$$\chi^{-1} = \omega_m^2 - \omega^2 + i\frac{\omega_m\omega}{Q_m} \quad (1)$$

where ω and ω_m represent the frequencies of tapplied vibration and the mth order natural frequency, and Q_m is the quality factor of the resonator. This expression indicates that the responses at the frequency range below the fundamental resonance frequency are very close to the static response. Here, we can alternatively analyze the optical response of the laser sensor to a static load. Since the beat frequency is originated from the round-trip phase difference between the two polarization modes, the response in terms of beat-frequency shift to a localized perturbation is highly dependent on the local intensity, which can be expressed by

$$\Delta v = \frac{c}{n_0\lambda} \int_{-\infty}^{+\infty} \Delta B(z)I(z)dz \quad (2)$$

where ΔB is the induced local birefringence which is in proportion to the applied load, and I represents the local intensity normalized by $\int_{-\infty}^{+\infty} I(z)dz = 1$. The intensity profile of an ideal DFB laser can be

expressed by

$$I(z) = \kappa \cdot e^{-2\kappa|z|} \tag{3}$$

where κ (m^{-1}) represents the grating coupling coefficient. The intensity profile presents a peak at the phase shift point where $z=0$, yielding a maximum sensitivity at this point. Equation (3) indicates that the sensitivity is proportional to the grating coupling strength.

Figure 3 shows the measured longitudinal sensitivity profile by scanning the loading position with a glass rod of 5.1 gram rolling along the fiber length via quasi-point contact by use of the setup described in [12]. The fiber laser is placed on an inclined plane with the angle $\theta=17°$ between the horizontal and inclined planes. Meanwhile, an ordinary singlemode fiber is placed parallel to the fiber laser as a support fiber to keep the balance of the rod. Therefore, the normal force subjected to the laser is $N=mg\cos\theta/2=2.39\times10^{-2}$ N. In this measurement as well as the accelerometer implementation in the following text, the load is applied in accordance with principle axis of the fiber to maximize the sensitivity. The measured curve presents a single peak in between the two gratings, in accordance with the measured intensity profile in [13]. The maximum sensitivity can be calculated as 72 MHz/gram. Considering the grating coupling coefficient can be as high as 1000 m^{-1} in our experiment, the laser is expected to have a sensitivity as high as 100 MHz/gram. However, due

to the limitation in fiber gain, we have to leave a certain separation between the two gratings to achieve laser oscillation. This deviation can be equivalently treated as a DFB laser with lower grating coupling strength or longer cavity length.

The frequency noise of a fiber grating laser with the single polarization mode output has been investigated in [14−16], and the dominate noise source is the nonequilibrium thermal fluctuation over the gain medium, yielding a $1/f$ profile over the low and intermediate frequencies (which are exactly the frequency ranges of interest). The two polarization modes of the present laser share the same cavity and have a high correlation degree. As a result, the noise level of the beat signal also presents the $1/f$ profile but is much lower than each polarization mode [15]. The noise spectral density can be simply expressed by $S(f)=C/f$, where C represents the noise strength. Figure 4 shows the measured spectral density profile of the frequency noise. The C value is estimated as $C=2.2\times10^{5}$ (Hz^{2}) via exponential fit. Generally speaking, active fibers with higher concentration can produce a higher noise level. Note that the noise level is also related to the intensity profile. It can be approximately considered proportional to the item $\sqrt{\int I^{2}(z)dz}$. Substituting the intensity profile depicted by (3), the noise level of a dual-polarization laser is proportional to $\sqrt{\kappa}$. This fiber is selected for accelerometer fabrication because it produces relatively low noise level among the individual

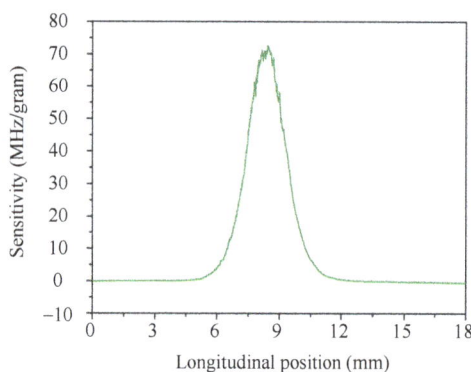

Fig. 3 Measured sensitivities of the fiber grating laser to a static transverse load along the fiber length.

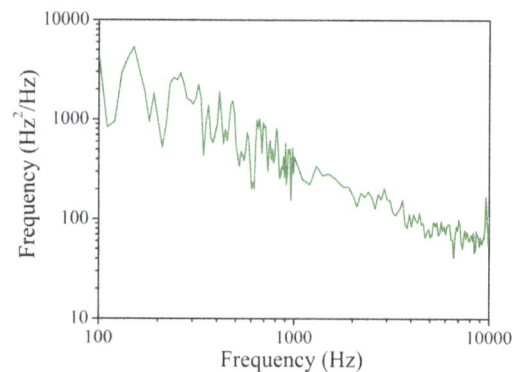

Fig. 4 Measured frequency noise of the beat signal.

products and can provide sufficient photosensitivity for grating inscription. In addition, the noise level is irrelevant with the power and the wavelength of the pump laser.

Regardless of the frequency demodulation capability, the detection capability of the laser sensor can be characterized by the ratio of sensitivity over noise, which is proportional to $\sqrt{\kappa}$. That means the higher grating coupling strength (or shorter effective cavity length) can result in lower detection limit. That is why we intend to shorten the laser cavity in the implementation of the laser accelerometer.

4. Implementation and characterization of the accelerometer

Figure 5 shows the experimental setup of the fiber laser accelerometer. A 10-gram proof mass is fixed onto the laser cavity at the intensity maximum. The contact length between the mass and the laser cavity is about 2 mm. In this mass-spring like structure, the fiber laser acts as a "spring". When the sensor package is subjected to vertical vibration, the proof mass periodically squeezes the fiber and consequently causes a beat-frequency shift. The amplitude, frequency, and phase of the applied vibration can be optically interrogated by reading out the beat frequency shift. A dummy fiber is placed parallel to the lasing fiber to keep balance of the structure. The fiber grating laser is pumped with a 980 nm laser diode via a wavelength division multiplexer (WDM). An in-line polarizer is used to maximize the intensity of the beat signal. A reference sensor is also attached on the exciter for calibration. Compared with our previous work based on the cantilever based transducer [17], the present structure has higher transducing efficiency. This configuration can be highly compact, compared to transducers for the single-frequency fiber laser accelerometers like the ones described in [8].

Figure 6 shows the measured beat-frequency deviation as a function of acceleration amplitude at

480 Hz. The accelerometer demonstrates a linear response versus vibration strength. Figure 6 inset shows a typical sensor output in terms of beat-frequency variation measured by the signal analyzer. The applied acceleration is 6.34 mg.

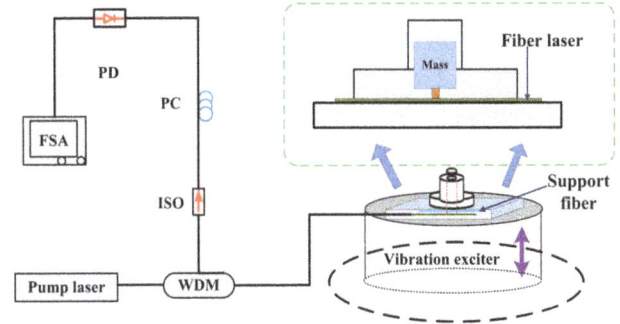

Fig. 5 Experimental setup of the fiber laser accelerometer (WDM: wavelength division multiplexer; ISO: isolator; PC: polarization controller; PD: photodetector; VSA: vector signal analyzer).

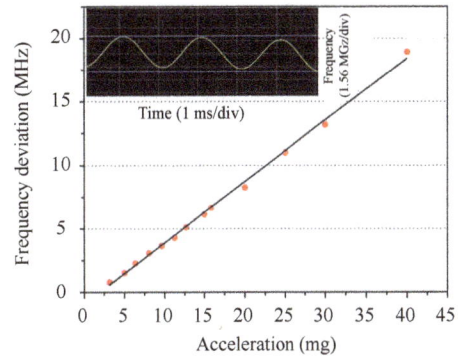

Fig. 6 Measured frequency deviation as a function of acceleration amplitude at 480 Hz (inset: typical output signal under variation with a frequency of 480 Hz and an amplitude of 6.34 mg).

Figure 7 shows the average sensitivity below 600 Hz is estimated as 290 MHz/g (1 g= 9.8 m²/s). The fundamental natural frequency is about 680 Hz as a result of resonance effect. The sensor presents a broad and flat frequency range from 60 Hz to 600 Hz as the operating bandwidth. This natural frequency is much lower than our calculated result, due to the out-of-phase vibration arising from the imperfection of the transducer fabrication. The minimal detectable signal is fundamentally limited by the frequency noise of the beat signal in Fig. 4. For example, the frequency noise at 200 Hz is about

971 Hz^2/Hz (31.2 Hz/Hz$^{1/2}$). Considering the sensitivity of 290 MHz/g and the sensitivity-to-noise ratio (SNR) of 139.3 dB/g/Hz$^{1/2}$, the minimal detectable acceleration is estimated as about 107 ng/Hz$^{1/2}$ at 200 Hz.

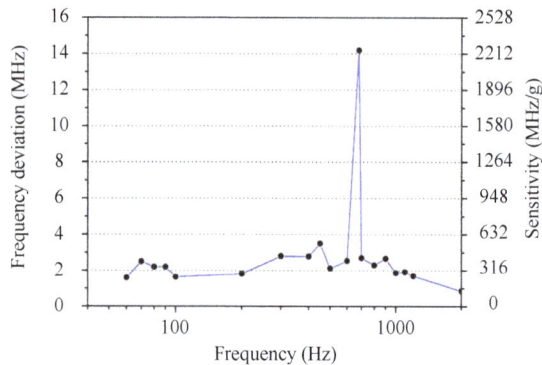

Fig. 7 Measured frequency response and related sensitivities with an acceleration amplitude of 6.34 mg.

5. Conclusions

In summary, we have demonstrated a high-resolution accelerometer based on a dual-polariztion fiber grating laser with a device weight only tens of grams. A resolution of 107 ng/Hz$^{1/2}$ at 200 Hz has been attained with a spring-mass like transducer scheme. Theoretical and experimental investigations reveal that an ideal DFB fiber laser with higher grating coupling strengths (or shorter effective cavity length) can enable higher detection capability to the applied load. The further performance improvement relies on both the fabrication of an active fiber with both high UV photosensitivity and optical gain and the improvement of the transducer fabrication.

Acknowledgment

This work is supported by the National Natural Science Foundation of China (Nos. 61235005 and 11474133), Guangdong Natural Science Foundation (No. S2013030013302), and the Planned Science & Technology Project of Guangzhou (Nos. 2012J510028 and 2014J2200003). L. Jin is supported by the Department of Education, Guangdong Province (No. Yq2013021) and by Open Fund of the Guangdong Provincial Key Laboratory of Fiber Laser Materials and Applied Techniques (South China University of Technology).

References

[1] X. Wu, X. Wang, S. Li, S. Huang, Q. Ge, and B. Yu, "Cantilever fiber-optic accelerometer based on modal interferometer," *Photonics Technology Letters*, 2015, 27(15): 1632−1635.

[2] Y. Liu, W. Peng, X. Zhang, Y. Liang, Z. Gong, and M. Han, "Fiber-optic anemometer based on distributed Bragg reflector fiber laser technology," *Photonics Technology Letters*, 2013, 25(13): 1246−1249.

[3] N. Zeng, C. Z. Shi, M. Zhang, L. W. Wang, Y. B. Liao, and S. R. Lai, "A 3-component fiber-optic accelerometer for well logging," *Optics Communications*, 2004, 234(1−6): 153−162.

[4] G. A. Cranch and P. J. Nash, "High-responsivity fiber-optic flexural disk accelerometers," *Journal of Lightwave Technology*, 2000, 18(9): 1233−1243.

[5] G. Chen, X. Zhang, G. Brambilla, and T. P. Newson, "Theoretical and experimental demonstrations of a microfiber-based flexural disc accelerometer," *Optics Letters*, 2011, 36(18): 3669−3671.

[6] Q. Liu, X. Qiao, Z. Jia, H. Fu, H. Gao, and D. Yu, "Large frequency range and high sensitivity fiber Bragg grating accelerometer based on double diaphragms," *IEEE Sensors Journal*, 2014, 14(5): 1499−1504.

[7] Q. Liu, X. Qiao, J. Zhao, A. Zhen, H. Gao, and M. Shao, "Novel fiber Bragg grating accelerometer based on diaphragm," *IEEE Sensors Journal*, 2012, 12(12): 3000−3004.

[8] G. H. Ames and J. M. Maguire, "Erbium fiber laser accelerometer," *IEEE Sensors Journal*, 2007, 7(4): 557−561.

[9] B. Guan, L. Jin, Y. Zhang, and H. Tam, "Polarimetric heterodyning fiber grating laser sensors," *Journal of Lightwave Technology*, 2012, 30(8): 1097−1111.

[10] K. Yu, C. Lai, C. Wu, Y. Zhao, C. Lu, and H. Tam, "A high-frequency accelerometer based on distributed Bragg reflector fiber laser," *Photonics Technology Letters*, 2014, 26(14): 1418−1421.

[11] A. G. Krause, M. Winger, T. D. Blasius, Q. Lin, and O. Painter, "A microchip optomechanical accelerometer," *Nature Photonics*, 2012, 6(11): 768‒772.

[12] F. Zhou, L. Jin, Y. Liang, L. Cheng, and B. Guan, "Spatial sensitivity characterization of dual-polarization fiber grating laser sensors," *Journal of Lightwave Technology*, 2015, 33(19): 4151‒4155.

[13] S. Foster and A. Tikhomirov, "Experimental and theoretical characterization of the mode profile of single-mode DFB fiber lasers," *Quantum Electronics*, 2005, 41(6): 762‒766.

[14] O. Llopis, P. H. Merrer, H. Brahimi, K. Saleh, and P. Lacroix, "Phase noise measurement of a narrow linewidth CW laser using delay line approaches," *Optics Letters*, 2011, 36(14): 2713–2715.

[15] S. Foster, G. A. Cranch, and A. Tikhomirov, "Experimental evidence for the thermal origin of $1/f$ frequency noise in erbium-doped fiber lasers," *Physical Review A*, 2009, 79(5): 1744‒1747.

[16] S. Foster, "Fundamental limits on 1/f frequency noise in rare-earth-metal-doped fiber lasers due to spontaneous emission," *Physical Review A*, 2008, 78(1): 013820.

[17] B. Guan, X. Sun, and Y. Tan, "Dual polarization fiber grating laser accelerometer," in *Proc. SPIE*, vol. 7653, pp. 76530Z-1‒76530Z-4, 2010.

9

Design and Optimization of Photonic Crystal Fiber for Liquid Sensing Applications

Md. Faizul Huq ARIF[1], Kawsar AHMED[1*], Sayed ASADUZZAMAN[1], and Md. Abul Kalam AZAD[2]

[1]*Department of Information and Communication Technology (ICT), Mawlana Bhashani Science and Technology University (MBSTU), Tangail-1902, Bangladesh*

[2]*Department of Material and Metallurgical Engineering (MME), Bangladesh University of Engineering and Technology University (BUET), Dhaka-1000, Bangladesh*

*Corresponding author: Kawsar AHMED   E-mail: kawsar.ict@mbstu.ac.bd and kawsar_it08050@yahoo.com

Abstract: This paper proposes a hexagonal photonic crystal fiber (H-PCF) structure with high relative sensitivity for liquid sensing; in which both core and cladding are microstructures. Numerical investigation is carried out by employing the full vectorial finite element method (FEM). The analysis has been done in four stages of the proposed structure. The investigation shows that the proposed structure achieves higher relative sensitivity by increasing the diameter of the innermost ring air holes in the cladding. Moreover, placing a single channel instead of using a group of tiny channels increases the relative sensitivity effectively. Investigating the effects of different parameters, the optimized structure shows significantly higher relative sensitivity with a low confinement loss.

Keywords: Photonic crystal fiber (PCF); liquid sensor; microstructure core; sensitivity; confinement loss

1. Introduction

The field of fiber optics is no longer limited into telecommunication and medical science only; it has been developing in an incredible pace with large dimensions of applications. Fiber optic technologies have made a revolutionary change after the invention of photonic crystal fiber (PCF). PCF is a new class of optical fiber which is one of the recent inventions in the field of fiber optics. PCF can be used as a transmission media as well as optical functional devices. In contrast to the conventional optical fiber, PCFs have additional design features, such as air-hole diameter, pitch size, and number of rings, which offer to overcome many limitations of conventional fiber.

Due to the well-known advantages, such as enhanced design freedom, low cost, short-time detection, small size, robustness, and high sensitivity and flexibility PCFs have received considerable attention in developing optodevices and sensors. Photonic crystal fibers (PCFs) have been attracted a great deal of attention for its incredible performance and large variety of applications. PCF can be used as filters [1], switches [2, 3], electro-optical modulators [4, 5], polarization converters [6], sensors [7−17], etc. PCF based sensors are smart applications in fiber optic technology which have been investigating and developing since last decade. A wide range of sensing applications of PCF are available, such as temperature sensors [7], refractive index (R.I) sensors [8], chemical sensors [9], mechanical sensors [10], pressure sensors [11], gas sensors [12,13], stress sensors [14], pH sensors [15], liquid sensors [16], biosensors [17], and so on. An ideal candidate of optical sensors is index guiding PCF. The sensing mechanism of index guiding PCF is evanescent interaction between the optical field and the analyte to be sensed. The evanescent field based PCF sensors have been developing rapidly for chemical and biomedical applications due to their attractive features.

Highly sensitive chemical (liquid and gas) sensors are playing an important role in the industrial processes [18] especially for detecting toxic and flammable chemicals (e.g., toxic gasses or liquids) to overcome the safety issues. So it has become one of the key challenges to enhance the performance of liquid and gas sensors. In recent years, researchers are keeping much interest on the development of photonic crystal fiber (PCF) based sensors for environmental and safety monitoring [19, 20] issues. Photonic crystal fiber based liquid and gas sensors through the evanescent field show excellent performance in terms of sensitivity, because core of the PCF directly interacts with the material to be analyzed.

PCF technologies allow for the accurate tuning of fiber through changing the air hole shape, size, and their position. A wide variety of PCF based sensing techniques have been reported by changing different geometric parameters of the PCF to gain sensitivity at a maximum and confinement loss at a minimum satisfactory level in liquid and gas sensing applications. J. Park *et al.* [21] enhanced relative sensitivity for chemical sensing, using a hexagonal PCF with a hollow high indexed ring defect. In the hollow core PCF, the direct interaction between light and the analyte in the hollow channel is higher than the index-guided PCFs. Recently, the idea of filling

core or cladding holes with various liquids or gases has been attracted much to the researchers. Cordeiro *et al.* [22] proposed a microstructure core PCF infiltrated with liquid analyte which enhanced the evanescent field. This concept introduced the sensing potentiality with infiltrated microstructure core. PCF of microstructure core offers to sense low indexed material because of the highly interaction of evanescent fields with the analyst to be sensed. A large number of published papers investigated and enhanced the performance of PCF based gas and liquid sensors with microstructure core [23−28].

In recent study, higher sensitivity and lower confinement loss of microstructure core PCF for liquid sensing have been attempted by using octagonal cladding structure [24, 25]. Reference [25] suggested 5-ring octagonal PCF for higher sensitivity and lower confinement loss; but in practical manufacturing octagonal structure requires extra more capillaries than the hexagonal structure. Keeping large number of capillaries will make high cost to fabricate. In this point of view, liquid sensing using a single infiltrated channel may also reduce the complexity of the core. To the best of our knowledge, no studies have been done in analyzing the sensitivity performance of PCF with a liquid filled core of a single channel.

In this research work, we have proposed and optimized simple evanescent hexagonal structure of PCF (H-PCF) with microstructure core and cladding for liquid sensing, which shows high relative sensitivity as well as low confinement loss. We have also explained the effect of single infiltrated channel replacing the microstructure core by proposing another structure of PCF, which achieved more enhancements of relative sensitivity and simplicity in design. We have not used any defect around the hollow core; though one of the previous articles [21] enhanced relative sensitivity by using a ring defect around the core. The relative sensitivity and confinement loss against different liquids (water, ethanol, and benzyne) have been investigated and compared. Although we have chosen water, ethanol, and benzyne as the targeted chemical species for characterization of our structures but these structures and the mechanism can be applied for all fluids and gases based on the absorption line of the targeted sample.

2. Design principle

Figure 1 shows the transverse cross sectional view of the four stages of our proposed PCF structure. The proposed PCF contains only four layers of air holes in the cladding. The distance between center and center of two adjacent air holes (pitch distance) has been denoted by Λ. The diameters of air holes in the innermost ring, second ring, third ring, and outermost ring are d_1, d_2, d_3, and d_4, respectively. In PCF$_1$, the diameter of all air holes is equal, where $d_1=d_2=d_3=d_4$.

In our numerical investigation, we found that the outermost ring holes diameter has greater impact on the confinement loss, and then we have come into PCF$_2$. In PCF$_2$, $d_1=d_2=d_3<d_4$. Another result of our numerical investigation shows that larger diameter of the innermost ring holes enhances the sensitivity and we have turned into PCF$_3$. In PCF$_3$, optimized values of air holes diameter have been kept as $d_2=d_3<d_1=d_4$. However, we have turned into PCF$_4$ and achieved higher sensitivity by replacing the group of tiny holes with a single hollow core filled with same analyte to be detected. The hollow core area is same as the area covered by supplementary tiny holes. In the PCF$_1$, PCF$_2$, and PCF$_3$, the core is designed with some tiny holes in circular form which are filled with various liquid samples: water, ethanol, and benzyne for this study. These supplementary core holes are arranged with the hole to hole pitch distance denoted by a. Figure 2 visualizes the enlarged view of core of PCF$_1$, PCF$_2$, PCF$_3$, and the replacement of hollow channel instead of using a group of tiny channels in PCF$_4$. Diameter of the hollow channel is $D_2=1.70\,\mu m$, which is same

as the diameter of the region of supplementary holes in the core ($D_1=D_2$).

Figure 3 shows the computational region of the proposed PCF$_3$ and PCF$_4$, which is divided into homogeneous triangular pieces forming a mesh. Each of the PCFs has two orthogonal sides of the computational region which are assigned with two artificial boundary conditions: perfect electric conductor (PEC) and perfect magnetic conductor (PMC). Perfectly matched layer (PML) is used as a boundary condition. Thickness of the PML is fixed to 10% of the radius of the proposed PCFs for efficient calculation of confinement loss [29].

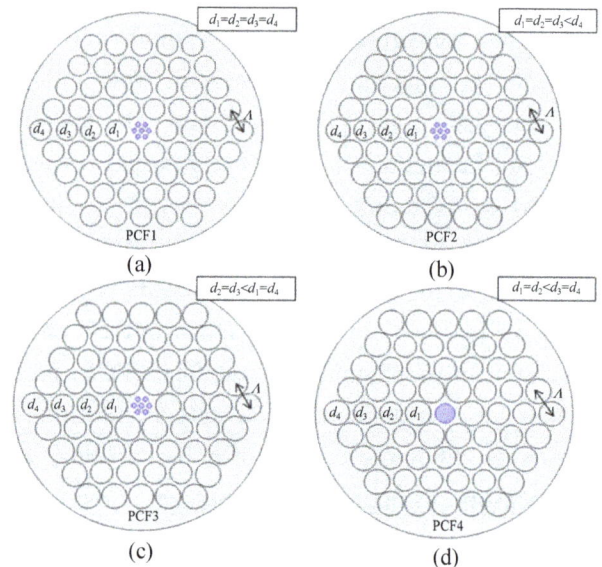

Fig. 1 Transverse cross sectional view of (a) PCF$_1$, (b) PCF$_2$, (c) PCF$_3$, and (d) PCF$_4$.

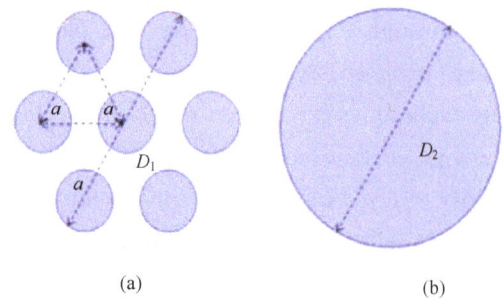

Fig. 2 Enlarged view of core region of (a) PCF$_1$, PCF$_2$, and PCF$_3$ and (b) PCF$_4$.

3. Principles of operation

PCFs act as a waveguide, and in this wave guide,

and the targeted analyte and light interact with each other. We have analyzed the evanescent field distribution of the proposed PCFs. Using the finite element method (FEM), the properties of propagating mode of the proposed PCFs is numerically investigated. We have considered circular perfectly matched layer (PML) as a boundary condition. The cross sections of the proposed PCFs are divided into homogeneous triangular subspaces using mesh analysis shown in Fig. 3. The liquid filled air holes' region is then divided into many sub-domains which are either triangular or quadrilateral in shape. Using FEM, Maxwell's equations are solved by accounting neighboring subspaces. As the wave propagates through z direction, the modal analysis has been performed in the x-y plane of the PCF structure. The following vectorial wave equation can be derived from the Maxwell's equation [30].

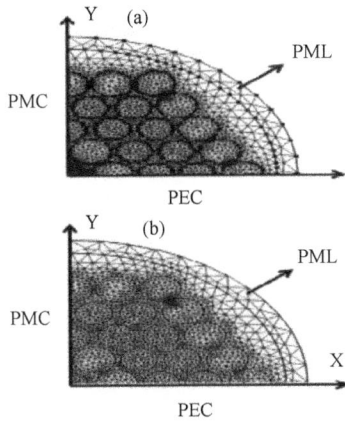

Fig. 3 FEM meshes and boundary conditions for computation of (a) PCF$_3$ and (b) PCF$_4$.

$$\nabla \times (\mathbf{S}^{-1}\nabla \times E) - k_0^2 n^2 \mathbf{S}E = 0 \qquad (1)$$

where $\mathbf{S}$ represents the PML matrix of 3×3 and $\mathbf{S}^{-1}$ is the inverse of $\mathbf{S}$ matrix. The symbol E denotes the electric field vector, n is the refractive index of the domain, K_0 is the wave number in free space, and λ is the operating wavelength. The propagating constant β is represented by the following equation.

$$\beta = n_{eff}K_0 \qquad (2)$$

Due to the finite number of air holes in the cladding part, there may cause leakage of light. The leakage of light from core to exterior materials results in confinement loss (dB/m) which can be obtained from the imaginary part of n_{eff} by using the following equation [24].

$$L_c = 8.868 \times K_0 I_m [n_{eff}](\text{dB/m}). \qquad (3)$$

However, this leakage of light energy can be omitted by using an infinite number of air holes. But in practical, the number of air holes is finite.

The relative sensitivity coefficient measures the interaction between the light and the analyte to be sensed. This interaction is measured through the absorption coefficient at a particular wavelength. According to the Beer-Lambert law, light is attenuated by the intensity of absorption of evanescent wave [31]

$$I(\lambda) = I_0(\lambda)\exp[-r\alpha_m l_c] \qquad (4)$$

The absorbance of the sample to be detected is defined by the following equation [27]:

$$A = \lg\left(\frac{I}{I_0}\right) = r\alpha_m l_c \qquad (5)$$

where I and I_0 are the input and output intensities, respectively, and c is the concentration of absorbing material. The length of the channel is l. The function of absorption coefficient is $\alpha_m(\lambda)$ and r is the relative sensitivity coefficient, which can be defined by the following equation [22]

$$r = \frac{n_r}{n_{eff}}f \qquad (6)$$

where n_r refers to the refractive index of the sample to be sensed, and n_{eff} is the effective index of the guided mode. f is the fraction of total power located in the core, and it is also known as a power distribution function [28] by using Poynting's theorem which can be expressed as the following equation:

$$f = \frac{\int_{holes}\text{Re}(E_x H_y - E_y H_x)dxdy}{\int_{total}\text{Re}(E_x H_y - E_y H_x)dxdy} \qquad (7)$$

where E_x and H_x are transverse electric field and

magnetic field respectively; E_y and H_x are longitudinal electric field and magnetic field respectively. Using FEM the mode field pattern and effective index are obtained. During the simulation, we have considered the material dispersion of silica background using the Sellmeier equation [32].

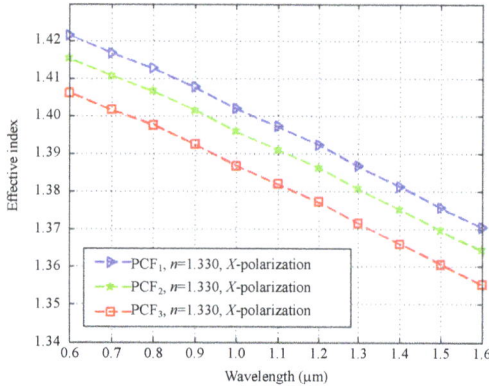

Fig. 4 Effective index curves of the fundamental mode for the X-polarization with Λ=2.4 μm, d=0.45 μm. PCF$_1$: $d_1=d_2=d_3=d_4$=1.9 μm; PCF$_2$: $d_1=d_2=d_3$=1.9 μm and d_4= 2.15 μm; PCF$_3$: $d_2=d_3$=1.9 μm and $d_1=d_4$=2.15 μm.

4. Results and discussion

This section describes the numerical analysis of propagation characteristics in fundamental mode and some higher order modes of the proposed PCFs. Three liquid analytes, water, ethanol, and benzyne, have been selected for filling the supplementary core holes. Here, it has been considered X-polarization of fundamental mode for this investigation. The initial analysis has been performed by assuming the geometric parameters of PCF$_1$: $d_1=d_2=d_3=d_4$=1.9 μm; PCF$_2$: $d_1=d_2=d_3$=1.9 μm and d_4=2.15 μm; PCF$_3$: $d_2=d_3$=1.9 μm and $d_1=d_4$=2.15 μm. The supplementary holes pitch ratio is d/a=0.70. The center-to-center air holes distance is Λ=2.4 μm at the cladding, which has been kept fixed for all of the proposed PCFs. The simulation has been performed at a wide range of wavelength from 0.6 μm to 1.6 μm. The simulation process has been down using COMSOL Multiphysics 4.2 by selecting a fine mode of mesh size. The convergence error seems very low of proposed PCFs about 3.55×10^{-5}% and 3.50×10^{-5}% for PCF$_3$ and PCF$_4$, respectively.

Initially, Fig. 4 shows the effective index profile of PCF$_1$, PCF$_2$, and PCF$_3$. It is clear from Fig. 4 that the effective indices decrease linearly with an increase in wavelength. It can be evidently seen that the PCF$_1$ shows higher effective index values among the first three proposed PCFs.

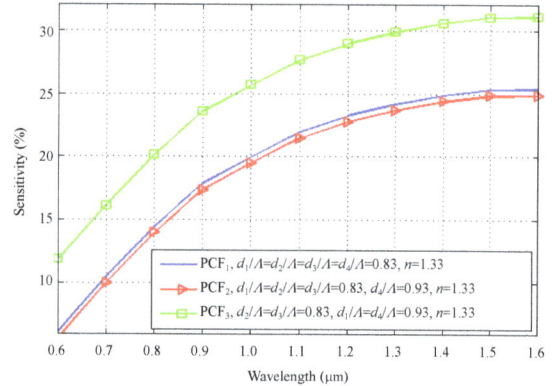

(a)

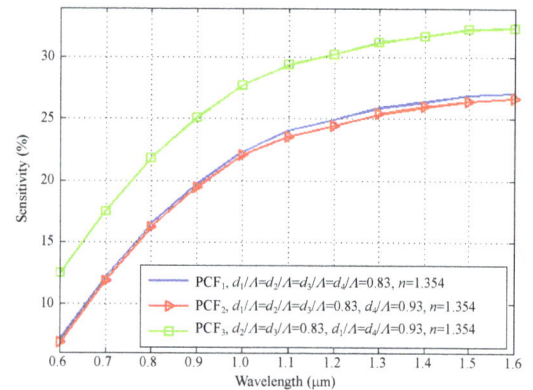

(b)

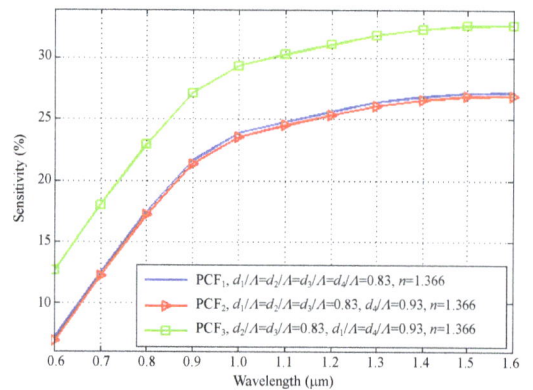

(c)

Fig. 5 Comparison of the relative sensitivity of PCF$_1$, PCF$_2$, and PCF$_3$ for (a) water, (b) ethanol, and (c) benzyne, where Λ=2.4 μm, d=0.45 μm. PCF$_1$: $d_1=d_2=d_3=d_4$=1.9 μm; PCF$_2$: $d_1=d_2=d_3$=1.9 μm, and d_4=2.15 μm; PCF$_3$: $d_2=d_3$=1.9 μm, and $d_1=d_4$=2.15 μm.

Figure 5 presents the relative sensitivity curves of PCF$_1$, PCF$_2$, and PCF$_3$ for the three analytes as a function of wavelength. There is no significant change in sensitivity for PCF$_1$ and PCF$_2$ in all wavelengths. Therefore, no significant impacts on sensitivity have been observed with increasing diameters of outer rings holes. However, the relative sensitivity of PCF$_3$ is greatly enhanced. At the wavelength λ=1.33 µm, for water, ethanol, and benzyne, the calculated sensitivity of PCF$_3$ is 30%, 32.5%, and 33.67%, respectively and the confinement loss is 3.25×10^{-10} dB/m, 2.95×10^{-10} dB/m, and 2.31×10^{-10} dB/m, respectively. The reason behind the enhanced sensitivity of PCF$_3$ is that the increment of the inner ring holes diameter leads them closer to the core area and the fraction of evanescent field penetrates to the holes increase and relative sensitivity of the PCF$_3$ increases consequently. It is also clear that higher index material shows higher relative sensitivity.

Figure 6 illustrates the relative sensitivity performance of PCF$_3$ varying the diameter (d) of the supplementary holes in the core region. According to this inquiry, the sensitivity increases with the increment of the diameter of supplementary holes. From Fig. 6, we have found the highest relative sensitivity when d=0.55 µm. For this value of the supplementary holes diameter, PCF$_3$ shows relative sensitivity 48.50% and 47.78%, and confinement loss 1.28×10^{-10} dB/m and 5.37×10^{-11} dB/m for ethanol and water, respectively, at the wavelength λ=1.33 µm.

To achieve much more relative sensitivity, we have proposed PCF$_4$ replacing a single hollow channel instead of using supplementary tiny holes. In PCF$_4$, the diameter of the hollow channel is D_2=1.70 µm. Figure 7 depicts the comparative performance of sensitivity of the last two proposed PCFs: PCF$_3$ and PCF$_4$ for all types of analytes used in this study. According to Fig. 7, PCF$_4$ shows great enhancement of relative sensitivity. At the

wavelength λ=1.33 µm, PCF$_4$ exhibits the relative sensitivity 50%, 55.83%, and 59.07%, confinement loss 4.25×10^{-10} dB/m, 8.72×10^{-10} dB/m, and 2.56×10^{-10} dB/m for water, ethanol, and benzyne, respectively.

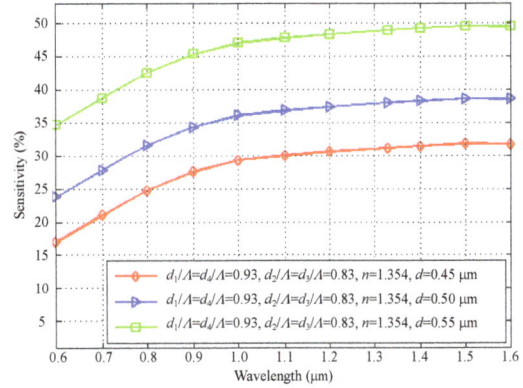

(a)

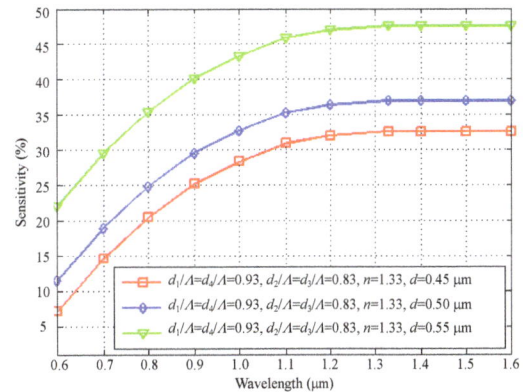

(b)

Fig. 6 Comparison of relative sensitivity of PCF$_3$ as a function of operating wavelength for (a) ethanol (b) water; where d=0.45 µm, d=0.50 µm, d=0.55 µm, and rest of the parameters are fixed as before.

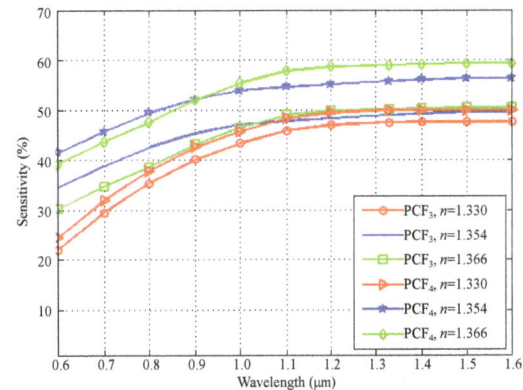

Fig. 7 Relative sensitivity versus wavelength for PCF$_3$ and PCF$_4$ with Λ=2.4 µm, d=0.55µm, D_2=1.70 µm, d_2=d_3=1.9 µm, and d_1=d_4=2.15 µm.

Figure 8 presents the confinement loss curves of PCF$_3$ and PCF$_4$. With the investigation of Fig. 8, it can be seen that PCF$_4$ exhibits better performance in terms of confinement loss for all types of analytes used in this study. Therefore, it can be said that the light mode is more confined in the core region for the proposed PCF$_4$ compared with the first three proposed PCF structures. This may be linked to the fact that the electromagnetic interaction between the propagated light and analyte is higher which causes an increase in relative sensitivity. In addition, from Fig. 8, it can be found that lower confinement losses are achieved with higher indexed liquids. According to the overall discussion, PCF$_4$ shows higher sensitivity and lower confinement loss than PCF$_3$.

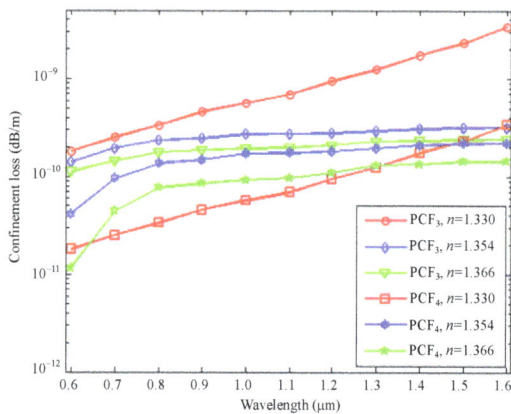

Fig. 8 Confinement loss versus wavelength for PCF$_3$ and PCF$_4$ with Λ=2.4 µm, d=0.55 µm, D_2=1.70 µm, d_2=d_3=1.9 µm, and d_1= d_4=2.15 µm.

In order to support the numerical results reported in the figures, the electric field distribution of the proposed PCF$_3$ and PCF$_4$ has been illustrated in Fig. 9, where the operating wavelength is set to λ=1.33 µm and the core holes are filled with ethanol. It can be clearly seen that the fundamental mode of our optimized PCFs (PCF$_3$ and PCF$_4$) is tightly confined in the core region.

From the overall discussion, we have found that our designed structures show better performance in relative sensitivity with more design simplicity than the prior structures [24, 25] for liquid sensing. Table 1 shows the comparative performance analysis

between prior PCFs and proposed PCFs for liquid sensing at the wavelength λ=1.33 µm. Tables 2 and 3 represent the sensitivity dependency on diameters and global parameters variations, respectively. Although, diameters are varied after fabrication but it has no global effects on result of the proposed structure.

Table 1 Comparison between proposed PCFs and prior PCFs for liquid sensing applications at λ=1.33 µm.

PCFs	Sensitivity (%)		Structural shape description
	Ethanol (n=1.354)	Benzyne (n=1.366)	
Ref. [24]	21.55	22.50	Octagonal shape: 3 rings
Ref. [25]	46.87	47.35	Octagonal shape: 5 rings
Proposed PCF$_3$	48.50	50.08	Hexagonal shape: 4 rings
Proposed PCF$_4$	55.83	59.07	Hexagonal shape: 4 rings

Table 2 Comparison of sensitivity for optimum design parameters and also for fiber's global diameter variations of order ±1%–±3% around the optimum value at λ=1.33 µm.

Only diameter variations (%)	Relative sensitivity (%)	
	PCF$_3$	PCF$_4$
−3	44.05	48.46
−1	46.98	53.29
Optimum	48.50	55.83
+1	50.09	57.84
+3	53.31	61.13

Through the experimental point of view, the fabrication feasibility of the proposed PCFs is an important part. It seems that the fabrication process of the micro cored region may not be easy. However, due to the technological advancement, the fabrication of our recommended PCFs is possible. Micro core must be filled with the analyte without damaging the fiber's integrity. Now, several techniques are available for filling the PCF holes with analytes. Huang *et al.* [33] proposed a unique method for selectively filling the all cladding holes as well as micro core holes. The fabrication of PCF with liquid filled core or cladding can be accomplished with the same method [34, 35]. Now, applying the sol-gel technique [36] any kind of complexity of fabrication of microstructure optical fiber can be removed. In this regard, our proposed

PCF structures can be fabricated with the currently available nanotechnology. Selective filling technique [37] can be used for fill the analytes (gas or liquids) at the core.

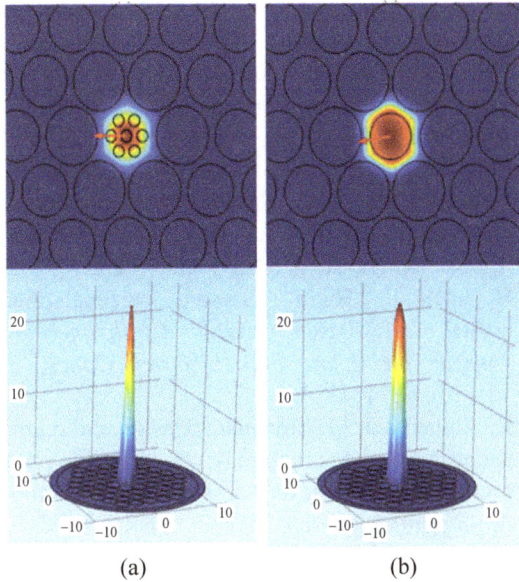

(a)   (b)

Fig. 9 2d and 3d views of modal Intensity distribution of (a) PCF_3 and (b) PCF_4 for X-polarized mode at the wavelength λ=1.33 μm where $n_{\text{eff-x}}$=1.3753 for PCF_3 and $n_{\text{eff-x}}$=1.3455 for PCF_4.

Table 3 Comparison of sensitivity for optimum design parameters and also for fiber's global parameters variations of order ±1%–±3% around the optimum value at λ=1.33 μm.

All parameters variation (%)	Relative sensitivity (%)	
	PCF3	PCF4
−3	48.72	56.46
−1	48.58	56.03
Optimum	48.50	55.83
+1	48.41	55.53
+3	48.23	55.03

5. Conclusions

In this study, the enhancement of the performance of the PCF based liquid sensor has been done by our recommended two structures of PCF, which are based on microstructure core and hollow core, and infiltrated with the liquid to be sensed. All of the proposed structures with microstructure core and liquid core have better

guiding capability and the manufacturing of this type of structure is possible with the current nanofabrication techniques [34−38]. Our proposed PCF provided higher relative sensitivity with tighter confinement of optical field than the prior PCF structures. Therefore, our proposed PCFs can successfully overcome the critical trade-off between confinement loss and sensitivity, and it is assumed that our proposed structures of PCF offer great potentiality for toxic chemical and gas detection in industrial safety purposes.

References

[1] M. Arjmand and R. Talebzadeh, "Optical filter based on photonic crystal resonant cavity," *Optoelectronics and Advanced Materials-Rapid Communications*, 2015, 9(1−2): 32−35.

[2] K. Fasihi, "High-contrast all-optical controllable switching and routing in nonlinear photonic crystals," *Journal of Lightwave Technology*, 2014, 32(18): 3126−3131.

[3] K. Cui, Q. Zhao, X. Feng, Y. Huang, Y. Li, D. Wang, et al., "Thermo-optic switch based on transmission-dip shifting in a double-slot photonic crystal waveguide," *Applied Physics Letters*, 2012, 100(20): 201102-1−201102-4.

[4] J. M. Brosi, C. Koos, L. C. Andreani, M. Waldow, J. Leuthold, and W. Freude, "High-speed low-voltage electro-optic modulator with a polymer-infiltrated silicon photonic crystal waveguide," *Optics Express*, 2008, 16(6): 4177−4191.

[5] Y. Gao, R. J. Shiue, X. Gan, L. Li, C. Peng, I. Meric, et al., "High-speed electro-optic modulator integrated with graphene-boron nitride heterostructure and photonic crystal nanocavity," *Nano Letters*, 2015, 15(3): 2001−2005.

[6] H. Xuan, J. Ma, W. Jin, and W. Jin, "Polarization converters in highly birefringent microfibers," *Optics Express*, 2014, 22(3): 3648−3660.

[7] Y. H. Chang, Y. Y. Jhu, and C. J. Wu, "Temperature dependence of defect mode in a defective photonic crystal," *Optics Communications*, 2012, 285(6): 1501−1504.

[8] Y. Liu and H. W. M. Salemink, "All-optical on-chip sensor for high refractive index sensing in photonic crystals," *Europhysics Letters*, 2014, 107(3):

1160−1170.

[9] S. Zheng, Y. Zhu, and S. Krishnaswamy, "Nanofilm-coated photonic crystal fiber long-period gratings with modal transition for high chemical sensitivity and selectivity," *Proc. SPIE*, 2012, 8346(14): 1844−1864.

[10] C. Lee and J. Thillaigovindan, "Optical nanomechanical sensor using a silicon photonic crystal cantilever embedded with a nanocavity resonator," *Applied Optics*, 2009, 48(10): 1797−1803

[11] S. Olyaee and A. A. Dehghani, "Ultrasensitive pressure sensor based on point defect resonant cavity in photonic crystal," *Sensor Letters*, 2013, 11(10): 1854−1859.

[12] Y. N. Zhang, Y. Zhao, and Q. Wang, "Multi-component gas sensing based on slotted photonic crystal waveguide with liquid infiltration," *Sensors and Actuators B: Chemical*, 2013, 184(8): 179−188.

[13] M. Morshed, M. F. H. Arif, S. Asaduzzaman, and K. Ahmed, "Design and characterization of photonic crystal fiber for sensing applications," *European Scientific Journal*, 2015, 11(12): 228−235.

[14] T. W. Lu and P. T. Lee, "Ultra-high sensitivity optical stress sensor based on double-layered photonic crystal microcavity," *Optics Express*, 2009, 17(3): 1518−1526.

[15] P. Hu, X. Dong, W. C. Wong, L. H. Chen, K. Ni, and C. C. Chan, "Photonic crystal fiber interferometric pH sensor based on polyvinyl alcohol/polyacrylic acid hydrogel coating," *Applied Optics*, 2015, 54(10): 2647−2652.

[16] W. C. Lai, S. Chakravarty, Y. Zou, and R. T. Chen, "Multiplexed detection of xylene and trichloroethylene in water by photonic crystal absorption spectroscopy," *Optics Letters*, 2013, 38(19): 3799−3802.

[17] E. K. Akowuah, T. Gorman, H. Ademgil, S. Haxha, G. K. Robinson, and J. V. Oliver, "Numerical analysis of a photonic crystal fiber for biosensing applications," *IEEE Journal of Quantum Electronics*, 2012, 48(11): 1403−1410.

[18] M. B. Pushkarsky, M. E. Webber, O. Baghdassarian, L. R. Narasimhan, and C. K. N. Patel, "Laser-based photoacoustic ammonia sensors for industrial applications," *Applied Physics B*, 2002, 75(2-3): 391-396.

[19] G. Whitenett, G. Stewart, K. Atherton, B. Culshaw, and W. Johnstone, "Optical fibre instrumentation for environmental monitoring applications," *Journal of Optics A: Pure and Applied Optics*, 2003, 5(5): S140−S145.

[20] J. P. Carvalho, H. Lehmann, H. Bartelt, F. Magalhaes, R. Amezcua-Correa, J. L. Santos, *et al.*, "Remote system for detection of low-levels of methane based on photonic crystal fibres and wavelength modulation spectroscopy," *Journal of Sensors*, 2009, 2009(2): 1−10.

[21] J. Park, S. Lee, S. Kim, and K. Oh, "Enhancement of chemical sensing capability in a photonic crystal fiber with a hollow high index ring defect at the center," *Optics Express*, 2011, 19(3): 1921−1929.

[22] C. M. Cordeiro, M. A. Franco, G. Chesini, E. C. Barretto, R. Lwin, C. B. Cruz, *et al.*, "Microstructured-core optical fibre for evanescent sensing applications," *Optics Express*, 2006, 14(26): 13056−13066.

[23] M. Morshed, H. M. Imarn, T. K. Roy, M. S. Uddinand, and S. A. Razzak, "Microstructure core photonic crystal fiber for gas sensing applications," *Applied Optics*, 2015, 54(29): 8637−8643.

[24] H. Ademgil, "Highly sensitive octagonal photonic crystal fiber based sensor," *Optik−International Journal for Light and Electron Optics*, 2014, 125(20): 6274−6278.

[25] K. Ahmed and M. Morshed, "Design and numerical analysis of microstructured-core octagonal photonic crystal fiber for sensing applications," *Sensing and Bio-Sensing Research*, 2016, 7: 1−6.

[26] S. Asaduzzaman, K. Ahmed, M. F. H. Arif, and M. Morshed, "Proposal of simple structure photonic crystal fiber for lower indexed chemical sensing," in *18th International Conference on Computer and Information Technology*, MIST, Bangladesh, 2015.

[27] S. Asaduzzaman, K. Ahmed, and M. F. H. Arif, "Numerical analysis of O-PCF structure for sensing applications with high relative sensitivity," in *2nd International Conference on Electrical Information and Communication Technology*, KUET, Bangladesh, 2015.

[28] S. Asaduzzaman, K. Ahmed, M., M. F. H. Arif, and M. Morshed, "Application of microarray-core based modified photonic crystal fiber in chemical sensing," in *International Conference on Electrical and Electronic Engineering*, RUET, Bangladesh, 2015.

[29] M. Morshed, M. I. Hasan, and S. M. A. Razzak, "Enhancement of the sensitivity of gas sensor based on microstructure optical fiber," *Photonic Sensors*, 2015, 5(4): 312−320.

[30] S. Selleri, L. Vincetti, A. Cucinotta, and M. Zoboli, "Complex FEM modal solver of optical waveguides with PML boundary conditions," *Optical and Quantum Electronics*, 2001, *33*(4−5): 359−371.

[31] B. Q. Wu, Y. Lu, C. J. Hao, L. C. Duan, N. N. Luan, Z. Q. Zhao, *et al.*, "December. hollow-core photonic crystal fiber based on C2H2 and NH3 gas sensor," *Applied Mechanics and Materials*, 2013, 411: 1577−1580.

[32] G. Ghosh, "Sellmeier coefficients and dispersion of

thermo-optic coefficients for some optical glasses," *Applied Optics*, 1997, 36(7): 1540‒1546.

[33] Y. Huang, Y. Xu, and A. Yariv, "Fabrication of functional microstructured optical fibers through a selective-filling technique," *Applied Physics Letters*, 2004, 85(22): 5182‒5184.

[34] M. Luo, Y. G. Liu, Z. Wang, T. Han, Z. Wu, J. Guo, *et al.*, "Twin-resonance-coupling and high sensitivity sensing characteristics of a selectively fluid-filled microstructured optical fiber," *Optics Express*, 2013, 21(25): 30911‒30917.

[35] R. M. Gerosa, D. H. Spadoti, C. J. de Matos, L. D. S. Menezes, and M. A. Franco, "Efficient and short-range light coupling to index-matched liquid-filled

hole in a solid-core photonic crystal fiber," *Optics Express*, 2011, 19(24): 24687‒24698.

[36] R. T. Bise and D. J. Trevor, "Sol-gel derived microstructured fiber: fabrication and characterization," in *Optical Fiber Communications Conference*, Anaheim, U.S.A., 2005.

[37] Y. Huang, Y. Xu, and A. Yariv, "Fabrication of functional microstructured optical fibers through a selective-filling technique," *Applied Physics Letters*, 2004, 85(22): 5182‒5184.

[38] F. M. Cox, A. Argyros, and M. C. J. Large, "Liquid-filled hollow core microstructured polymer optical fiber," *Optics Express*, 2006, 14(9): 4135‒4140.

10

A Portable Analog Lock-In Amplifier for Accurate Phase Measurement and Application in High-Precision Optical Oxygen Concentration Detection

Xi CHEN[1], Jun CHANG[1*], Fupeng WANG[1], Zongliang WANG[2], Wei WEI[1], Yuanyuan LIU[1], and Zengguang QIN[1]

[1]*School of Information Science and Engineering and Shandong Provincial Key Laboratory of Laser Technology and Application, Shandong University, Jinan, 250100, China*
[2]*School of Physics Science and Information Technology and Shandong Key Laboratory of Optical Communication Science and Technology, Liaocheng University, Liaocheng, 252059, China*

*Corresponding author: Jun CHANG   E-mail: changjun@sdu.edu.cn

Abstract: A portable analog lock-in amplifier capable of accurate phase detection is proposed in this paper. The proposed lock-in amplifier, which uses the dual-channel orthometric signals as the references to build the *xy* coordinate system, can detect the relative phase between the input and *x*-axis based on trigonometric function. The sensitivity of the phase measurement reaches 0.014 degree, and a detection precision of 0.1 degree is achieved. At the same time, the performance of the lock-in amplifier is verified in the high precision optical oxygen concentration detection. Experimental results reveal that the portable analog lock-in amplifier is accurate for phase detection applications. In the oxygen sensing experiments, 0.058% oxygen concentration resulted in 0.1 degree phase shift detected by the lock-in amplifier precisely. In addition, the lock-in amplifier is small and economical compared with the commercial lock-in equipments, so it can be easily integrated in many portable devices for industrial applications.

Keywords: Portable analog lock-in amplifier; phase shift measurement; high accuracy; optical oxygen detection

1. Introduction

Nowadays, the accurate detection of phase shift is of extreme importance in many different research fields such as medicine, chemistry, environment and marine analysis, molecular biotechnology, bioprocess control, and industrial production monitoring. But high noise level environments make the right acquisition of accurate phase measurements difficult, especially when the output signal is very small compared to the noise level [1]. In these cases, the ordinary low pass filter is not an option, and some special methods for extracting the signal information should be considered. In this direction, an interesting possibility is the lock-in amplifier, which uses the phase sensitive detection (PSD) technique to take out the data signal at a specific reference frequency and reject noise signals at different frequencies without affecting the measurement significantly [2, 3].

However, the commercial lock-in system proposed in the literature, which are expensive, heavy, and power-hungry, is typically appropriate for multi-frequency operation and not considered suitable used in the portable sensing system that utilizes single-supply battery cells. Therefore, the dedicated lock-in amplifier for phase measurement has been developed for measurement problems. On the other hand, another advantage of a dedicated lock-in amplifier is that generally the circuit can be much smaller than a complete instrument and can be placed much closer to the sensor. This can dramatically decrease parasitic effects, noise, and disturbance levels in the system [4–6].

To the best of authors' knowledge, there are no cheap portable commercial lock-in amplifiers [7]. To acquire the phase information precisely, it demands that the PSD block operates extremely accurately and stably, because a little deviation of this block will lead to invalid results of the whole system [8]. With respect to a previously published solution, a circuit that permits a single-supply portable analog lock-in amplifier properly processes phase delay information buried in high noise levels, showing recovery equations and the detail for phase measurement [9, 10]. Furthermore, the exhaustive experimental verification carried out here improves the previously presented results. The conducted experimental measurements for phase, performed through the AD630 from analog devices, and the fabricated printed circuit board (PCB), have confirmed the correct functionality of the designed amplifier, as well as the system capability to reveal very small signal coming from resistive sensors with satisfactory performances in terms of both sensitivity and resolution improvements. Furthermore, the exhaustive experimental verification of the lock-in amplifier is verified in the high precision optical oxygen concentration detection.

2. Basic principle of phase measurement with a lock-in amplifier

Figure 1 shows the basic principle of phase measurement using lock-in techniques. The theory of the phase measurement with a lock-in amplifier can be explained by the mathematical model of tangent in the rectangular coordinate system. As depicted in Fig. 1, the system oscillator generates an original signal, with the frequency ω and amplitude A_s. After phase delay block, the input signal is generated with the phase θ, but the frequency and amplitude are not changed. Reference channel path leads the reference signal to be multiplied with the signal. Theoretically, the duty ratio has a great influence on the phase measurement, and it is decided to use the square wave signal V_r as the reference signal, because the duty ratio of square wave is more stable than the sinusoidal wave. As depicted in Fig. 1, the relative reference signals can be easily synchronized with the input signal because they are all produced by the system oscillator. The chopper circuit changes the original wave to the square wave synchronously to provide the reference signal for the lock-in amplifier. As the phase measurement system adopts the dual-channel orthometric reference architecture, two orthometric square waves are prepared through part of 90 degree move-phase. In the practical system, it is more secure to keep phase stable between the input and reference signals, because the functions of the oscillator, chopper circuit, and 90° phase shift are all achieved by the microcontroller. In this way, the next block, a multiplier, where it is multiplied by the reference signal, generates a periodic signal, whose direct current (DC) component is proportional to the amplitude of alternating current (AC) component input signal, and depends on the mentioned phase difference.

The signal generated by the multiplier may be easily extracted by means of a suitable low-pass filter, which represents the final block of the complete system, gives a DC output proportional to the input signal amplitude, and cancels noise coupled at frequencies different with a different ω.

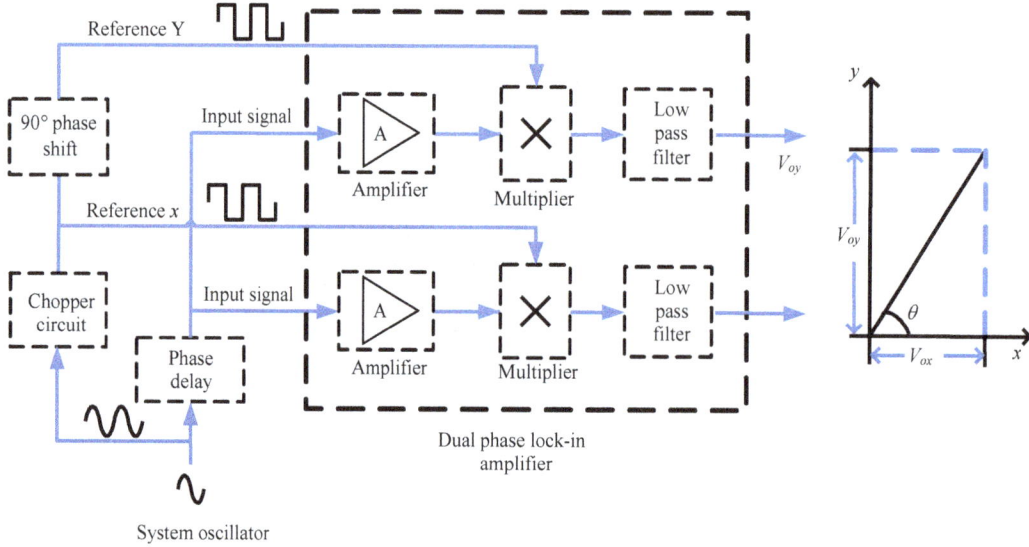

Fig. 1 Complete block scheme of the proposed lock-in amplifier architecture.

According to the preceding part of the text, the input signal is

$$V_s = A_s \sin(\omega t + \theta). \tag{1}$$

The reference signal (square wave) in the x axis is

$$V_{\text{refx}} = \frac{4}{\pi} A_r \sum_{n=1}^{\infty} \frac{(-1)^{n+1}}{2n-1} \cos\left[(2n-1)\omega t\right] \tag{2}$$

where A_r is the amplitude of the reference signal. Therefore, applied to the circuit of Fig. 1, it can be seen that the frequencies are equal, and using an ideal amplifier with the amplification n, the signal the output of the multiplier becomes

$$V_p = V_s \times V_{\text{refx}}$$

$$= A_s \sin(\omega t + \theta) \times \frac{4}{\pi} A_r \sum_{n=1}^{\infty} \frac{(-1)^{n+1}}{2n-1} \cos\left[(2n-1)\omega t\right]$$

$$= \frac{2A_s A_r}{\pi} \sum_{n=1}^{\infty} \frac{(-1)^{n+1}}{2n-1} \cos\left[(2n-2)\omega t - \theta\right] +$$

$$\frac{2A_s A_r}{\pi} \sum_{n=1}^{\infty} \frac{(-1)^{n+1}}{2n-1} \cos\left[2n\omega t + \theta\right]. \tag{3}$$

Then, after the low pass filter, whose cut-off frequency has to be very small, the sum-frequency component is removed resulting in a DC output:

$$V_{ox} = A_1 \cdot \frac{2A_s A_r}{\pi} \cos\theta \tag{4}$$

where A_1 is the magnification of the circuit. The same principle also applies to the y axis. After the low pass filter, the DC output in the y axis is

$$V_{oy} = A_1 \cdot \frac{2A_s A_r}{\pi} \sin\theta. \tag{5}$$

Thus, although the phase shift θ changes, it is possible to recover the phase shift according to (6):

$$\theta = \arctan\left(\frac{V_{oy}}{V_{ox}}\right). \tag{6}$$

3. Design of lock-in amplifier for phase measurement

The proposed lock-in amplifier for phase measurement is composed of two phase detection channels. The orthometric square waves are the reference signals separately. For one channel, it divides two parts, modulation and demodulation. An analog device, AD630, as the balanced modulator, is the most important part of the lock-in amplifier, which combines a flexible commutating architecture with the accuracy and temperature stability afforded by laser wafer trimmed thin-film resistors. It is a complete high precision function with the preamplifier, multiplier, and chopper circuit. Theoretically, if the duty ratio of the reference signals is not 50%, the output DC signal will be mixed with extra DC component caused by the unbalanced duty ratio. In the whole lock-in amplifier block, there is a calibration circuit, which has wide bandwidth, making sure the output signals are screened from the incorrect duty ratio. In particular, Fig. 2 depicts the measured signals from the output of AD630, when a clean 16 kHz sine wave is the input signal, and the reference square signals are "in-quadrature" and "in-phase", respectively. The figure is collected by the oscilloscope (Tektronix TBS 1102B-EDU). Through a suitable low-pass filter, the modulator output periodic wave is filtered so as to extract a DC voltage signal that gives information about AC input signal amplitude. In particular, the output of "quadrature" mixer: (a) is symmetric with respect to the zero, while the output of "in-phase" mixer; (b) is above zero, and a non-zero DC signal can be detected corresponding to the "in-phase" mixer inputs.

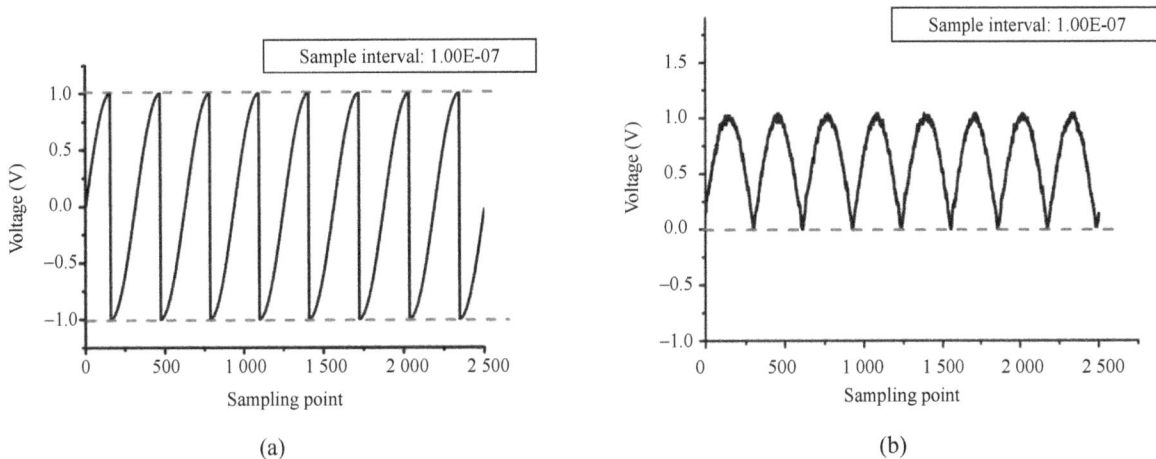

Fig. 2 Measurement results at the balance modulator output for: (a) "in-quadrature" and (b) "in-phase" inputs.

The demodulator is divided into two parts: the integrator and the low-pass filter. Theoretically, if the frequency of the input signal is much larger than the central frequency of the integrator, the amplitude of the output signal from the integrator will decrease to close to zero. Conversely, the amplitude with lower frequency will be enlarged. So the integrator with low central frequency in this system is regarded as a low-pass filter, which reduces the influence of sum-frequency and keeps the phase DC component. Here, the central frequency of the integrator is set as 32 Hz. A 2nd order filter also has been implemented after the integrator. The filter with 8 Hz cut-off frequency removes the high frequency component remaining from the integrator and acquires the purer phase DC information.

4. Laboratory experiments and discussion

4.1 Lock-in amplifier performance test

Referring to the system block scheme above, which presents two AC inputs and one DC output in one channel lock-in amplifier, the measured value (U_o) is proportional to the input signal amplitude (U_i) only when the calibration output is zero, as previously described, the same as the other lock-in amplifier channel.

The electrical characterization of the designed lock-in amplifier has been performed on the fabricated PCB. In order to both perform the designed system analysis and achieve preliminary experimental result, the fabricated PCB has been tested, utilizing high accuracy laboratory instrumentation.

According to (1) and (4), the V_{ox} can be expressed as

$$V_{ox} = \sqrt{(\frac{2A_1 A_s A_r}{\pi})^2 - V_{oy}^2}$$

$$\text{if}: \quad A = \frac{2A_1 A_s A_r}{\pi}; \quad X = V_{ox}; \quad Y = V_{oy} \quad (7)$$

$$\therefore X = \sqrt{A^2 - Y^2}$$

where A is the constant for the certain input and reference signal. Thus, the phase shift becomes as

$$\theta = \arctan \frac{Y}{\sqrt{A^2 - Y^2}}. \quad (8)$$

Then the derivation of (8) is given by

$$\theta' = \left(\arctan \frac{Y}{\sqrt{A^2 - Y^2}} \right)' = \frac{1}{\sqrt{A^2 - Y^2}}. \quad (9)$$

Based on (9), the θ' increases with the growth of the factor Y in the range of factor Y bigger than 0. So the change rate of phase θ is least when the Y is close to zero. Therefore, the minimum sensitivity of the phase measurement system can be caught at the tiny degree. As we all know, the tiny phase measurement function can be changed approximately as

$$\tan \theta \approx \sin \theta = \frac{Y}{A}. \quad (10)$$

The ARM7 (LPC1758) used here, as the core component of signal acquisition, contains a signal 12-bit successive approximation analog-to-digital converter (ADC) and has a working voltage range of 0 V – 3.0 V and resolution of 0.73 mV. For this condition, the sensitivity of lock-in amplifier for phase measurement is 0.014 degree [10].

Furthermore, the designed analog lock-in system has been proved to recover with success also very small noisy signals without performance degradation. More in details, measurement results are reported in Fig. 3, showing the amplitude measurement and the phase measurement for the proposed lock-in amplifier. The figures of Part (1) express the linearity between the measured output DC voltage and different amplitudes of the applied input AC signal, even pure and noisy. Considering the limit voltage of AD in the signal acquisition block, the system sensitivity has been set to about 50 [V/V] here. As shown in the figure, the correlation coefficients of the three groups are all equal to 0.999, extremely close to 1, indicating the output DC presents great linear increasing property with the growth of the amplitudes. The relative error of the worst measurements is 0.4%, even the SNRs differ greatly. So no matter how the SNRs deteriorate, the system has shown the capability to reveal accurately with success an input AC voltage signal with different SNR levels.

To test the precision of the lock-in amplifier in phase measurement, some experiments have been implemented. An alignment of phase shift has been set factitiously between the input signal and the reference signal. As depicted in Figs. 3(a) – 3(c), the figures of Part (2) show the relationship between the system measurement and the set phase value, both pure signal and noisy signal, as setting the left part. The measurement also has very good linearity, and the slope of each function is extremely close to 1. The measuring error is below 0.1 degree. So the precision of the lock-in amplifier is at least 0.1 degree. In conclusion, the measurement of this system can reflect the real phase under test, even the quality of the input signal is not satisfying.

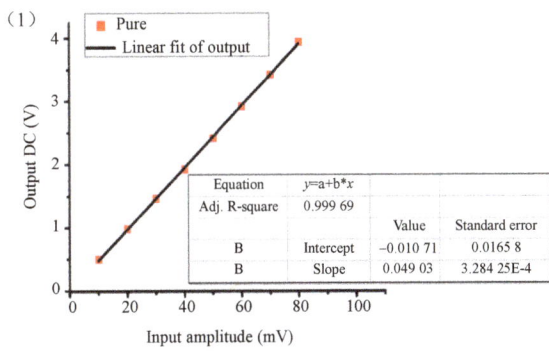

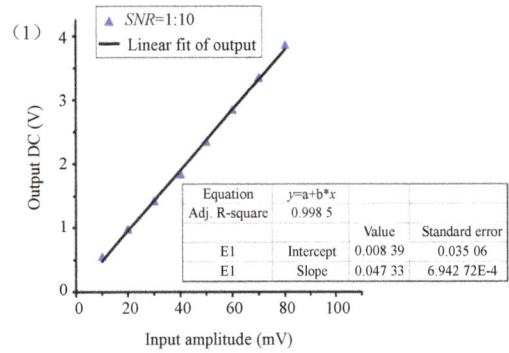

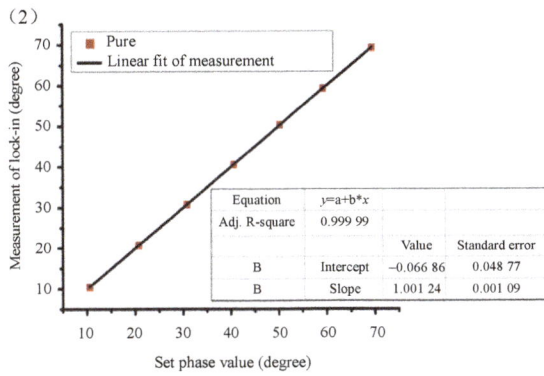

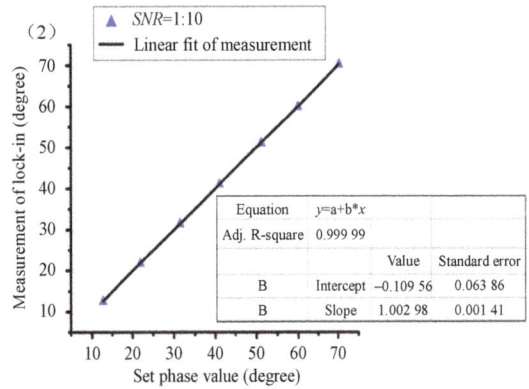

(a)

(b)

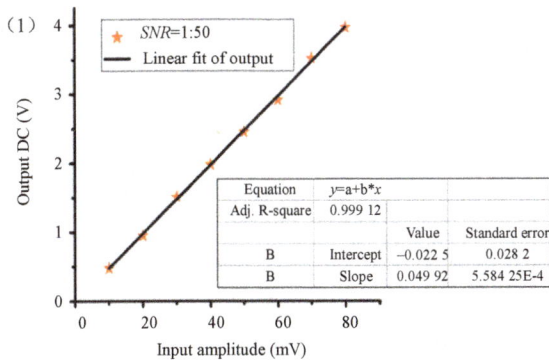

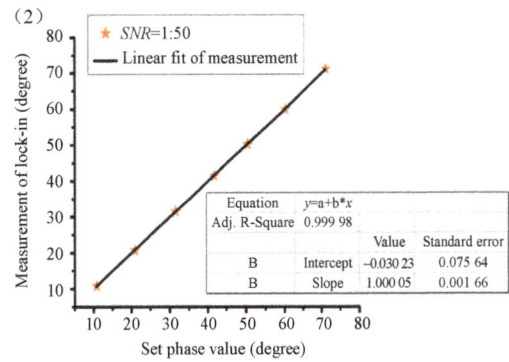

(c)

Fig. 3 Experimental measurements (dots) and the linearity (continuous line): (a, (1)) measurements for the applied input pure signal with different amplitudes, (a, (2)) measurements for the applied input pure signal with different phase measurements, (b) measurements for the noisy signal (SNR=1:10), and (c) measurements for the noisy signal (SNR=1:50).

4.2 Application in the high precision optical oxygen concentration detection system

The purpose of designing the lock-in amplifier for phase measurement is some practical applications. Oxygen sensing is of major importance in many fields, and phase shift measurement is always the crucial step in the fluorescence phase based oxygen sensor. The dynamic quenching of the luminescence of ruthenium complexes is a popular method for optical oxygen detection. Lifetime of luminescent complex, the key factor for concentration measurements, is closely connection with the phase shift measurements. Luminescence sensing requires a change in the spectral properties of the indicator in the presence of oxygen. Changes can occur in the form of luminescence intensity or lifetime of the luminescent complex [11–15]. The oxygen concentration is related to these parameters by the Stern-Volmer (SV) equation:

$$\frac{I_0}{I} = \frac{\tau_0}{\tau} = 1 + K_{SV}[O_2] \qquad (11)$$

where I and I_0 are the luminescence intensities in the presence and absence of the quencher, τ and τ_0 are the lifetimes of the luminophore in the presence and absence of the quencher, K_{SV} is the Stern-Volmer quenching constant, and $[O_2]$ is the concentration of the oxygen. Applying a sinusoidal modulation to the optical source results in a phase delay, φ, in the fluorescent emission that can be related to the lifetime by the equation:

$$\tan(\varphi) = 2\pi f \tau \qquad (12)$$

where f is the modulation frequency, which can be tuned to the suitable optimum sensor sensitivity that the luminescence lifetime can be detected accurately. Using these equations above, the phase delay information can easily be related with the oxygen concentration.

The fabricated system has been tested with a suitable experimental apparatus to detect the presence of O_2 into a closed chamber. Figure 4 shows the prototype PCB photo, utilized for the system testing (the designed lock-in amplifier delimited by the solid line). In the experimental measurements, we have utilized the commercial oxygen sensing film as the oxygen sensor and the

light-emitting diode (LED) as the optical source. The experimental setup scheme is reported in Fig. 5, where the system is allowed to properly select the target gas and detect its concentration. The microcontroller (LPC1758) generates the sine wave signal to excite the LED, and the optical fiber carries blue light from LED to the oxygen sensing film to emit fluorescent light that travels back to a photo detector. Then the fluorescence signal produced on the film is send to lock-in amplifier. The voltage signal demodulated by the lock-in amplifier has been acquired through the signal acquisition block based on the microcontroller (LPC1758) and calculated to be the phase shift signal related to the oxygen concentration by a computer. The achieved experimental results have demonstrated the correct functionality of the system and satisfactory performance in terms of resolution enhancements.

Fig. 4 Photo of fabricated oxygen detection system: the designed lock-in amplifier is under the picture, delimited by the solid line.

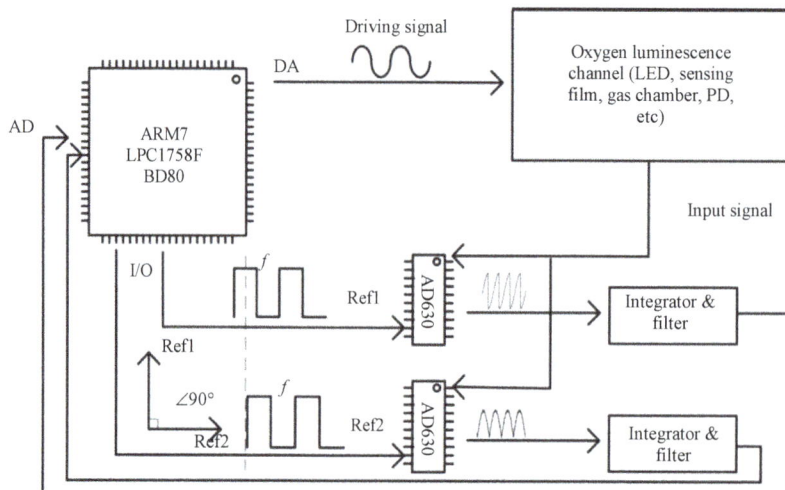

Fig. 5 Sketch of the experimental setup utilized for O_2.

By means of a straightforward analysis of the experimental result, it is possible to evaluate the proposed lock-in amplifier performance in terms of the resolution improvement. All experiments were performed at a temperature of $(24 \pm 1)\,℃$ every 10 min and humidity of 23%. The 0.99%, 3%, 5%, and 8% oxygen concentrations were tested separately. At the beginning of the test, we injected the dry pure nitrogen into a closed chamber for 20 minutes. And then, a kind of oxygen was fluxed into the chamber. After about 10 minutes, the chamber was cleaned again by pure nitrogen. Then another kind of oxygen could be tested. For facilitating the measurements, the system was set an offset phase at the reference coordinate. Figure 6(a) reveals the average phase measurements for different concentrations in 10 minutes, and the fitting function also has been calculated. As depicted in the figures, the phase shift grows with an increase in the concentration. When the concentration changed 3%, the minimum phase shift was 1.62 degree. So at the low concentration, the phase shift caused by the different concentration is available to separate the slight variant of oxygen concentration.

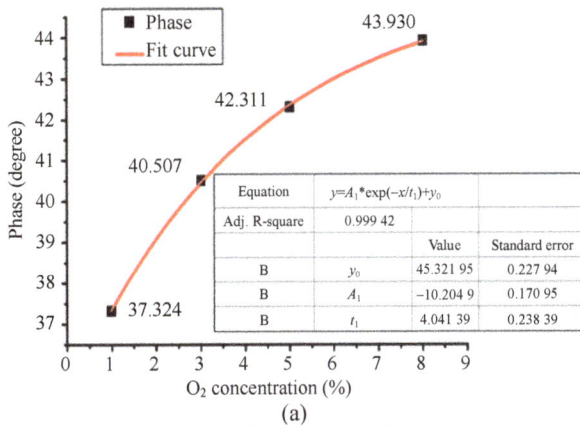

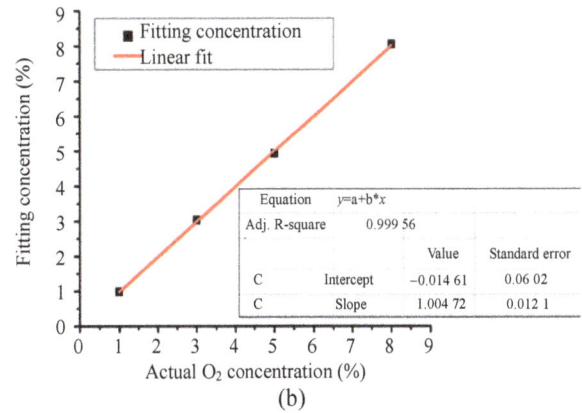

Fig. 6 Measurements linearity of different concentrations: (a) the phase measurements for different concentrations and (b) the fitting concentrations measured by the fit curve.

The solid line in Fig. 6(a) is the fit curve. The phase measurement shows an exponential function with the O_2 concentration. At 0.99% – 8% concentration, the phase response produced by the oxygen sensing film is more sensitive at the lower concentration. At this range of concentration, the sensitivity of the system is greater than 0.58% O_2 concentration for 1 degree. Then, we fluxed dry pure nitrogen into the chamber for about 30 minute to clean the chamber thoroughly. And another round test started. The four kinds of the standard gas with different concentrations were injected orderly, 10 minutes for everyone. Then the new group of phase measurements for each concentration was acquired by the system. Afterwards, these data were plugged into the fit equation in Fig. 6(a), so the fit concentrations have been calculated. Figure 6(b) depicts the relationship between the fit concentration and the actual O_2 concentration. As the figure depicts, the correlation coefficient is equal to 0.999, and the slope is 1.00, which means the fit measurements have perfect linearity and 1:1 corresponding relationship with the actual concentration, so the measurements of the system are credible and have good repeatability.

In Fig. 7, the measurement of 3% concentration was tested for a period of time. According to the instability of the gas flow, the phase measurements have a certain extent drift during the measuring time. In Fig. 7, the phase ranges from 40.495 degree to 40.528 degree. The dash line demonstrates the fluctuation range for the period of time. Then the system has 0.0263% O_2 concentration drift calculated by the fit equation shown in Fig. 6(a).

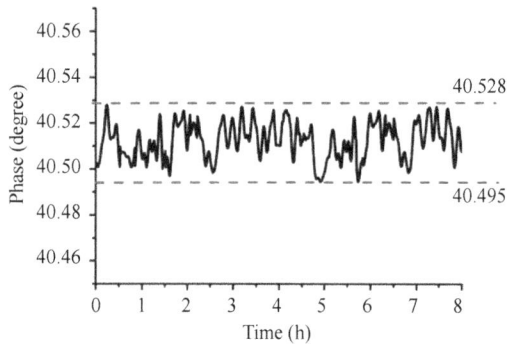

Fig. 7 Measurement of 3% concentration for a period of time.

5. Conclusions

In this work, a portable low-cost low-power portable analog lock-in amplifier suitable for sensor interface application was fabricated, providing continuous and accurate measurement of the noisy input signal. It can recover the signal buried in the noise and can measure phase shift well and truly. The sensitivity of the phase measured system reached 0.014 degree, and the precision was at least 0.1 degree. The main advantages of the proposed lock-in amplifier concern both its utilization in portable instrumentations, due to its small size characteristics, and its capability to perform accurately phase measurement between the input and reference signals. The designed system allows enhancing the *SNR*, so reducing the minimal resolution of the sensor front-end and extracting the phase information effectively and allowing the precision detection of target gas concentration in the application of the high precision optical oxygen concentration detection system. Besides, this system also can be utilized to another measurement system, such as the precision distance measurement, which the measurement can be indicated by the phase shift. The final goal of this project is to complete the integrator and the low-pass filter to get purer DC signal through lock-in amplifier and achieve higher *SNR* of the whole system. Furthermore, we need to continue to estimate the stability and repeatability of the lock-in block to realize the goal that could be utilized in the modern high-accuracy high precision portable measurement system for sensor application.

Acknowledgment

This work was supported by the National Natural Science Foundation of China (Nos. 61475085 & 61405105), the Science and Technology Development Project of Shandong Province (No. 2014GGX101007), and the Fundamental Research Funds of Shandong University (No. 2014YQ011).

References

[1] A. D. Marcellis, G. Ferri, A. D'Amico, and C. D. Natale, "A fully-analog lock-in amplifier with automatic phase alignment for accurate measurements of ppb gas concentrations," *Sensors Journal IEEE*, 2012, 12(5): 1377–1383.

[2] G. Graaf and R. F. Wolffenbuttel, "Lock-in amplifier techniques for low-frequency modulated sensor applications," *Conference Record – IEEE Instrumentation and Measurement Technology Conference*, 2012, 8443(8): 1745–1749.

[3] J. Aguirre, D. García-Romeo, N. Medrano, B. Calvo, and S. Celma, "Square-signal-based algorithm for analog lock-in amplifiers," *IEEE Transaction on Industrial Electronics*, 2014, 61(10): 5590–5598.

[4] M. Gabal, N. Medrano, B. Calvo, P. A. Martínez, S. Celma, and M. R. Valero, "A complete low voltage analog lock-in amplifier to recover sensor signals buried in noise for embedded applications," *Procedia Engineering*, 2010, 5(41): 74–77.

[5] A. Amico, A. Marcellis, C. D. Carlo, C. D. Natale, G. Ferri, E. Martinelli, *et al.*, "Low-voltage low-power integrated analog lock-in amplifier for gas sensor applications," *Sensors and Actuators B Chemical*, 2010, 144(2): 400–406.

[6] J. Aguirre, N. Medrano, B. Calvo, and S. Celma, "Lock-in amplifier for portable sensing systems," *Electronics Letters*, 2011, 47(21): 1172–1173.

[7] W. D. Walker, "Sub-microdegree phase measurement technique using lock-in amplifiers," in *IEEE International Frequency Control Symposium*, Hawaii, pp. 825–828, 2008.

[8] C. Azzolini, A. Magnanini, M. Tonelli, G. Chiorboli, and C. Morandi, "Integrated lock-in amplifier for contact-less interface to magnetically stimulated mechanical resonators," in *3rd International Conference on Design and Technology of Integrated Systems in Nanoscale Era*, Tozeur, pp. 1–6, 2008.

[9] A. D. Marcellis and G. Ferri, *Detection of small and noisy signals in sensor interfacing: the analog lock-in amplifier*. Netherlands: Springer, 2011: 181–204.

[10] H. M. Kalayeh, G. R. Paz-Pujalt, and J. P. Spoonhower, "System and method for remote quantitative detection of fluid leaks from a natural gas or oil pipeline, " US Patent 6822742, 2004.

[11] M. Quaranta, S. M. Borisov, and I. Klimant, "Indicators for optical oxygen sensors," *Bioanalytical Reviews*, 2012, 4(2–4): 115–157.

[12] A. K. McEvoy, C. M. Mcdonagh, and B. D. Maccraith, "Dissolved oxygen sensor based on fluorescence quenching of oxygen-sensitive ruthenium complexes immobilized in sol–gel-derived porous silica coatings," *Analyst*, 1996, 121(6): 785–788.

[13] M. T. Murtagh, D. E. Ackley, and M. R. Shahriari, "Development of a highly sensitive fiber optic O_2 sensor based on a phase modulation technique," *Electronics Letters*, 1996, 32(5): 477–479.

[14] C. Mcdonagh, C. Kolle, A. K. Mcevoy, D. L. Dowling, A. A. Cafolla, S. J. Cullen, *et al.*, "Phase fluorometric dissolved oxygen sensor," *Sensors and Actuators B Chemical*, 2001, 74(1–3): 124–13.

[15] Z. Limpouchová, and K. Procházka, *Theoretical principle of fluorescence spectroscopy*. Netherlands: Springer, 2016.

11

Harmful Intrusion Detection Algorithm of Optical Fiber Pre-Warning System Based on Correlation of Orthogonal Polarization Signals

Fukun BI[1], Chong FENG[1], Hongquan QU[1*], Tong ZHENG[1], and Chonglei WANG[2]

[1]School of Electrical and Information Engineering, North China University of Technology, Beijing, 100144, China

[2]Beijing Institute of Technology Department of Information and Electronic, Beijing, 100081, China

[*]Corresponding author: Hongquan QU   E-mail: dolly115dolly@163.com

Abstract: At present, advanced researches of optical fiber intrusion measurement are based on the constant false alarm rate (CFAR) algorithm. Although these conventional methods overcome the interference of non-stationary random signals, there are still a large number of false alarms in practical applications. This is because there is no specific study on orthogonal polarization signals of false alarm and intrusion. In order to further reduce false alarms, we analyze the correlation of optical fiber signals using birefringence of single-mode fiber. This paper proposes the harmful intrusion detection algorithm based on the correlation of two orthogonal polarization signals. The proposed method uses correlation coefficient to distinguish false alarms and intrusions, which can decrease false alarms. Experiments on real data, which are collected from the practical environment, demonstrate that the difference in correlation is a robust feature. Furthermore, the results show that the proposed algorithm can reduce the false alarms and ensure the detection performance when it is used in optical fiber pre-warning system (OFPS).

Keywords: Optical fiber; birefringence; orthogonal polarization signals; correlation

1. Introduction

The optical fiber pre-warning system (OFPS) can be extensively applied in the fields of border lines, key security areas, oil and gas pipelines, and other scenes [1–5]. By extracting intrusion signals in the buried optical fiber, the system realizes long-distance, wide-range intrusion detection, and real-time pre-warning. It is an effective way to monitor illegal entry, ensure communication safety, and prevent to damage pipelines. The optical fiber intrusion signal processing can be divided into two

steps, namely detection and recognition. The detection function is used to locate the intrusion. According to the detection results, we can extract the intrusion signal segment to identify the type of intrusion. Therefore, accurate detection of intrusion signal is an important prerequisite for the effective operation of OFPS.

Currently, phase-sensitive optical time domain reflectometer (Φ-OTDR) is a typical structure in OFPS [6]. It can be used to detect concurrent intrusions with every small resolution cell. Fixed threshold detection is adopted in Φ-OTDR initially, but its performance degrades dramatically in complicated application environment [7–11]. Furthermore, the constant false alarm rate (CFAR) methods are introduced to detect the intrusion signal of optic fiber [12–16]. CFAR improves the performance of OFPS in a certain extent because adaptive threshold of it, but there is still a mass of false alarms in detection results. A large quantity of false alarms will increase workload and reduce security personnel's work efficiency. To solve this problem, it is necessary to develop new algorithms to further decrease false alarms.

The proposed Φ-OTDR based on detection algorithm will improve the performance of traditional CFAR. At present, most CFAR algorithms used in OFPS are based on spatial characteristics of signals. They detect the amplitude change of synthesis of two orthogonal polarization signals. However, in previous researches, characteristics of these two signals and their relationships are neglected. According to the theory of birefringence of single-mode fiber [17–21], the signals collected by OFPS are analyzed in the paper. The results show that the difference between false alarm and intrusion is obvious and robust, thus it can be used to decrease false alarms. Moreover, by analyzing and discriminating false alarm signals and intrusion signals collected in practical applications, we find that false alarm signals and intrusion signals can be distinguished from the correlation coefficient of the two orthogonal polarization signals. According to this conclusion, this paper presents a harmful intrusion detection algorithm based on correlation of orthogonal polarization signals. Further experiments prove the feasibility and effectiveness of this algorithm. In the end, the OFPS is tested in Shangweidian Village of Mentougou District in Beijing. Experiments of multiple intrusions detection are implemented and the results show the proposed algorithm can significantly decrease false alarms compared with previous methods.

The remainder of this paper is organized as follows. Section 2 gives an introduction of the proposed method. Section 3 is devoted to the correlation analysis of optical fiber signals based on birefringence, including intrusion signals, noise, and false alarm signals. The OFPS detection algorithm based on correlation of orthogonal polarization signals is proposed in Section 4. In Section 5, experiments on real data demonstrate the effectiveness of the proposed method. The discussions and conclusions are provided in Section 6.

2. Analysis on optical fiber signals

In this section, we introduce birefringence of the single mode optical fiber first, then analyze three types of optical fiber signals to investigate the principle of the proposed algorithm.

2.1 Birefringence of single mode optical fiber

Single mode optical fiber is used in OFPS to extract intrusion signals. The principle is that external stresses can lead to birefringence through the photo elastic effect. A bent fiber is shown in Fig. 1. As seen, there are two orthogonal polarization modes HE_{11}^{X} and HE_{11}^{Y}. Birefringence is induced, when propagation constant U_x and U_y of polarization modes are not equal.

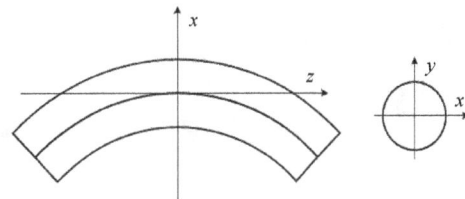

Fig. 1 Geometry of a bent fiber.

On the one hand, wavelengths of HE_{11}^{X} and HE_{11}^{Y} are equivalent when there is no bending caused by external stresses. In this situation, $U_x = U_y$, and polarization modes degenerate. Therefore, components of signal remain equal in the directions of x and y because of the same phase. On the other hand, the wavelengths of HE_{11}^{X} and HE_{11}^{Y} are no longer equal anymore because reflective indexes of x and y change differently when optical fiber is bent in direction of x axis as shown in Fig. 1. In this case $U_x \neq U_y$, and the two polarization modes do not degenerate. Accordingly, the components in x and y are unequal.

Actually, direction of external stresses is rarely identical to vector sum direction of two polarization modes composition, so changes of U_x and U_y are unequal. Hence, amplitude fluctuation induced by intrusion is obvious in one of orthogonal polarization signals and dim in the other.

2.2 Correlation analysis on optical fiber signals based on birefringence

This section analyzes three kinds of typical optical fiber signals, including intrusion signals, noises, and false alarm signals.

2.2.1 Correlation analysis on intrusion signals

Signal I and signal Q are two orthogonal polarized components of intrusion signals. They are depicted in Fig. 2. A 1-s long fragment of the signals marked with a rectangle are shown in Fig. 3. This kind of signal is detected as intrusion by CFAR. Obviously, waveform in Fig. 2(a) is significantly different from that in Fig. 2(b) which contains lots of peaks and valleys. The propagation constants of two polarization modes are distinct according to Section 2.1.

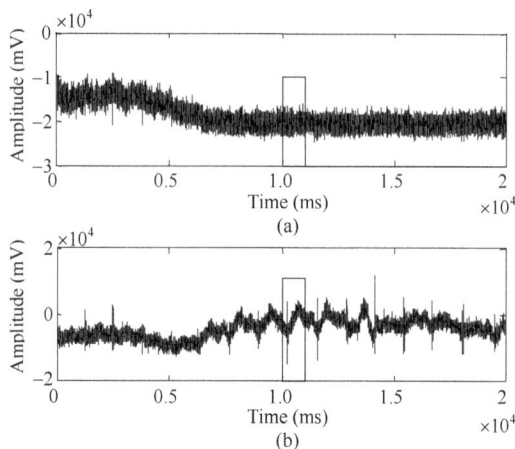

Fig. 2 Intrusion signals: (a) signal I and (b) signal Q.

In order to compare the correlation difference of three kinds of signals, following steps are taken:

(1) The signal is divided into 20 segments with each length of 1 s.

(2) The correlation coefficients of 20 groups of two signals are calculated.

(3) Fit curve of distribution function by making statistical analysis on the correlation coefficients.

(4) The mean and variance is computed based on the curve obtained in Step (3).

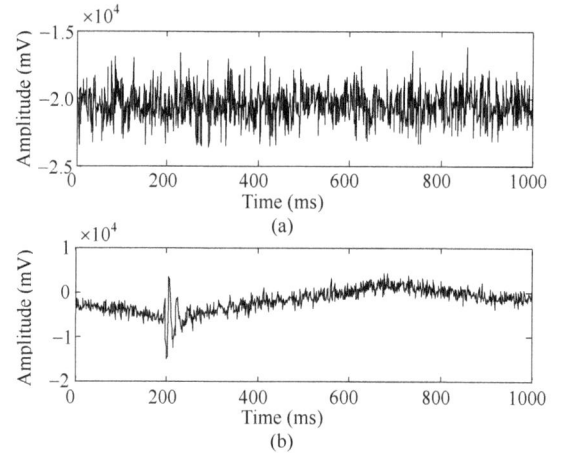

Fig. 3 Intrusion signals: (a) signal I and (b) signal Q.

The result of relevance analysis of intrusion signal is shown in Table 1.

Table 1: Statistic on intrusion signals.

Signal type	Mean	Variance
Intrusion	0.1662	0.0140

The correlation coefficient is Gaussian distributed. Its mean and variance are 0.1662 and 0.140, respectively.

2.2.2 Correlation analysis on noise

The noise is collected by OFPS when no external stresses are applied on optical fiber, and it cannot be detected by CFAR. Noise is shown in Fig. 4. Signal I and signal Q are two orthogonal polarized components, respectively. Segments of the signals marked by rectangle are shown in Fig. 5. There is no prominent peak in the wavelet, which means no intrusion.

Noise signals are analyzed in the same way as in Section 2.2.1. The result is shown in Table 2. The correlation coefficient follows Gaussian distribution, and its mean and variance of correlation coefficient is 0.2710 and 0.0111.

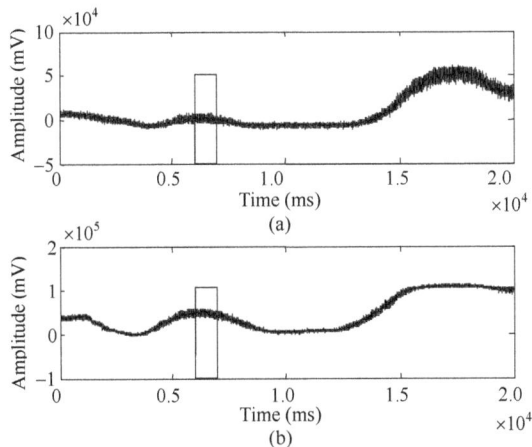

Fig. 4 Noise signals: (a) signal I and (b) signal Q.

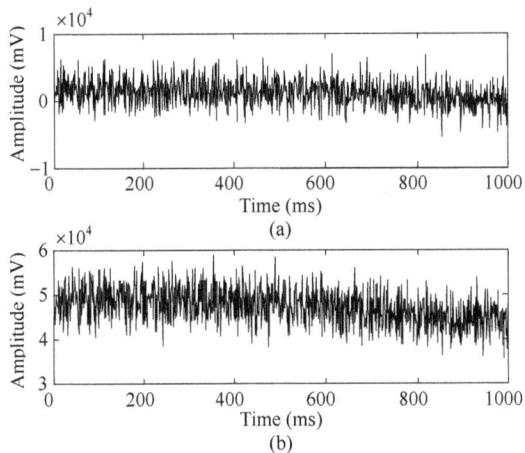

Fig. 5 Noise signals: (a) signal I and (b) signal Q.

Table 2: Statistic on noise.

Signal type	Mean	Variance
Noise	0.2710	0.0111

2.2.3 Correlation analysis on false alarm signals

Signal I and signal Q, orthogonal polarized components of false alarm signal, are shown in Fig. 6. Segments marked by rectangle are shown in Fig. 7. Peaks are found in both orthogonal polarization signals, and can be detected by CFAR. According to birefringence, bending induced by external stresses will lead to change in the one of orthogonal polarization signals, but the changes are synchronized and similar to false alarm signals in figures. Consequently, it is not the strain that leads to fluctuation in the false alarm signals. Statistics on alarm signals are shown in Table 3.

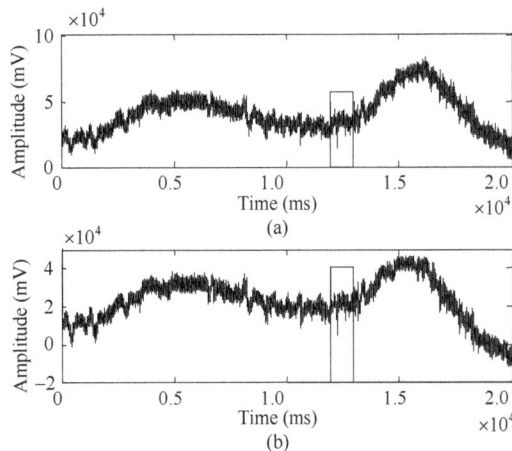

Fig. 6 False alarm signals: (a) signal I and (b) signal Q.

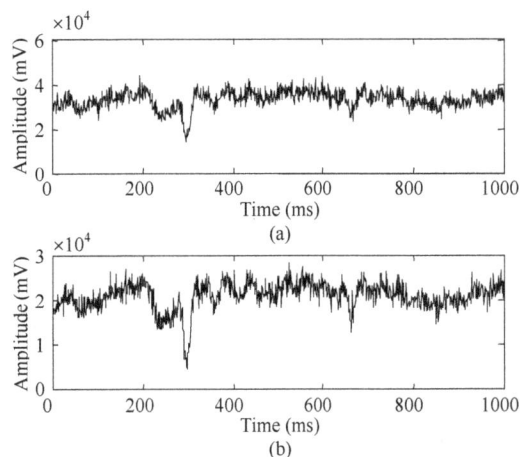

Fig. 7 False alarm signals: (a) signal I and (b) signal Q.

Table 3: Statistic on false alarm signals.

Signal type	Mean	Variance
False alarm	0.7705	0.0163

Correlation coefficient mean of false alarms is 0.7705, and it is much greater than that of intrusion signals 0.1662 and that of noise 0.2710. Based on these analyses, the correlation coefficient of orthogonal polarization signals is an effective feature to distinguish intrusion and false alarm, and it can be used to exclude false alarms in the result obtained by CFAR.

3. Detection algorithm based on correlation of cross-polarization signals

According to the correlation analysis, this paper proposes a harmful intrusion detection algorithm based on correlation of orthogonal polarization signals. The main procedures of the proposed method are shown in Fig. 8.

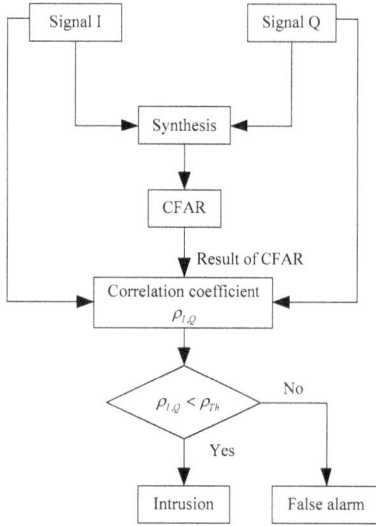

Fig. 8 Proposed algorithm.

(1) Synthesize signal I and signal Q collected by OFPS into D as (1):

$$D = I^2 + Q^2 \qquad (1)$$

(2) Apply CFAR to detect D to extract alarm signals.

(3) Calculate correlation coefficient between I and Q of alarm signal as (2), and compare it with presupposed threshold.

$$\rho_{I,Q} = \mathrm{Cov}(I,Q)\big/\sqrt{D(I) \cdot D(Q)} \qquad (2)$$

where $\rho_{I,Q}$ is correlation coefficient, $\mathrm{Cov}(I, Q)$ is covariance of I and Q, and $D(I)$ and $D(Q)$ are variance of I and Q, respectively. If $\rho_{I,Q}$ is less than ρ_{Th} which is the threshold, the alarm is judged as an intrusion otherwise it is a false alarm.

Hypothesis testing is used to determine the threshold. Firstly, all sample of false alarm signals are divided into segments of 1 s. Subsequently, their $\rho_{I,Q}$ of them are calculated. Then, fit all $\rho_{I,Q}$ values into Gaussian distribution. Last, calculate ρ_{Th} of hypothesis testing from probability distribution function based on false alarm rate which is set as 1×10^{-7}.

4. Experiment and analysis

4.1 Collection of real data

In order to validate correlation feature of optical fiber signals and performance of proposed algorithm, test data is collected in Shangweidian Village of

Mentougou District in Beijing. A 1000-m long optical fiber is buried 20 cm under the ground.

Signals are collected under three conditions as follows.

(1) Experiment site without any external intrusion interference.

(2) Testers jog at the 700 m to the end of optical fiber for 8 s.

(3) Testers dig the ground with a pickaxe at the 500 m to the end of optical fiber for 8 s.

4.2 Experiment to verify correlation feature of optical fiber signals

The signals collected by OFPS are analyzed in Section 2.2. Alarms in Condition (1) are considered as false alarms. In addition, alarms in non-intrusion positions in Conditions (2) and (3) are false alarms too. The results are shown in Table 4.

Table 4: Analysis on four types of measurements.

	Noise	Jogging	Digging	False alarm
Mean of correlation coefficient	0.2588	0.1714	0.1594	0. 7493
Variance of correlation coefficient	0.0133	0.0160	0.0122	0.0194

Accordingly, the correlation coefficient of the false alarm signal is much larger than two intrusion signals. This result agrees with analysis on signals collected by OFPS in Section 2.2. It proves that the correlation coefficient of the false alarm is greater than that of the intrusion signals. And this is a general characteristic in OFPS.

4.3 Experiments of the OFPS detection algorithm

Both CFAR and the proposed algorithm are used to process signals under three situations, including non-intrusion, jogging, and digging. Then, the performance of these two algorithms will be compared in detection probability and false alarm probability.

The detection results of signals without intrusion are shown in Figs. 9 and 10. Horizontal axis and vertical axis are situation and time, respectively. According to Fig. 9, it is clear that there are a lot of false alarms in the position of 300 m and other sporadic false alarms after CFAR detection. The results of the proposed algorithm are displayed in Fig. 10. Apparently, most false alarms are eliminated effectively.

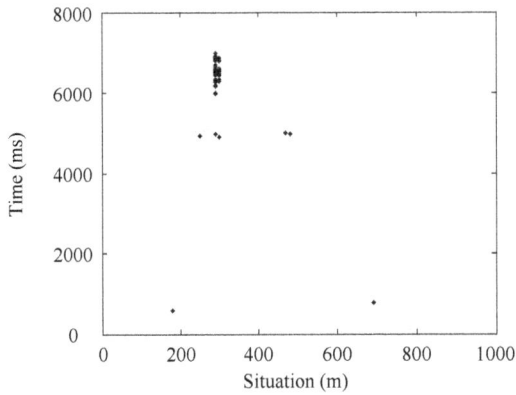

Fig. 9 Detection result of CFAR.

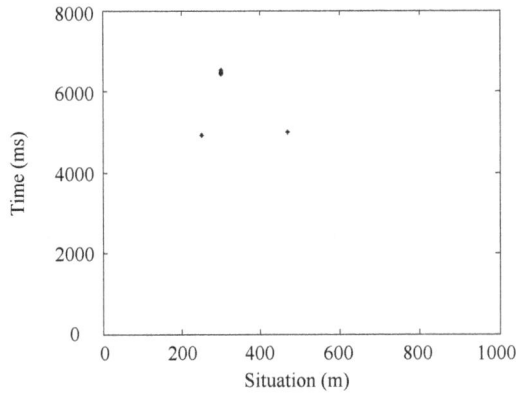

Fig. 10 Detection result of proposed method.

According to the detection outcome of jogging shown in Figs. 11 and 12, the proposed algorithm not only eliminates false alarm, but also detects intrusion in position of the 700 m to the end, which coincides with the experiment setup.

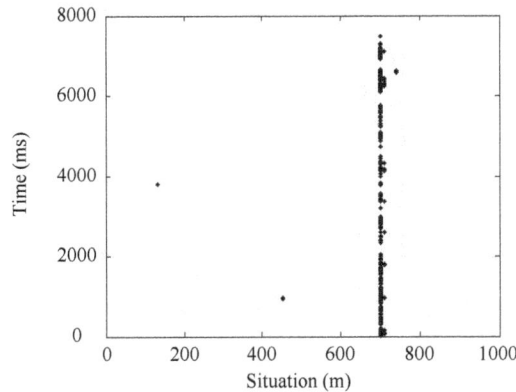

Fig. 11 Detection result of CFAR.

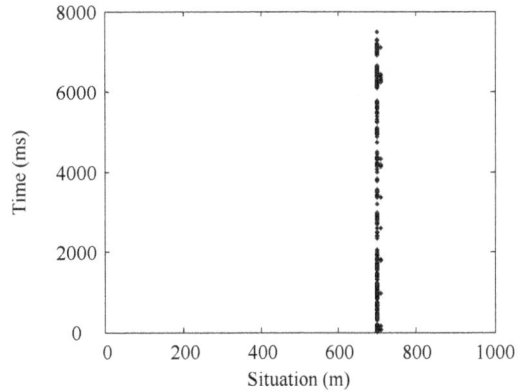

Fig. 12 Detection result of proposed method.

Figures. 13 and 14 show the detection results of the digging signal. As seen in the figure, intrusion at position of 500 m is effectively detected, and the duration of the alarm exactly agrees with the experimental setup. Significantly, false alarms in Fig. 13 are noticeably less than those in Fig. 14, so the proposed algorithm achieves better performance in reducing false alarms.

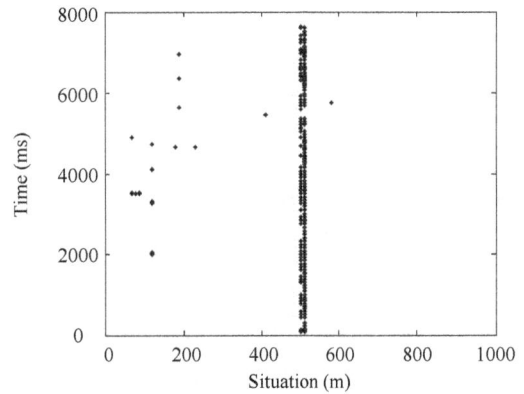

Fig. 13 Detection result of CFAR.

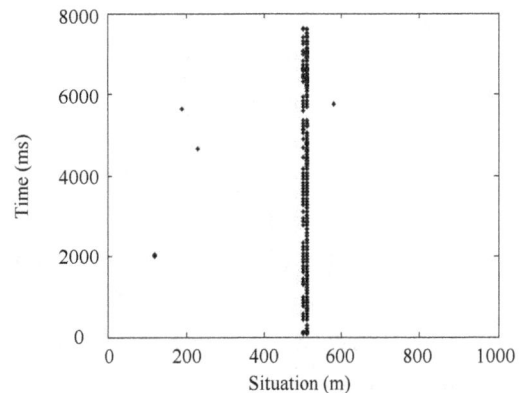

Fig. 14 Detection result of proposed method.

In order to demonstrate the availability of proposed algorithm, we count the number of both false alarms and alarms at the intrusion site. The results are given in Tables 5 and 6. As can be seen, the proposed algorithm performs better than CFAR. The detection performance of both methods is comparable, but the proposed algorithm reduces false alarms by 87.8%.

Table 5: Number of false alarms.

Type of signal	CFAR	Proposed method
No intrusion	137	10
Jogging	3	0
Digging	65	15

Table 6: Number of alarms in intrusion site.

Type of signal	CFAR	Proposed method
Jogging	1246	1246
Digging	471	471

By comparing detection results of three types of signals, it is clear that the proposed OFPS detection algorithm can remove the harmless interferences based on CFAR detection results without the degradation of the detection performance.

5. Discussions and conclusions

In this paper, false alarm in OFPS is studied. Based on the theory of birefringence of single mode optical fiber, we find that the correlation coefficient of two orthogonal polarization signals can be used to distinguish false alarm and intrusion. According to the results of statistical analysis on correlation coefficient of intrusion signals, noises, and false alarm signals, we find that correlation coefficient of false alarms signals is notably greater than that of intrusion signals. In addition, we show that the difference in correlation feature of optical signals commonly exists in OFPS.

We propose a harmful intrusion detection algorithm based on the correlation analysis of orthogonal polarization signals. Firstly, the synthesis of two cross-polarization signals is detected by CFAR to get alarms. Afterwards, the coefficient of correlation of alarm signals is calculated. Eventually, once the coefficient is lower than the presupposed

threshold, the alarm is judged as an intrusion, or it is a false alarm.

To test the algorithm, we design experiments on real data which is collected in Shangweidian Village of Mentougou District in Beijing. The threshold is determined through hypothesis testing from probability distribution function and false alarm rate is set as 1×10^{-7}. The results show that the proposed algorithm outperforms CFAR in terms of false alarms. Accuracy of OFPS can be significantly improved because false alarms are drastically reduced and intrusions are detected accurately as well.

Acknowledgement

This work was supported by the National Nature Science Foundation of China (Grant Nos. 61571014, 61601006); Beijing Nature Science Foundation (Grant No. 4172017); General project of science and technology program of Beijing Education Commission(Grant No. KM201610009004).

Reference

[1] Z. G. Qu, H. Feng, Z. M. Zeng, J. C. Zhuge, and S. J. Jin, "A SVM-based pipeline leakage detection and pre-warning system," *Measurement*, 2010, 43(4): 513–519.

[2] J. Kang and Z. H. Zou, "Time prediction model for pipeline leakage based on grey relational analysis," *Physics Procedia*, 2010, 25(2): 2019–2024.

[3] W. Liang, L. L. Lu, and L. B. Zhang, "Coupling relations and early-warning for 'equipment chain' in long-distance pipeline," *Mechanical Systems and Signal Processing*, 2013, 41(1–2): 335–347.

[4] W. Liang, L. B. Zhang, Q. Q. Xu, and C. Y. Yan, "Gas pipeline leakage detection based on acoustic technology," *Engineering Failure Analysis*, 2013, 31(6): 1–7.

[5] T. T. Zhang, Y. F. Tan, H. X. Yang, J. H. Zhao, and X. D. Zhang, "Locating gas pipeline leakage based on stimulus-response method," *Energy Procedia*, 2014, 61: 207–210.

[6] Q. Y. Lv, L. J. Li, H. B. Wang, Q. Li, and X. Zhong, "Influences of laser on fiber-optical distributed

disturbance sensor based on Φ-OTDR," *Infrared and Laser Engineering*, 2014, 43(12): 3918–3923.

[7] H. F. Martins, S. Martin-Lopez, P. Corredera, M. L. Filograno, O. Frazão, and M. Gonzáez-Herraez, "Coherent noise reduction in high visibility phase-sensitive optical time domain reflectometer for distributed sensing of ultrasonic waves," *Journal of Lightwave Technology*, 2013, 31(23): 3631–3637.

[8] Q. Li, C. X. Zhang, L. J. Li, and X. Zhong, "Localization mechanisms and location methods of the disturbance sensor based on phase-sensitive OTDR," *Optik–International Journal for Light and Electron Optics*, 2014, 125(9): 2099–2103.

[9] Q. Lin, C. X. Zhang, and C. S. Li, "Fiber-optic distributed sensor based on phase-sensitive OTDR and wavelet packet transform for multiple disturbances location," *Optik–International Journal for Light and Electron Optics*, 2014, 125(24): 7235–7238.

[10] A. R. Bahrampour and F. Maaoumi, "Resolution enhancement in long pulse OTDR for application in structural health monitoring," *Optical Fiber Technology*, 2010, 16(4): 240–249.

[11] L. D. Lu, Y. J. Song, X. J. Song, X. P. Zhang, and F. Zhu, "Frequency decision multiplexing OTDR with fast signal processing", *Optics & Laser Technology*, 2012, 44(7): 2206–2209.

[12] H. Q. Qu, T. Zheng, F. K. Bi, and L. P. Pang, "Vibration detection method for optical fiber pre-warning system," *IET Signal Process*, 2016, 10(6): 692–698.

[13] H. Q. Qu, T. Zheng, L. P. Pang, and X. L. Li, "A new two-dimension method to detect harmful intrusion vibrations for optical fiber pre-warning system," *Optik–International Journal for Light and Electron Optics*, 2016, 127(10): 4461–4469.

[14] R. L. Zhang, W. X. Sheng, and X. F. Ma, "Improved switching CFAR detector for non-homogeneous environments," *Signal Processing*, 2013, 93(1): 35–48.

[15] G. V. Weinberg, "Management of interference in Pareto CFAR processes using adaptive test cell analysis," *Signal Processing*, 2014, 104(104): 264–273.

[16] B. Shi, C. P. Hao, C. H. Hou, X. C. Ma, and C. Y. Peng, "Parametric Rao test for multichannel adaptive detection of range-spread target in partially homogeneous environments," *Signal Processing*, 2015, 108(108): 421–429.

[17] M. Karimi, T. Sun, and K. T. V. Grattan, "Design evaluation of a high birefringence single mode optical fiber-based sensor for lateral pressure monitoring applications," *IEEE Sensors Journal*, 2013, 13(11): 4459–4464.

[18] L. Palmieri, A. Galtarossa, and T. Geisler, "Distributed characterization of bending effects on the birefringence of single-mode optical fibers," *Optics Letters*, 2010, 35(14): 2481–2483.

[19] Z. Y. Li, C. Q. Wu, H. Dong, P. Shum, C. Y. Tian, and S. Zhao, "Stress distribution and induced birefringence analysis for pressure vector sensing based on single mode fibers," *Optics Express*, 2008, 16(6): 3955–3960.

[20] B. Y. Kim and S. S. Choi, "Backscattering measurement of bending-induced birefringence in single mode fibres," *Electronics Letters*, 1981, 17(5): 193–194.

[21] O. G. Leminger, "Stress birefringence in single-mode fibres," *Electronics Letters*, 1977, 13(12): 370–371.

Research Progress in the Key Device and Technology for Fiber Optic Sensor Network

Deming LIU, Qizhen SUN[*], Ping LU, Li XIA, and Chaotan SIMA

School of Optical and Electronic Information, Huazhong University of Science and Technology; National Engineering Laboratory for Next Generation Internet Access System, Wuhan, 430074, China

[*]Corresponding author: Qizhen SUN   E-mail: qzsun@hust.edu.cn

Abstract: The recent research progress in the key device and technology of the fiber optic sensor network (FOSN) is introduced in this paper. An architecture of the sensor optical passive network (SPON), by employing hybrid wavelength division multiplexing/time division multiplexing (WDM/TDM) techniques similar to the fiber communication passive optical network (PON), is proposed. The network topology scheme of a hybrid TDM/WDM/FDM (frequency division multiplexing) three-dimension fiber optic sensing system for achieving ultra-large capacity, long distance, and high resolution sensing performance is performed and analyzed. As the most important device of the FOSN, several kinds of light source are developed, including the wideband multi-wavelength fiber laser operating at C band, switchable and tunable 2 μm multi-wavelength fiber lasers, ultra-fast mode-locked fiber laser, as well as the optical wideband chaos source, which have very good application prospects in the FOSN. Meanwhile, intelligent management techniques for the FOSN including wideband spectrum demodulation of the sensing signals and real-time fault monitoring of fiber links are presented. Moreover, several typical applications of the FOSN are also discussed, such as the fiber optic gas sensing network, fiber optic acoustic sensing network, and strain/dynamic strain sensing network.

Keywords: Fiber optic sensor network; fiber optic sensor laser; intelligent management; fiber optic gas sensing; fiber optic acoustic sensing; fiber optics strain sensing

1. Introduction

Due to the distinct advantages of light weight, small size, high sensitivity, immunity to electromagnetic interference, and ease to network, fiber optic sensors have been intensively studied for measuring numerous parameters including temperature, strain, refractive index, vibration, displacement, etc. in the past 40 years. Along with the bloom of optical fiber sensors and the tidal wave of Internet of Things (IOT), the fiber optic sensing network is becoming an inevitable tendency for the sensing industry [1]. Especially for the application areas requiring multipoint and large area measurement, such as structural health monitoring and geophysical surveying, the fiber optic sensing networks are ideal solutions. For a typical fiber optic sensing network, the fiber sensors can be multiplexed by specific schemes, including time division multiplexing (TDM), wavelength division multiplexing (WDM), frequency division multiplexing (FDM), coherent division multiplexing (CDM), and space division multiplexing (SDM) [2–6]. Recently, following the development trend of the large capacity, long distance, and high resolution of the fiber optic sensing network, the hybrid of above multiplexing schemes are intensively studied. By exploring high-performance light sources and demodulation modules, the information of each sensor unit can be interrogated, enabling the corresponding sensing parameters to be detected.

Stable and compact light sources, with wide bandwidth, multiple and selective operating wavelengths, ultra-short pulse width, high output power, or other special properties, are essential for the fiber optic sensing network to support specific multiplexing schemes. As a promising candidate of specialized light sources, fiber lasers have attracted considerable attention due to their overwhelming advantages of compatibility toward fiber, flexibility in wavelength control, high beam quality, low pump-power requirements, and reliability in harsh environment [7].

In addition, the intelligent management especially the survivability of the fiber optic sensor network, is also important. For example, the fiber optic sensor networks usually work in harsh environment with large temperature difference, high pressure, strong magnetic field, etc. Thus, the fiber fracture could happen occasionally [8], while it may cost much time and manpower to locate the fiber breakpoint in the sensor network with the large coverage area and complex structure [9]. Therefore, automatic and precision fiber fault locating techniques should be developed.

In this paper, we review the efforts to advance the fields of fiber optic sensing networks. (1) Firstly, a brief description of various multiplexing schemes provides some insight into the advantages and remaining challenges of them. Then, we expect to build up a highly adaptive fiber optic sensor network (FOSN) with large capacity and multi-parameters/functions/mechanisms integration, by employing hybrid WDM/TDM multiplexing techniques and matrix topology architecture similar to the fiber communication passive optical network (PON). Consequently, on account of a selected hybrid TDM/WDM/FDM scheme, a novel sensing network with linear typology along a single fiber is established, mainly focusing on ultra-large capacity, long distance, and high resolution sensing performance. (2) Secondly, as the most important device of the FOSN, several kinds of fiber laser, including the wideband multi-wavelength fiber laser operating at C band, switchable and tunable 2 μm multi-wavelength fiber lasers, ultra-fast mode-locked fiber laser, and optical wideband chaos source, are explored to meet the requirements of the FOSN. (3) Thirdly, intelligent management techniques of the FOSN, including the wide bandwidth, high resolution, and high speed optical spectrum demodulation, as well as the real-time and precisely fault monitoring of fiber link are performed. (4) Finally, several typical applications of the FOSN are discussed, including the fiber optic gas sensing network, fiber optic acoustic sensing network, and quasi-distributed static/dynamic strain sensing network.

2. Networking technology of the fiber optic sensors

Accompanied by the development of fiber optic communication technologies, fiber optic sensing technologies are greatly improved. At present, the reported FOSN techniques are mainly focused on the various sensor multiplexing schemes, including WDM, TDM, SDM, FDM, and CDM, or the combinations of them.

2.1 Single division multiplexing schemes

TDM is the simplest and earliest reported multiplexing scheme, in which the fiber sensors in the sensing link are usually arranged by the cascaded way with a delay line inserted between two adjacent fiber sensors to separate them in time domain. Differed by the time delay to receive the sensing signal, different fiber sensors can be distinguished and located. Many researchers have been focused on increasing the capacity and the spatial resolution of the TDM scheme. For instance, the capacity of the fiber Bragg grating (FBG) based TDM scheme has increased from tens to hundreds points in the past decades [10–12]. Recently, Q. Z. Sun and C. Y. Hu separately reported the TDM technique based on ultra-weak FBGs and demonstrated an FOSN along a single fiber with over 1000 sensors [13, 14]. However, limited by the topological structure and time slot based locating method, the spatial resolution of the TDM scheme is still limited.

The WDM scheme has been very mature in the wavelength encoded fiber sensors. Optical fiber sensors always occupy a particular bandwidth of the light spectrum. By using a broadband light source and allocating different bandwidths to the fiber sensors, different fiber sensors can be distinguished through the wavelengths of sensing signals. For the FBG sensor network, the capacity of the WDM sensing network is usually limited to only several dozen, decided by the wavelength range of the light source and bandwidth allocated to each FBG [15, 16].

The most important and typical technique of FDM is the frequency modulation continuous wave technique (FMCW), which is also called the optical frequency domain reflector (OFDR). By employing the narrow linewidth light source and high speed acquisition setup, the OFDR can realize a very high spatial resolution and large capacity. Kazumasa Takada reported a high spatial resolution of $850\,\mu m$ with a fiber optic frequency encoder in 1992 [17]. Brooks A. Chiders achieved a test with 3000 FBGs distributed on four 8-meter fibers. However, the system cost is relatively high, and the application in

the long-distance detection is limited [18].

The CDM technology multiplexes the interferometric fiber sensors like the Mach-Zehnder interferometer, Michelson interferometer, and the Fabry-Perot (FP) interferometer. This technique was first proposed by Brooks in 1985 [19]. However, it requires a compensating interferometer for each sensor to realize demodulation, which cannot satisfy the requirement for large capacity applications.

Besides, sensors on different fibers can be multiplexed through the optical switch and optical coupler in the SDM scheme. Y. J. Rao came up with a spatially multiplexed fiber sensing system and realized 32 FBGs multiplexed [20]. The topological structure for SDM is quite complex, and the speed is limited by the optical switch.

2.2 Hybrid schemes of different multiplexing formats

From the above discussion, single multiplexing schemes cannot satisfy the great demand of large capacity and high spatial resolution. Combination of different multiplexing schemes together is an effective way to improve the performances of the FOSN [21, 22]. Here we mainly discuss two hybrid multiplexing schemes for flexible or ultra-large capacity sensor units accessing to the sensor network.

2.2.1 Hybrid sensor passive optic network (HSPON) with flexible access [23, 24]

Most of us are familiar with the topology of fiber to the X (FTTX), which consists of the optical line terminal (OLT), the optical distribution network (ODN), and the active optical network units (ONU). The downstream is broadcasted from OLT to every ONU, while the upstream is directionally transmitted from certain ONU to OLT. According to this configuration, we proposed the architecture of fiber optic sensor PON (SPON), as displayed in Fig. 1. The SPON is also composed of three main parts, including OLT, ODN, and optic sensor units (OSU). However, it is different with the optical communication PON. Firstly, the transmission data from OLT are always analog and is broadcasted to every OSU from ODN. Secondly, the OSU is always passive due to the characteristics of the optical sensor. Hence, in order to make the SPON work effectively, the OSU should be of self-feedback, i.e. it can reflect the modulated signal which carries the sensing information to the OLT through the ODN, and then the modulated signal can be demodulated at OLT to achieve the sensing parameters. In this network, the OSUs can be multi-functions, multi-structures, and also multi-points, provided that the OLT can provide appropriate light source and signal demodulation processing. Therefore, the SPON is of large capacity, good adaptability, high extendibility, and great flexibility as the fiber communication PON [23].

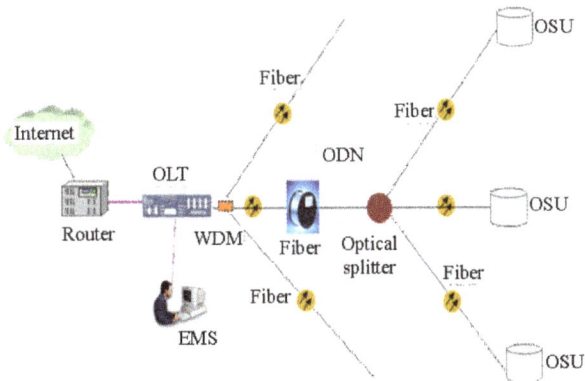

Fig. 1 Schematic diagram of the SPON architecture [23].

Figure 2 presents a typical hybrid TDM/WDM multiplexing SPON (HSPON). The TDM is realized by employing fiber delay lines of different lengths in different links, and the WDM is achieved by the WDM device connected to every OSU. The OLT generates high-power and wideband pulse light, which is transmitted to each OSU through the TDM/WDM distribution network, and then receives the sensing optical signal from OSUs. Assuming that the number of the wavelength channels is m, the number of time-domain channels is n, and the total number of the sensing units is $m \times n$. m is determined by the spectrum bandwidth of super luminescent diode (SLD), and the channel interval of WDM and etc., while n is limited by the power of the light

source and the optical link loss. According to the performance of the commercial devices, the capacity of the HSPON can reach $8 \times 32 = 256$ [24]. In this configuration, OSU can be interferometer based vibration sensor, optical spectrum absorption based gas sensor, and Faraday effect based circulator sensor, etc. Therefore, the HSPON can be widely utilized in many application fields such as multi-zone monitoring, multi-parameter detection, and multi-function monitoring [23].

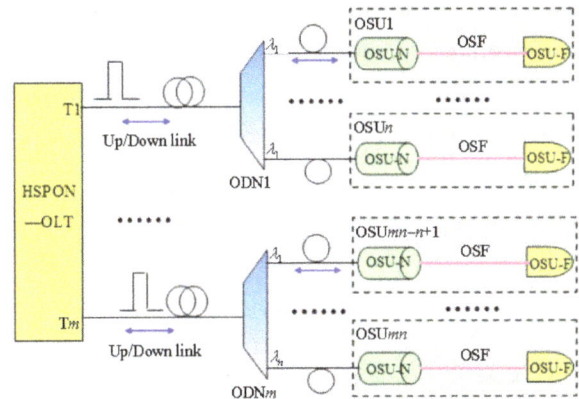

Fig. 2 Schematic diagram of the configuration of the HSPON [24].

2.2.2 Fiber optic microstructure based hybrid WDM/FDM/TDM sensor network with ultra-high capacity access

We further proposed a fiber optic microstructure based hybrid WDM/FDM/TDM sensor network for further expanding the capacity and improving the practicality. The structure and spectral characteristics of the microstructure are presented as Figs. 3(a) and 3(b), respectively [25]. Actually, it is an FP interferometer composed of two closely spaced and identical ultra-short FBGs with ultra-weak reflectivity. Then, the typical spectrum of the microstructure is a low reflective Bragg reflection spectrum modulated with certain resonant periods which is defined as the frequency of the microstructure. It is obvious that the microstructures can be encoded with different wavelengths, frequencies, and time, which are realized by choosing different central wavelengths of FBGs, different spatial distances between the FBGs pairs, and delay fiber with a certain length. The measurand affected on the microstructure can be demodulated from the wavelength shift of the spectrum. Hence, it has a great potential for large capacity multiplexing by using hybrid WDM/FDM/TDM along a single fiber. As shown in Fig. 3, the microstructures are divided into several time groups through the delay fiber. In every time group, the microstructures are hybrid WDM and FDM. Defining the multiplexing numbers of the WDM, FDM, and TDM as M, N, and Q, respectively, the total network capacity should be $M*N*Q$, which could be significantly enhanced compared with other multiplexing schemes.

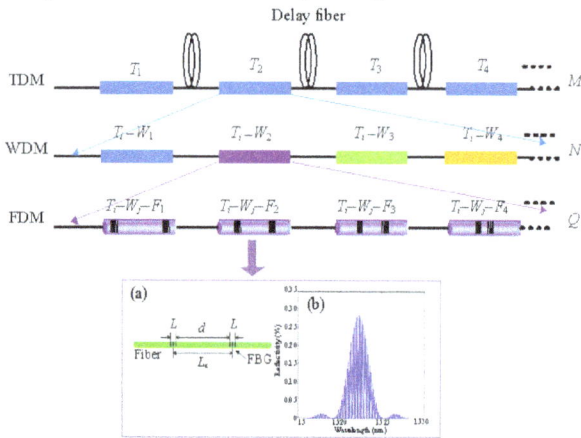

Fig. 3 Schematic diagram of the hybrid WDM/FDM/TDM fiber sensor network [inset: (a) configuration and (b) reflective spectrum of the microstructure].

Assuming that the scanning bandwidth of the system is set as 80 nm from 1510 nm to 1590 nm, the working bandwidth budget of the sensors is optimized

to 2.5 nm. Hence, the number of WDM (M) will be more than 32. Meanwhile, and there is a trade-off between the TDM and FDM channels, due that more microstructures with FDM in one time group means higher total reflectivity of the group and then will limit the TDM number. Therefore, the optimized capacity of hybrid TDM and FDM, i.e. $N*Q$, could reach to 1000 when the reflectivity of the microstructures is 0.04% [13]. Consequently, the total capacity of the three-dimensional hybrid multiplexing FOSN could be greatly improved to an ultra-large amount of 32000.

3. Specialized light sources for FOSN

3.1 Multi-wavelength fiber lasers operating at C band

Multi-wavelength fiber lasers possess versatile applications in dense WDM systems for expanding the capacity of the sensing network, as well as developing the remote sensing. Appropriate multichannel all-fiber filters play crucial roles in the implementation of multi-wavelength lasing. So far, a lot of all-fiber filters, such as equivalent Lyot birefringent fiber filters [26], fiber Sagnac loop mirrors [27], Mach-Zehnder interferometers [28], Fabry-Pérot filters [29], cascaded long-period fiber gratings [30], and chirped or sampled fiber gratings [31], have been selected as intra-cavity comb filters to successfully achieve stable multi-wavelength lasing oscillations at room temperature.

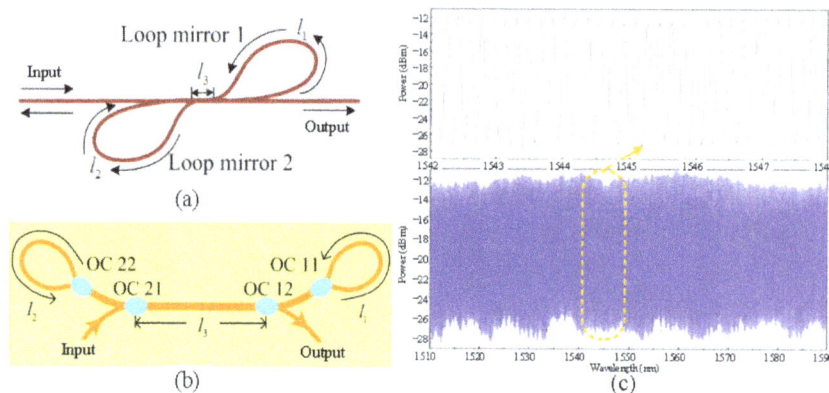

Fig. 4 Characteristics of the MFP: (a) configuration diagram, (b) equivalent model, and (c) comb reflection spectrum [35].

Recently, several kinds of micro/nanofiber (MNF) based fiber lasers have been demonstrated for the miniaturization of fiber-optic components or devices [32–34]. Especially, the microfiber Fabry-Pérot (MFP) filter acting as a comb filter paves another way to implement multi-wavelength lasing. Figures 4(a) and 4(b) show the configuration diagram and equivalent model of the MFP, respectively, where the coupling regions can be regarded as optical couplers (OCs) owing to the evanescent coupling between adjacent microfibers. Benefiting from the all fiber and compact size of the Sagnac loop mirror structure, the MFP presents excellent comb filtering spectrum as depicted in Fig. 4(c), with high extinction ratio of about 15 dB, narrow 3 dB linewidth of each channel of only 55 pm, and the channel flatness better than 2 dB within the broad bandwidth from 1510 nm to 1590 nm [35].

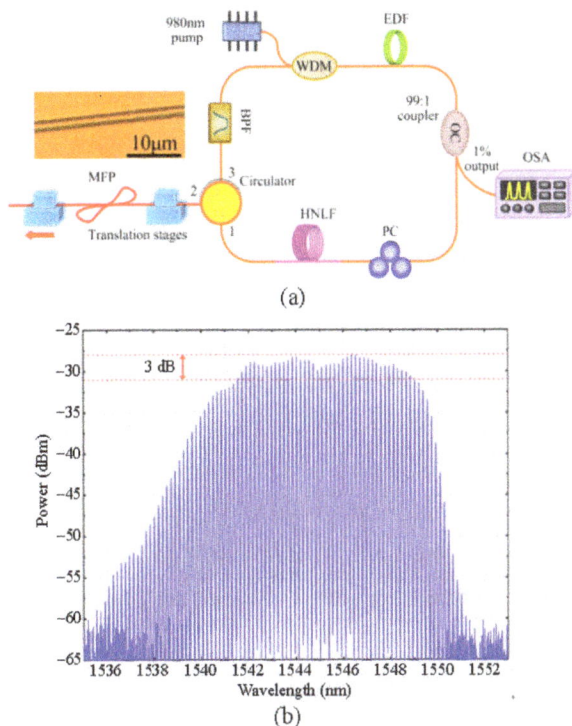

(a)

(b)

Fig. 5 Characteristics of the MFP based fiber laser: (a) schematic of the multi-wavelength fiber laser [inset: micrograph of the microfiber and (b) typical multi-wavelength lasing output [35, 36].

The corresponding multi-wavelength fiber laser is schematically depicted in Fig. 5(a), where the MFP is served as the multi-wavelength selector and

highly non-linear fiber (HNLF) is utilized to suppress the mode competition induced by the erbium doped fiber (EDF). A typical multi-wavelength lasing output is illustrated in Fig. 5(b). Up to 42 simultaneously lasing wavelengths with a wavelength spacing of 0.18 nm are obtained. The 3 dB linewidth of each channel is about 25 pm, and the side-mode suppression ratio (SMSR) is more than 35 dB [35]. Further, by moving the translation stages to mechanically adjust the cavity length of the MFP, the wavelength-spacing as well as the channels-numbers of the multi-wavelength laser can be continuously tuned [36].

3.2 Switchable and tunable multi-wavelength fiber lasers operating at the 2 μm region

Recently, interest of operating wavelength among fiber lasers has been transferred from C band and L band to the 2 μm region, for the reason that infrared fiber lasers demonstrate intriguing advantages in atmospheric transmission, hydrocarbon gas sensing, laser lidar, and eye safety [37, 38].

Additionally, switchable and tunable multi-wavelength fiber lasers mainly employ overlap cavities or polarization hole burning (PHB) to control or adjust the properties of wavelength and therefore to achieve wavelength switching. According to the two basic theories, various methods have been proposed and reported, such as cascaded FBGs [39], highly birefringent (Hi-Bi) fiber loop mirrors [40], Sagnac loop mirror [41], fiber Bragg gratings (FBGs) written in Hi-Bi fibers [42, 43], and acoustic waves [44]. In the following, two typical kinds of switchable and tunable multi-wavelength fiber laser operating at the 2 μm region, which adopt the above methods in the Tm-Ho co-doped fiber laser (THCDFL), are introduced.

3.2.1 Design of the Sagnac loop as the comb filter [45]

Figure 6 illustrates the simple configuration of multi-wavelength lasing based on the Sagnac comb filter and Tm-Ho co-doped fiber (THCDF) coiling round the plates of three-loop polarization controller (PC), where the comb filter is a Sagnac loop mirror composed of a piece of polarization maintaining fiber (PMF) and a 3 dB coupler. The polarization hole burning can be obtained without polarization maintaining-Tm-Ho co-doped fiber (PM-THCDF) and avoids the splice loss between the PM-THCDF and single mode fiber. The comb filter which consists of PMF and PC is effective to suppress the wavelength competition at room temperature. Tunable multi-wavelength lasing of different positions can be achieved by simply adjusting PC.

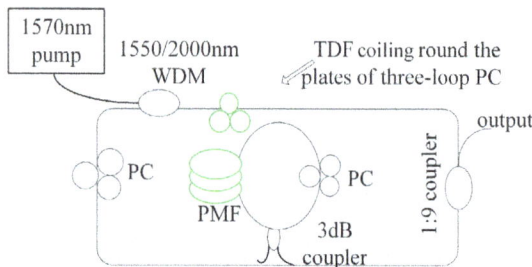

Fig. 6 Schematic of the proposed multi-wavelength THCDF laser based on the Sagnac comb filter [45].

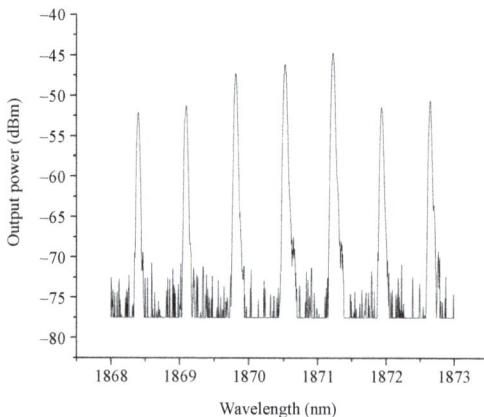

Fig. 7 Spectra of the multi-wavelength lasers with PMF [45].

Figure 7 presents the spectrum of the multi-wavelength lasers with PMF. It can be seen that stable seven wavelengths with the smallest wavelength spacing of 0.65 nm are obtained. The shortest and longest wavelengths are 1868.4 nm and 1872.7 nm, respectively. The SMSR is about 35 dB with the maximum peak power of −53 dBm at around 1871.2 nm. The number of the wavelength

and the channel spacing can be tuned by adjusting the angle of PC, for changing the birefringence and phase difference in the Sagnac loop.

3.2.2 Design of the parallel cavities based on 3×3 coupler [46]

The configuration of the switchable and tunable dual-wavelength fiber laser at the 2 μm region is schematically shown in Fig. 8. The laser topology is based on the parallel connection of FBGs using a 3×3 coupler which acts as two individual cavities, so that the dual wavelengths are tunable and switchable by adjusting the central wavelength of FBGs and the cavity losses.

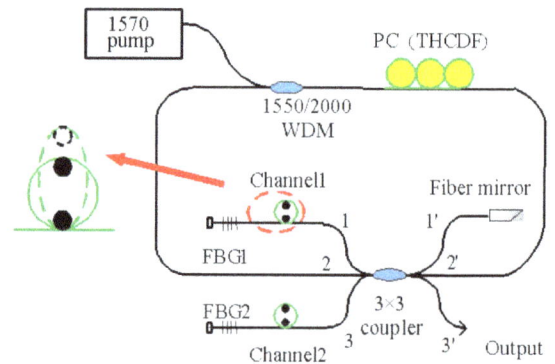

Fig. 8 Configuration of the switchable and tunable dual-wavelength fiber laser around 2 μm [46].

The dual wavelengths enjoy the same gain medium by the parallel connection of FBGs. Introducing lower losses for one wavelength will cost some gain for another wavelength, and thus, the output laser can be switched from one wavelength lasing to another or dual-wavelength lasing together by adjusting the variable attenuator (VA) and pump power. Figure 9 shows the laser operating in switching mode.

The dual-wavelength of this laser is dependent on the central wavelengths of the FBGs. Changing the central wavelengths of the FBGs by strain or temperature will tune the wavelength of output laser. Figure 10 shows the tunability of the laser by strain tuning the FBG1 and FBG2 individually. Tuning ranges of 6.5 nm (1890.6 nm − 1897.1 nm) and 4 nm (1919.4 nm − 1923.4 nm) are obtained, respectively.

The period and the reflectivity of FBGs will be changed when tuning the central wavelengths of the FBGs by strain.

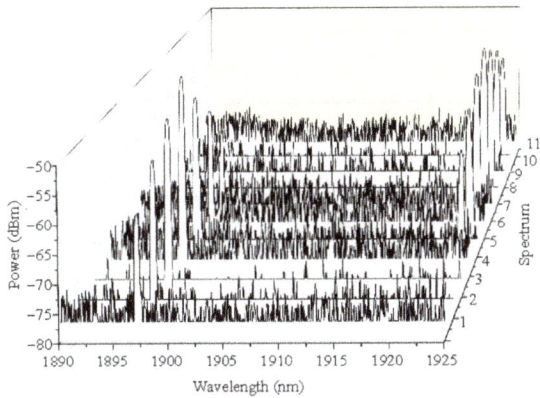

Fig. 9 Output spectra when laser operates in the switching mode [46].

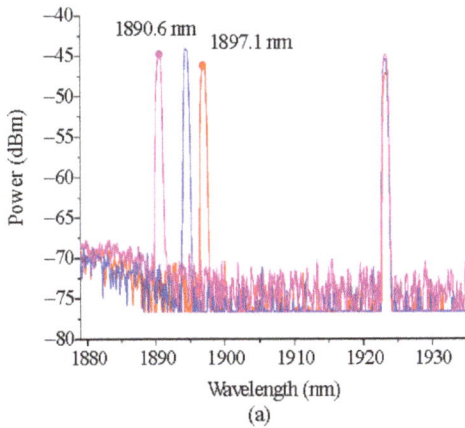

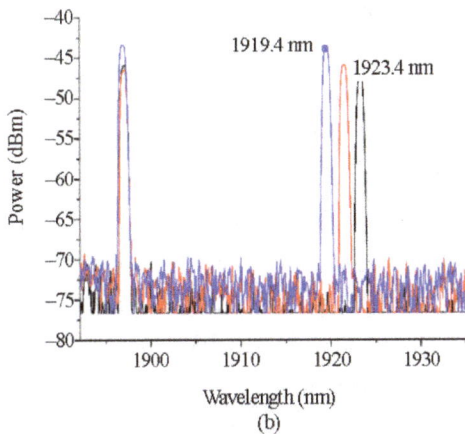

Fig. 10 Output spectrum of the dual-wavelength fiber laser when strain tuning the (a) FBG1 and (b) FBG2 [46].

3.3 Ultra-fast mode-locked fiber lasers

Mode-locked fiber laser has the potential for a spectrally bright pulsed broadband source that can be used for interrogating in the fiber sensor network. The light source is made from relatively low-cost fiber-pigtailed off-the-shelf components and does not require an optical modulator to achieve ultra-short light pulses [47]. Thus far, single-wavelength ultrafast fiber lasers have been investigated theoretically and demonstrated experimentally [48–52].

Multi-wavelength mode-locked fiber lasers have also been widely explored by using actively and passively mode-locked mechanisms in different wavelength regimes (1 μm, 1.5 μm, and 2 μm) [53–57]. Compared with the actively mode-locked fiber lasers, passively mode-locking fiber lasers are technically easy to operate and compact in structure as well as has low jitter in performance. Recently, mode-locked fiber lasers have attracted a great deal of interest due to their compact size and high flexibility for both pulse width and repetition rate. Therefore, this kind of fiber lasers possesses a flexible application in the TDM-based sensing network. Especially, the ultrashort pulse width contributes to the improvement in the spatial resolution.

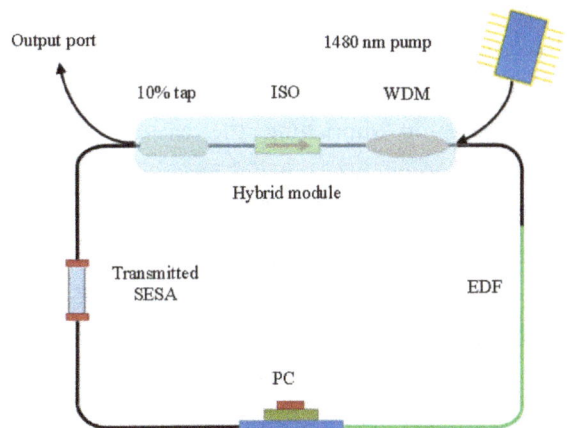

Fig. 11 Schematic of the passively mode-locked fiber laser [58].

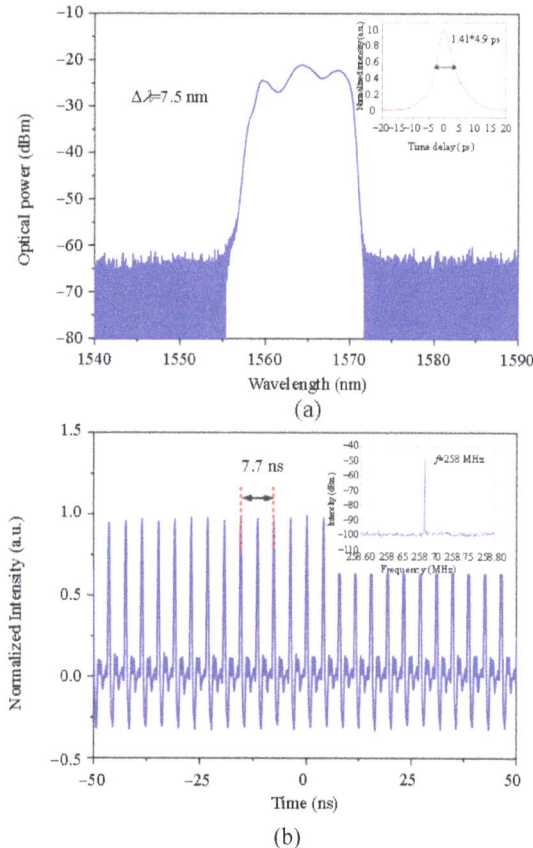

Fig. 12 Characteristics of the passively mode-locked fiber laser:
(a) optical spectrum of the 2nd-order harmonic mode-locked
DSs [inset: the corresponding autocorrelation trace] and
(b) corresponding oscilloscope trace [inset: the RF spectrum] [58].

High repetition rate dissipative soliton (DS) pulses are generated in the normal-dispersion erbium-doped fiber lasers mode-locked by a transmitted semiconductor saturable absorber (SESA), as schematically depicted in Fig. 11. The wavelength division multiplexer/isolator/tap hybrid module (WDM/isolator/tap hybrid module) plays a crucial role in shortening the cavity length for achieving a high fundamental repetition rate. As a result, through tuning the pump power and appropriately adjusting the intra-cavity polarization state, 2nd-order harmonic mode-locked DS pulses are obtained as illustrated in Fig. 12. The rectangular optical spectrum of these DS pulses is shown in Fig. 12(a) with a 3-dB bandwidth of 7.5 nm, and the autocorrelation trace depicted in the inset declares a pulse width of 4.9 ps if a Gaussian pulse shape is assumed. Thus, the time-bandwidth product is 4.6,

indicating that these DS pulses are strongly chirped. Figure 12(b) and its inset show the oscilloscope trace and radio frequency (RF) spectrum, respectively. The harmonic repetition rate is fixed at 258 MHz, and two pulses coexist in a cavity round-trip interval of 7.7 ns, which agree well with the cavity length. The average output power of fundamental mode-locking is around 6.6 mW with the pulse energy of 25.5 pJ. The strongly chirped pulses could be potentially compressed to femtosecond pulses through chirp compensation, which provides a possibility for achieving ultrashort pulsed light source with high repetition rate [58].

3.4 Optical broadband chaos

Optical chaos has been widely applied to fiber fault monitoring [59–61] and fiber sensing [62–66]. The study of optical chaos has lasted for nearly half a century. By introducing an optical feedback into a semi-conductor laser, the laser output is no longer stable, but it shows the property of small random fluctuation in intensity, wavelength, and phase, i.e. chaotic state. With the noise-like property, the correlation of optical chaos has the characteristic of Dirac function, making it extremely suitable for applications in fiber fault locating. Moreover, as the amplitude of the correlation is determined by the intensity of the optical chaos, intensity demodulation in fiber sensing networks can also be achieved at the same time.

The experimental setup of an optical broadband chaos is shown in Fig. 13. A semiconductor optical amplifier (SOA) ring structure with an isolator (ISO) acts as the chaotic light source. The SOA has the parameters of central wavelength (1500 nm), optical bandwidth (74 nm), saturation output power (14 mW), and the small signal gain (13 dB). The 80:20 optical coupler (OC1) provides 20% feedback and 80% output. The polarization controller (PC) is used to adjust the polarization of the light into the SOA, and the erbium doped fiber amplifier (EDFA) can enhance the power of output chaos. A 99:1 optical coupler (OC2) provides 1% transmission light as the reference signal and 99% transmission light as the detected signal. The circulator assures that the return signal can be detected. Two photon detectors (PDs) with 1 GHz bandwidth are used in the correlation detection [61, 65].

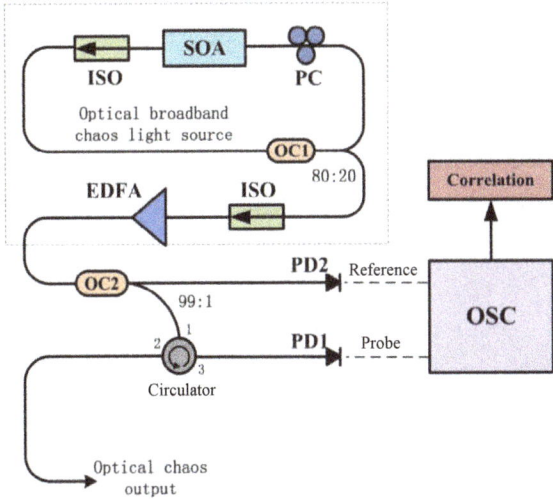

Fig. 13 Schematic diagram of the proposed optical broadband chaos [65].

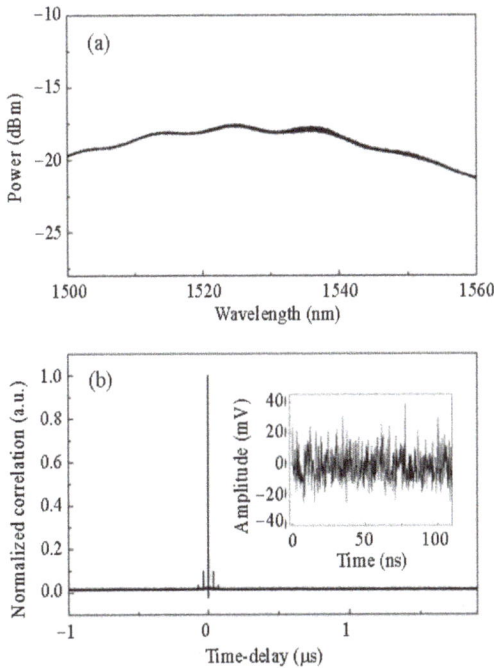

Fig. 14 Characteristics of the broadband chaotic source: (a) the optical spectrum and (b) the autocorrelation curve [inset: the time series] [65].

An optical spectrum analyzer (Yokogawa AQ6370C) and a radio frequency (RF) spectrum analyzer (Agilent 40GHz E4447A) are used to observe the chaotic output light. A real-time oscilloscope (OSC) with 12.5 GHz bandwidth and 50 GSa/s sampling rate is used to record the reference signal and reflected signal. At last, the data recorded in the OSC are processed by the cross correlation method. The characteristics of our proposed optical broadband chaos are demonstrated in Fig. 14(a) with a high quality of autocorrelation curve in Fig. 14(b) (inset: the time series).

4. Intelligent management for FOSN

4.1 Demodulation techniques for FOSN

The demodulation technique is critical for the fiber sensing network to extract the sensing parameters. As the optical spectrum includes comprehensive information such as the wavelength, intensity, and repetition period, the wide bandwidth, high resolution, and high speed optical spectrum scanning are the most important. The most common way to demodulate the spectrum is based on the charge coupled device (CCD) or tunable FP filter. The CCD module has a relatively high acquisition speed up to 3.5 kHz and is suitable for a low light detection. But the physical resolution is limited by the diffraction part to only 20 pm and the integration time limits the speed of the demodulation [67, 68]. The tunable FP filter can be configured flexibly and achieve a high resolution up to 1 pm. However, the FP filter is easy to be affected by temperature fluctuation [69, 70]. Another ultra-high resolution demodulation scheme is the coherence spectrum analysis [71], while the total processing time for the narrow wavelength band will be as long as tens of seconds, which is not suitable for dynamic sensing. In the following, we will present a flexible, wide bandwidth, high resolution, and high speed demodulation scheme for the hybrid multiplexed FOSN.

Figure 15 depicts the composition of the demodulation module. Light from the broadband light source is modulated into pulse by the acoustic optical modulator (AOM) and then transmits to the sensing fiber through the circulator. The light pulse reflected from the microstructure is first amplified by the EDFA, then passes through the tunable FP filter, and finally detected by the high speed avalanche photodiode (APD) detector with the rise time of only 300 ps [72, 73]. To guarantee the synchronization of the AOM, the tunable FP filter, and the analog-to-digital (AD) data acquisition, all of them are driven by the same field programmable gate array (FPGA) card [73]. A symmetrical triangular signal composed of multiple steps is utilized to control the tunable FP filter, which could ensure stable transformation of the driving signal. As a result, the light intensity of different time groups corresponding to certain driving signals, i.e.

wavelength, can be obtained. For accurate demodulation, a tunable laser is employed to calibrate the function of the driving signal-wavelength curve in advance. In addition, a compensating weak FBG isolated from the strain is adopted to eliminate the spectral drift of the FP filter induced by the environment temperature variation through the real-time wavelength calibration, which is beneficial for high accurate and stable demodulation. Based on the above demodulation principle, there is trade-off between the scanning bandwidth and the wavelength resolution, due that the ratio of the bandwidth to the resolution equals to the number of the sampling points. Therefore, the demodulation performance could be flexibly controlled according to the application requirement.

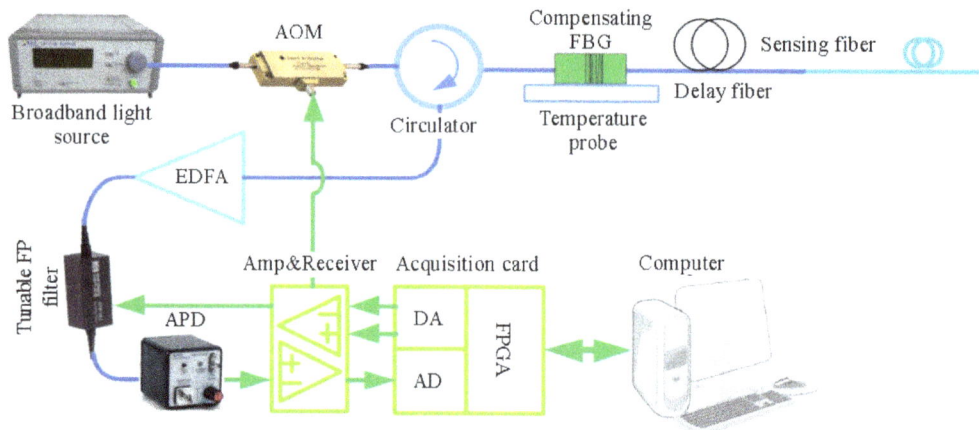

Fig. 15 Schematic diagram of the demodulation module for the hybrid fiber sensor network.

An NI Data acquisition platform is utilized to build the demodulation module. Taking the hybrid WDM/FDM/TDM fiber sensor network discussed in Section 2.2.2 as an example, the demodulation performance is investigated. Figure 16(a) presents the wavelength tracing results of a tunable laser from 1511 nm to 1585 nm, which demonstrates that the scanning bandwidth of the demodulation module is beyond 75 nm. To test the wavelength resolution, we heat the sensing microstructure from 24.2 ℃ to

24.3 ℃, and then it is clear that the peak wavelength changes from 1540.151 nm to 1540.152 nm as shown in Fig. 16(b), which proves that the wavelength resolution is higher than 1 pm. Figure 16(c) displays the driving signal of the FP filter with each step taking 0.5 μs. For the occasion of the 20 nm bandwidth and 5 pm resolution, the sample points could be decreased to 4000. Then, each sweeping cycle will take only 2 ms, corresponding to a high sweeping speed of 500 Hz. In general, a flexible demodulation platform with the maximum bandwidth over 75 nm, high wavelength resolution up to 1 pm, and fast sweeping speed of 500 Hz can be realized for hybrid multiplexed FOSN.

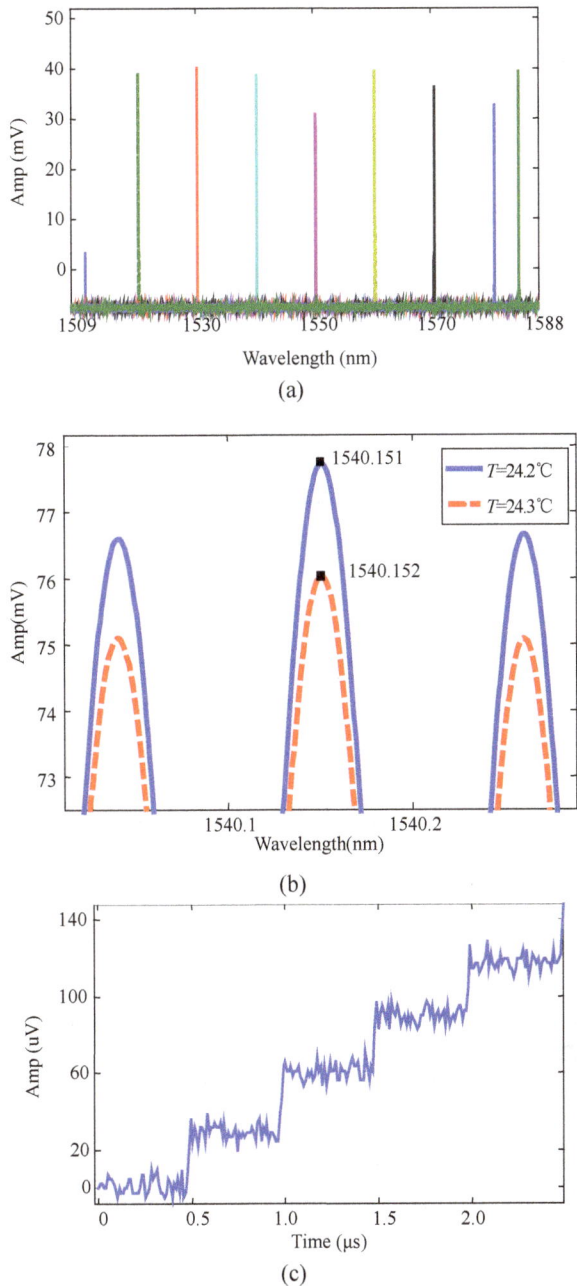

Fig. 16 Demonstration of the demodulation performances: (a) wavelength scanning of a tunable laser, (b) recovered spectrum for 1pm wavelength shift, and (c) fast driving signal of the FP filter.

4.2 Fault monitoring of the fiber transmission link in FOSN

The survivability of the FOSN in harsh environment can not be ignored in practical applications, where the most important for intelligent management is the real-time fault monitoring and locating in the fiber transmission link of the FOSN. Specifically, the detection of faults in fiber links at different wavelengths is urgently required to guarantee the reliable operation of the WDM FOSN. Conventional optical time domain reflectometry (OTDR) has difficulty in diagnosing the faults in a WDM-PON due to its fixed wavelength probe light [74]. Although a wavelength tunable sequence OTDR has been recently reported, without an expensive highspeed sequence generator and a wideband modulator, its resolution is still limited to tens of meters [75].

In order to locate simultaneously and precisely the faults among all fiber links in an FOSN, we provide a simple solution based on optical wideband chaos. Take a large capacity FOSN with hybrid TDM and WDM as the example. The schematic diagram of the large capacity quasi-distributed sensing network is shown in Fig. 17. Chirped fiber Bragg gratings (CFBG) based intensity demodulation is implemented by the programmed optical band pass filter (OBPF, Waveshaper, and FINISAR), which slices optical broadband chaos into multi-wavelength channels in consistent with the CFBG sensors [76].

A cross-correlation algorithm is applied to the chaotic reference signal and probe signal to interrogate the measurand sensing and the precise locating of each sensor. Meanwhile, the fiber breakpoint location can be also achieved through processing the reference signal and the Fresnel reflection signal of the fiber fault based on the same principle [61]. A proof-of-concept experiment is conducted with six CFBG sensors (S11, S12, S13, S21, S23, and S32) deployed in three sensing fiber lines. Three wavelength channels are filtered by the OBPF in consistent with the CFBG sensors, and an additional channel (testing channel) of 2.4-nm (300-GHz) bandwidth is reserved to provide sufficient optical power for fiber fault monitoring, as depicted in Fig. 18 [76].

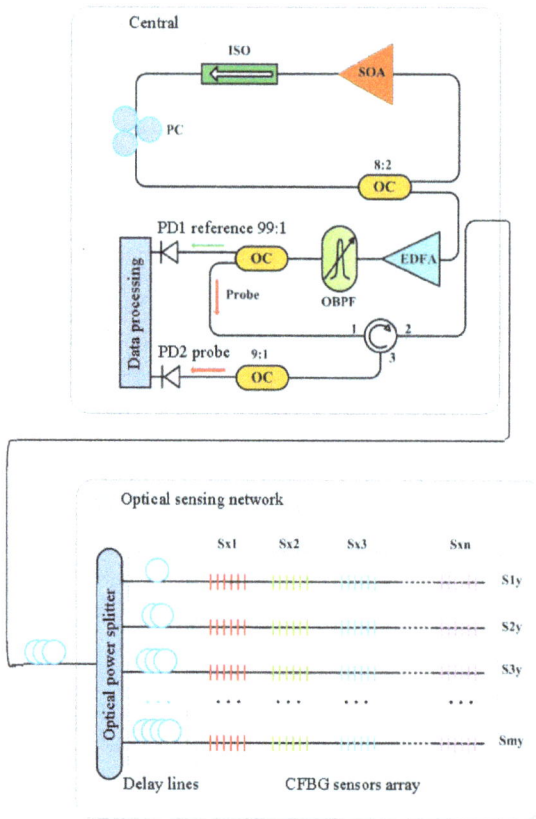

Fig. 17 Schematic diagram of the large capacity quasi-distributed sensing network [76].

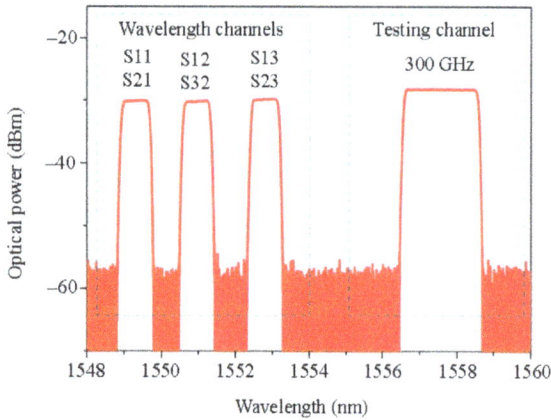

Fig. 18 Optical spectrum of the multichannel chaos, including three wavelength channels and a testing channel [76].

When strain is applied to a certain CFBG, the amplitude of the corresponding correlation peak will decrease due to a decrease in the reflection intensity. When a breakpoint occurs in the sensing network, an extra correlation peak will appear in the correlation spectrum, and its position indicates the location of the breaking point. Figure 19 presents the experimental results of simultaneously three fault points in different sensing lines, respectively. Through the peak searching algorithm, precise locating of the breakpoints can be pinpointed with a spatial resolution of 2.8 cm [76].

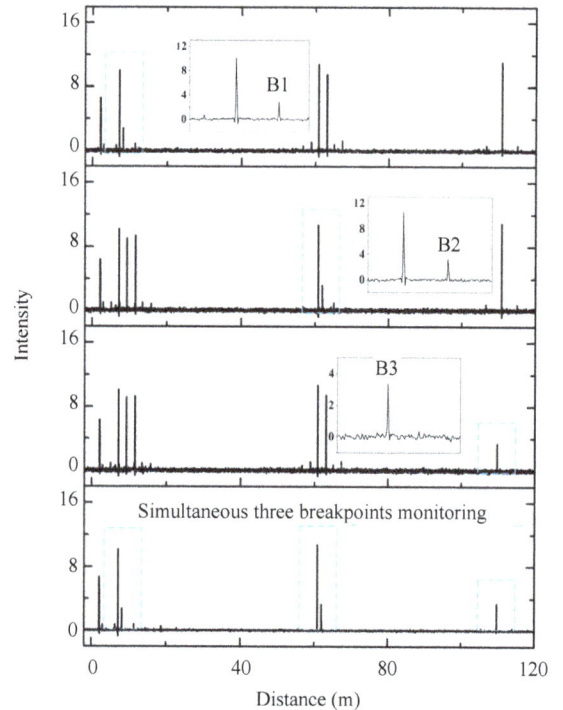

Fig. 19 Experimental results of the real-time fiber fault monitoring [76].

5. Fiber optic gas sensing structures and networks

Fiber optic gas sensors have intrinsic benefits over conventional dielectric and semiconductor sensors, as they are immune to electromagnetic interference, safe in dangerous or hazardous environments, and furthermore facilitate remote and distributed sensing. Numerous techniques are investigated for optical gas sensing. For instance, techniques based on spectrum absorbing including non-dispersive infrared (NDIR), tunable diode laser spectroscopy (TDLS), cavity ring-down spectroscopy (CRDS) as well as photo-acoustic detection (PAS) have been demonstrated [77]. In this section, research progress on special fiber optic structures designed for gas sensing, realizations of optical gas sensing systems by combination the fiber optic sensors and optical networks, and typical practical applications and demonstrations of fiber optic gas sensing networks are reviewed.

5.1 Engineered fiber optic gas sensing structures

Novel fiber structures such as surface processed fibers with sensitivity enhancement, fiber gratings, and photonic crystal fibers can be employed as gas sensors due to the evanescent field of the core mode. For gas absorption measurement, the evanescent field has to be extended to penetrate the external environment to provide interaction between the guided light and the external gas [78].

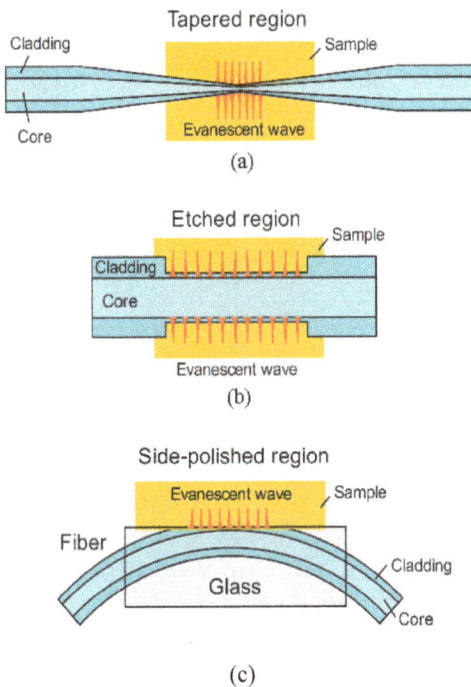

(a)

(b)

(c)

Fig. 20 Scheme diagrams of (a) tapered fiber, (b) etched fiber, and (c) side-polished fiber [78].

Specific fiber surface processing techniques have been adopted, including tapered fiber, etched fiber, and side-polished fiber, as illustrated in Fig. 20. These different types of sensor elements are used in fiber-cavity ring-down spectroscopy for gas sensing

[78].

Micro/nano fiber (MNF) with the sub-wavelength diameter is also a good candidate for gas sensing. By using a graphene coated micro fiber, an all-fiber ammonia gas sensor can be realized, as shown in Fig. 21. The microfiber is bilaterally stretched from a standard single mode fiber (SMF) by employing the flame heating and taper-drawing technology. Graphene is coated on the microfiber by pump light induced deposition, which improves gas sensitivity through its significant high specific surface area to attract gas molecules and thus achieve a resolution of 0.74 ppm in the ammonia gas detection [79].

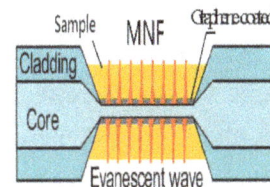

Fig. 21 Structure of graphene coated micro/nano fiber served as the ammonia gas sensor.

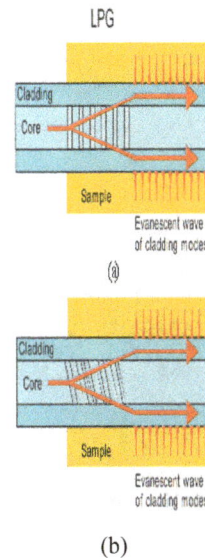

(a)

(b)

Fig. 22 Scheme diagrams of (a) LPG and (b) TFBG [78].

Fiber gratings have similar advantages along with the surface processed fiber sensor in stability and establishing fiber networks. Evanescent waves of cladding modes of the fiber grating interact with the surrounding medium, which leads to the change in the resonant wavelength. Most used fiber gratings are long period grating (LPG) and tilted FBG (TFBG) [78, 80], as presented in Fig. 22.

The holes in the photonic crystal fiber (PCF) can also be utilized as convenient repositories for gases. A significant overlap between the physical location of the holes and the guided modes leads to strong interaction between the two and allows great potential for gas sensing. The PCF is extremely potential for being a long path length, and low volume gas cell, which can bring big benefit in enhancing sensitivity from the Bear-Lambert law. Nevertheless, one of the main challenges of using these PCF gas cells is the effort to fill the PCF with the target gas [81]. Techniques to improve the sample filling time include increasing the pressure difference across the fiber and increasing the cell diameter or introducing holes to allow gas flow or diffusion along the waveguide's length [82], as

(a)

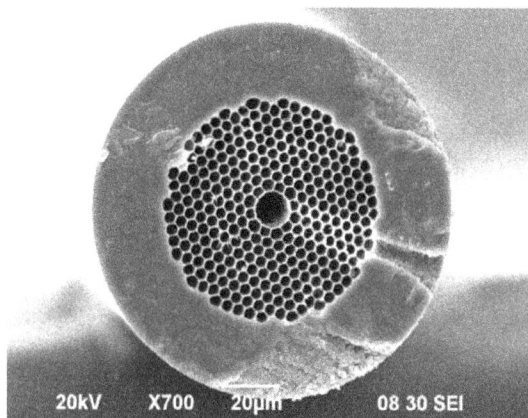

(b)

Fig. 23 PBGF images for gas detection: (a) side-view and (b) cross-section of a micro channel fabricated on the hollow core PBGFs (HC-PBGFs) [81].

illustrated in Fig. 23. For example, Parry's group has reported a methane sensor with 500 ppm detection limit by using a 10 μm core photonic band-gap fiber (PBGF) operated at 1.650 μm [83].

5.2 All fiber optic gas sensing networks

In many occasions, the multipoint gas sensing network is desired to obtain a large-scale gas distribution and reduce the cost per sensing point. This can be realized by the combination of fiber optic sensors and fiber optic networking technologies. In order to accurately and effectively detect the multipoint data, versatile multiplexing techniques have been developed, such as spatial SDM [84], TDM [85], frequency division multiple access (FDMA) [86], frequency modulated continuous wave (FMCW) [87], synthesis of optical coherence function (SOCF) [88], and WDM [89]. These techniques own unique features. For example, For FWCW, the laser source needs intensity modulation, and the complex system requires the voltage controlled oscillator and RF mixer. For SOCF, only one sensor can be measured each time, and the location resolution is limited by the coherent length of the laser [88].

Recently, based on TDLS and TDM techniques, commercial products named OptoSniff® as fiber optic gas sensing networks are presented by the OptoSci Ltd [90]. The overview of the system is schematically shown in Fig. 24, including the central control unit (CCU), fiber optic network composed of power splitters and optical multicore cables, and all-optical gas sensing cells. Due to the benefits of optical fiber networks and passive devices, it offers continuous, multipoint methane/natural gas leak detection and monitoring, from hundreds of sensing points (240 points) over a long distance (20 km). These products have been applied in British landfills and Japanese Tokyo gas service tunnels, monitoring methane and carbon dioxide, lasting for over 2 years. This is a typical case and major breakthrough for the practical applications of all-fiber optic gas sensing networks.

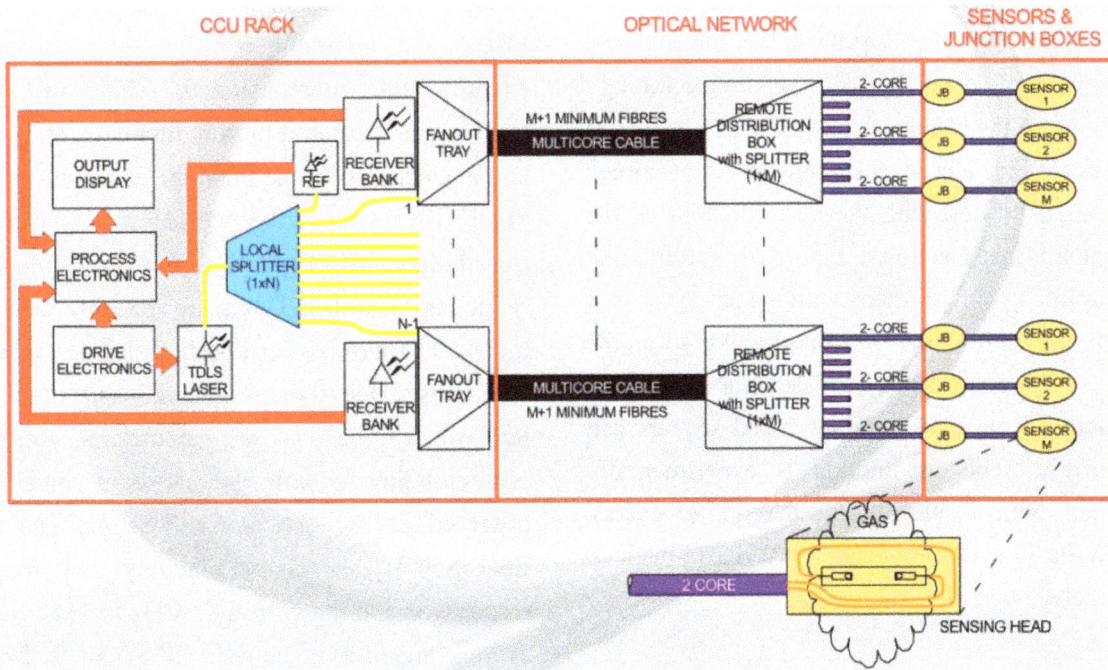

Fig. 24 Schematic diagram of the OptoSniff® multipoint fiber optic gas sensing system [90].

6. Fiber optic acoustic sensors for the sensing network

Acoustic sensors have been studied extensively for decades and play an important role in the modern society. The acoustic wave has a wide frequency spectrum range that extends from infrasound (<20 Hz) up to ultrasound (in the GHz-band), and different frequency bands have different sensing applications [91]. For the unique advantages of large bandwidth, high sensitivity, immunity to electromagnetic interference, remote detection, and multiplexing capability, fiber optic acoustic sensors can break the limitations of traditional electric acoustic sensors [92]. Currently, numerous fiber optic acoustic sensors have been researched, of which the configurations are based on fiber optic interferometers such as Mach-Zehnder, Michelson, Sagnac, Fabry-Pérot, or FBGs [93], and also some special structures such as fiber lasers [94], couplers [95], and tapers [96].

The low-frequency acoustic wave detection is of great significance because of its wide applications in

the fields of early warning of natural disasters such as earthquakes [97]. However, little work has been done on the research of the optical acoustic sensor for the low-frequency acoustic detection. In order to offer the fiber acoustic sensors with higher sensitivity, good low-frequency response, and easy networking, we have designed several kinds of fiber acoustic sensors, such as intensity modulated sensors with the single-mode/multimode fiber coupler, wavelength modulated acoustic sensor based on dual FBGs, and beat-frequency modulated sensor based on multi-longitudinal mode fiber laser.

6.1 Intensity modulated acoustic sensors [98]

Intensity modulated acoustic sensors can detect the acoustic signal by the changes in the optical intensity directly without the high cost and complex demodulation system, which can be easily used in the FOSN through TDM and SDM. One typical scheme is based on the non-standard fused single-mode/multimode fiber coupler and aluminum foil. For a fused 2×2 single-mode fiber coupler, the coupling ratio (CR) of two output ports is dependent on the distance (d) between centers of two optical fibers as shown in Fig. 25 [98]. When the acoustic pressure is applied on the foil, the deformation of foil will cause the d change of the fused coupler as shown in Fig. 26. Hence, the acoustic vibration can be measured by detecting the insertion loss of the non-standard fused coupler. Figure 25 implies that the "coupling cycle" shortens with a decreasing in d (from right to left of the horizontal axis), so as the number of coupling cycle increases, the sensitivity of the multi-cycle fused coupler based sensors can be improved. This assumption is experimentally conformed with result of 0.18 mW/Pa (1 cycle), 0.65 mW/Pa (3 cycles), and 1.71 mW/Pa (5 cycles), respectively.

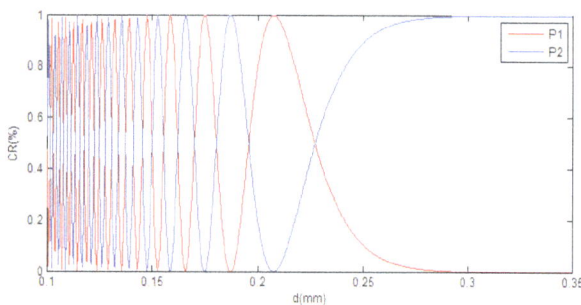

Fig. 25 Simulation of the relationship between CR and d during the fused process [98].

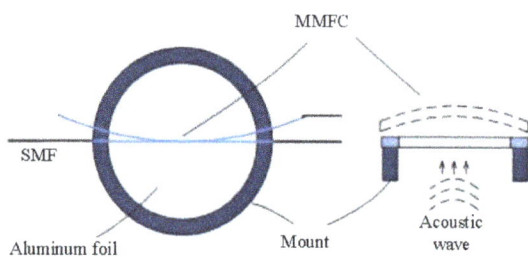

Fig. 26 Sensing mechanism of the fused coupler based acoustic sensor [98].

Moreover, the multimode fused coupler based sensor also has been studied which shows a sensitivity of 2.63 mW/Pa, with the operation frequency range from 20 Hz to 20 kHz and average SNR about 20 dB.

6.2 Wavelength modulated acoustic sensors [99]

Wavelength modulated acoustic sensors are suitable for WDM. An acoustic sensor based on dual FBGs and a titanium film is proposed and experimentally demonstrated, where the sensing head is manufactured by longitudinally sticking one end of each FBG to the titanium film as illustrated in Fig. 27 [99]. One of the two FBGs is employed for transmitting light, and the other is employed for reflecting. The dual FBGs are not only working as the signal transmission line but also as the sensing component. Due to the wavelength shift of the FBGs induced by the external acoustic vibration, the corresponding periodic fluctuation in power can be observed on the optical oscilloscope. The optical fiber sensor shows a relatively flat frequency response in the range from 100 Hz to 1 kHz with the average signal-to-noise ratio (SNR) above 22 dB. In addition, the maximum sound pressure sensitivity of the proposed sensor is found to be 90 μW/Pa within a sound pressure range 100.3 dB – 118.5 dB. The sensing system presents good stability and reliability, and has the advantage of direct self-demodulation.

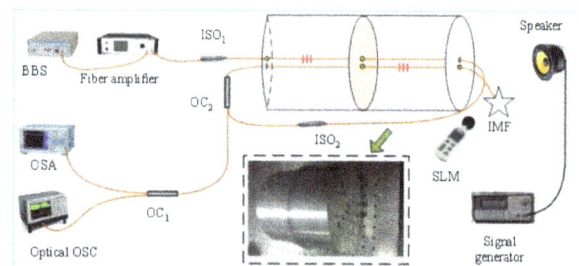

Fig. 27 Schematic diagram of the acoustic sensor based on the dual FBGs and a titanium film [99].

6.3 Beat-frequency modulated acoustic sensors [100]

The passively mode-locked technique which employs carbon nanotubes (CNT) is a widely used method for mode locking and frequency stabilization. The unique frequency stability can be also used for the acoustic pressure measurement [100]. The polymer membrane is employed beneath the pre-strained Er^+ doped fiber (EDF) to convert the sound pressure disturbance into axial strain, alter the cavity length, and induce shift of the longitudinal modes beat as displayed in Fig. 28. Hence, the acoustic pressure measurement can be carried out by detecting the shift of beat frequency. The sensor shows comparable strain and sound

pressure sensitivity of 0.5 kHz/µε and 147.2 Hz/Pa, respectively. The sensor can be an alternative to measure acoustic pressure with good stability, as well as a valuable candidate for the FOSN.

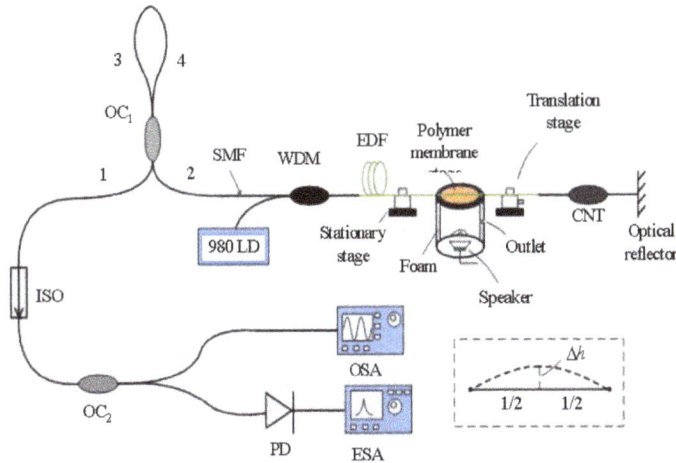

Fig. 28 Schematic diagram of the passively mode-locked multi-longitudinal mode fiber laser sensor [100].

7. Fiber optic static/dynamic strain sensing network

Fiber sensor networks have been widely used in health monitoring of civil and mechanical structures due to their advantages over electrical sensing devices, such as electric insulation and immunity to electromagnetic interference. Meanwhile, the research and application of optical chaos have attracted more and more attention of researchers recently. In the following, we mainly introduce an FBG sensor network for static/dynamic strain sensing. By adopting the optical wideband chaos described in Section 3.4 as the light source, the location of sensing FBGs can be distinguished by the different peak positions along the correlation spectrum, which means that the identical FBGs can be adopted in the quasi-distributed sensing network, which will increase the network multiplexing capabilities [65].

The WDM based multi-point static strain sensing setup is constructed with a 100-GHz-spaced WDM device with 2 channels (CH33 and CH34)

and the corresponding sensing grating at each branch, as shown in Fig. 29. The WDM device acts as a band pass filter. The reflection spectrum of the corresponding sensing grating is chosen to match the transmission spectrum of WDM. In the initial, the reflection intensity will get almost 100% reflection after the combined filtering from the WDM and the grating, if we ignore the insertion loss of WDM. When the strain is applied, the wavelength of the grating will shift towards the longer wavelength. Thus, the wavelength spacing between the WDM and the sensing grating will lead to a decrease in the reflection intensity. As we know, when the intensity of optical chaotic reflection changes, the amplitude of peak in the correlation spectrum will also change after the cross correlation calculation. In this way, strain is demodulated from the amplitude of the correlation peaks. In addition, the position of correlation peaks suggests the location of the gratings, thus strain sensing and locating can be achieved at the same time. The correlation spectrum with and without strain applied is shown in Figs. 30(a) and 30(b) [65].

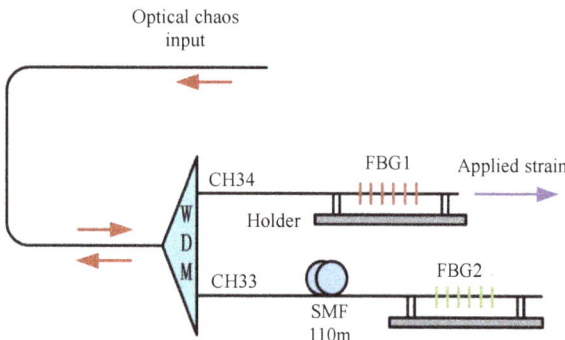

Fig. 29 Schematic of the static strain sensing setup [65].

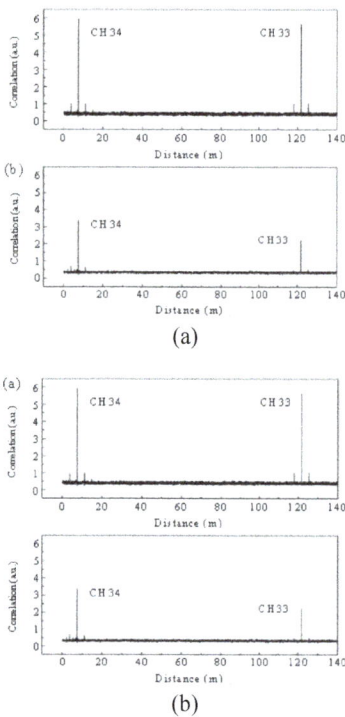

Fig. 30 Correlation spectra of (a) free of strain and (b) strain applied [65].

The dynamic sensing setup is shown in Fig. 31 [65]. The dynamic sensing unit is a section of no-core fiber (NCF) with the length of 29.2 mm, which is spliced to the lead-in SMF and lead-out SMF. NCF is a special fiber which has the same diameter of 125 μm as SMF, but only contains solid cladding and coating made of pure fused silica materials and polymer materials, respectively. The tiny bending of NCF leads to an increase in transmission loss induced by the leaking modes. When NCF vibrates, the light intensity will be modulated by the vibration [101].

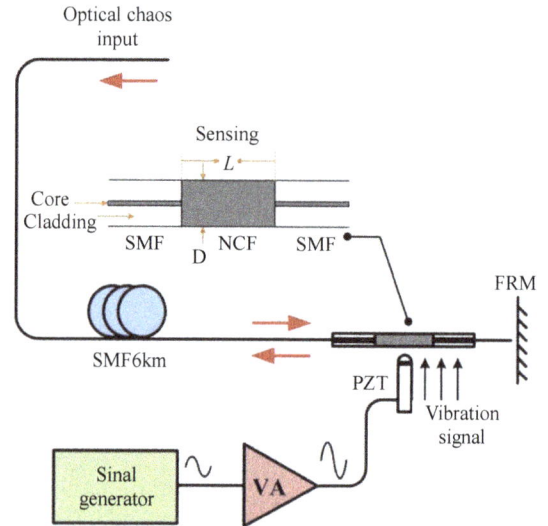

Fig. 31 Schematic of the dynamic strain sensing setup [65].

The locating of sensing point can be accomplished easily from the position of the correlation peaks, similar to that of the static sensing application. For the demodulation of the vibration signal, the reflected signal and the reference signal can be divided into segments in time serial, and the corresponding segments are processed with cross correlation calculation. The correlation peak amplitude of each segment is determined by the reflection light intensity at different recording time slots, which fluctuates with the vibration. Thus, each correlation peak can be seen as a sampling point of vibration. By splicing the correlation curves in time order, we get the fluctuation curve of light intensity from the envelope of correlation peaks, as shown in Fig. 32. Then by conducting the Fourier transformation, the vibration signal can be demodulated [65].

Further, this system possesses a possibility of multi-point sensing along one fiber line. Different points can be distinguished by different positions of correlation peaks. The vibration frequency at each point can be detected by the amplitude variation of the corresponding correlation peak. The wideband chaos also can guarantee the usage of multiple multiplexing methods such as TDM and WDM together in the FOSN and can be extended to many other sensing parameters, e.g. temperature, refractive index, acceleration, and liquid level, as long as the intensity of chaotic reflection can be modulated by these sensing parameters [65].

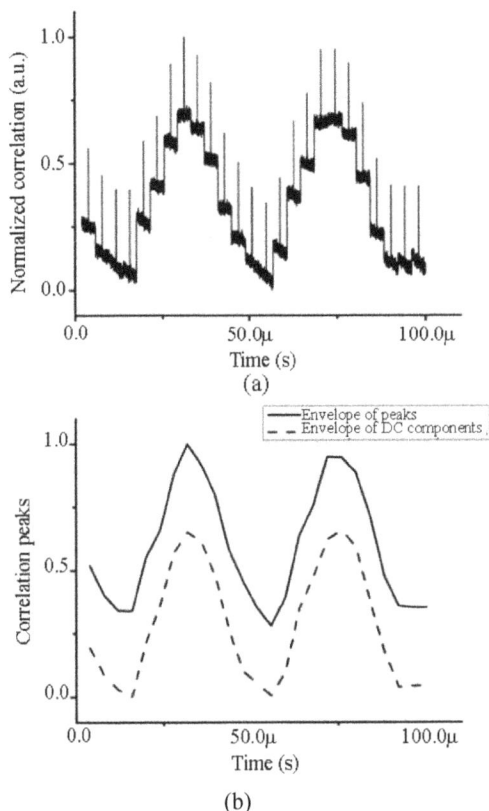

Fig. 32 Vibration demodulation scheme: (a) sampling multiple correlation peak profile and (b) envelope of multiple correlation peaks and direct current (DC) components [65].

8. Conclusions

So far, many efforts to improve the performance of the FOSN have been proposed and realized. Benefiting from the high efficient utilization of the signal resources, the topology architecture of the SPON and hybrid multiplexing based FOSN have shown the advantages of large capacity, flexible access, and integrated management. Further, special fiber lasers as well as the optical chaos have demonstrated the superiorities of high adaptability, good integration, and low cost to meet the requirements of the FOSN. Also, the flexible optical spectrum demodulation technique for sensing signals, as well as the precisely fault monitoring

technique for the fiber link have been provided to enhance the sensing performance and survivability of the FOSN. Based on the above key technologies and devices, the FOSN exhibits good adaptiveness for practical application fields such as gas sensing, acoustic sensing, and strain/dynamic strain sensing. The study on the key device and technology of the FOSN will provide the fundamental support and future outlook for fiber sensing.

Acknowledgment

These works are supported by a grant from the Sub-Project of the Major Program of the National Natural Science Foundation of China (No. 61290315), the National Natural Science Foundation of China (No. 61275083, 61275004, and 61404056), the National Key Foundation of Exploring Scientific Instrument of China (No. 2013YQ16048707), and the Fundamental Research Funds for the Central Universities (HUST: No. 2014CG002, and 2014QNRC005). Much appreciation should be given to the students, Zhinlin Xu, Yiyang Luo, Fan Ai , Wei Yang, Enci Chen, Shun Wang ,Shui Zhao, Li Liu, Hao Liao, Xin Fu, Shun Wang, Wei Yang, Wang Yang, and Mingren Su.

References

[1] S. Huang, W. Lin, and M. Chen, "Time-division multiplexing of polarization-insensitive fiber-optic Michelson interferometric sensors," *Optics Letters*, 1995, 20(11): 1244–1246.

[2] D. T. Jenstrom and C. L. Chen, "A fiber optic microbend tactile sensor array," *Sensors and Actuators*, 1989, 20(3): 239–248.

[3] G. Stewart, C. Tandy, D. Moodie, M. A. Morante, and F. Dong, "Design of a fibre optic multi-point sensor

for gas detection," *Sensors and Actuators B: Chemical*, 1998, 51(98): 227–232.

[4] C. Zhou, Y. Rao, and J. Jiang, "A coarse wavelength-division-multiplexed extrinsic fiber Fabry-Perot sensor system," in *Proc. SPIE*, vol. 5634, pp. 219–224, 2005.

[5] Z. Wang, F. Shen, L. Song, X. Wang, and A. Wang, "Multiplexed fiber Fabry-Perot interferometer sensors based on ultrashort Bragg gratings," *IEEE Photonics Technology Letters*, 2007, 19(8): 622–624.

[6] B. J. Vakoc, M. J. F. Digonnet, and G. S. Kino, "A novel fiber-optic sensor array based on the Sagnac interferometer," *Journal of Lightwave Technology*, 1999, 17(11): 2316–2326.

[7] H. Fu, D. Chen, and Z. Cai, "Fiber sensor systems based on fiber laser and microwave photonic technologies," *Sensors*, 2012, 12(5): 5395–5419.

[8] R. A. Perez-Herrera, M. Fernandez-Vallejo, and M. Lopez-Amo, "Robust fiber-optic sensor networks," *Photonic Sensors*, 2012, 2(4): 366–380.

[9] G. R. Kirikera, O. Balogun, and S. Krishnaswamy, "Adaptive fiber Bragg grating sensor network for structural health monitoring: applications to impact monitoring," *Structure Health Monitoring*, 2011, 10(1): 5–16.

[10] C. Chan, W. Jin, H. L. Ho, and M. Suleyman Demokan, "Performance analysis of a time-division-multiplexed fiber Bragg grating sensor array by use of a tunable laser source," *IEEE Journal of Selected Topics in Quantum Electronics*, 2000, 6(5): 741–749.

[11] D. Cooper, T. Coroy, and P. Smith, "Time-division multiplexing of large serial fiber-optic Bragg grating sensor arrays," *Applied Optics*, 2001, 40(16): 2643–2654.

[12] Y. Dai, Y. Liu, J. Leng, G. Deng, and A. Asundi, "A novel time-division multiplexing fiber Bragg grating sensor interrogator for structural health monitoring," *Optics and Lasers in Engineering*, 2009, 47(10): 1028–1033.

[13] M. Zhang, Q. Sun, Z. Wang, X. Li, H. Liu, and D. M. Liu, "A large capacity sensing network with identical weak fiber Bragg gratings multiplexing," *Optics Communications*, 2012, 285(13): 3082–3087.

[14] C. Hu, H. Wen, and W. Bai, "A novel interrogation system for large scale sensing network with identical ultra-weak fiber Bragg gratings," *Journal of Lightwave Technology*, 2014, 32(7): 1406–1411.

[15] G. D. Lloyd, L. Everall, K. Sugden, and I. Bennion, "Resonant cavity time-division-multiplexed fiber Bragg grating sensor interrogator," *IEEE Photonics Technology Letters*, 2004, 16(10): 2323–2325.

[16] Y. Yu, L. Lui, H. Y. Tam, and W. Chung, "Fiber-laser based wavelength division multiplexed fiber Bragg grating sensor system," *IEEE Photonics Technology Letters*, 2001, 13(7): 702–704.

[17] K. Takada, "High-resolution OFDR with incorporated fiber-optic frequency encoder," *IEEE Photonics Technology Letters*, 1992, 4(9): 1069–1072.

[18] B. A. Childers, M. E. Froggatt, S. G. Allison, T. C. Moore, D. A. Hare, C. F. Batten, *et al.*, "Use of 3000 Bragg grating strain sensors distributed on four 8-m optical fibers during static load tests of a composite structure," in *Proc. SPIE*, vol. 4332, pp. 133–142, 2001.

[19] J. L. Brooks, R. H. Wentworth, R. C. Youngquist, M. Tur, Y. K. Byoung, and H. J. Shaw, "Coherence multiplexing of fiber-optic interferometric sensors," *Journal of Lightwave Technology*, 1985, 3(5): 1062–1072.

[20] Y. Rao, K. Kalli, G. Brady, D, J, Webb, D. A. Jackson, L. Zhang, *et al.*, "Spatially-multiplexed fibre-optic Bragg grating strain and temperature sensor system based on interferometric wavelength-shift detection," *Electronics Letters*, 1995, 31(12): 1009–1010.

[21] Y. Rao, D. A. Jackson, L. Zhang, and I. Bennion, "Strain sensing of modern composite materials with a spatial/wavelength-division multiplexed fiber grating network," *Optics Letters*, 1996, 21(9): 683–685.

[22] Y. Rao, A. B. L. Ribeiro, D. A. Jackson, L. Zhang, and I. Bennion, "Simultaneous spatial, time and wavelength division multiplexed in-fibre grating sensing network," *Optics Communications*, 1996, 125(1): 53–58.

[23] Q. Sun, X. Li, M. Zhang, Q. Liu, H. Liu, and D. Liu, "High capacity fiber optic sensor networks using hybrid multiplexing techniques and their applications," in *Proc. SPIE*, vol. 9044, pp. 90440L-1–90440L-10, 2013.

[24] X. Li, Q. Sun, J. Wo, M. Zhang, and D. Liu, "Hybrid TDM/WDM based fiber-optic sensor network for perimeter intrusion detection," *Journal of Lightwave Technology*, 2012, 30(8): 1113–1120.

[25] X. Li, Q. Sun, D. Liu, R. Liang, J. Zhang, J. Wo, *et al.*, "Simultaneous wavelength and frequency encoded microstructure based quasi-distributed temperature sensor," *Optics Express*, 2012, 20(11): 12076–12084.

[26] Z. Zhang, L. Zhan, K. Xu, J. Wu, Y. Xia, and J. Lin,

"Multiwavelength fiber laser with fine adjustment, based on nonlinear polarization rotation and birefringence fiber filter," *Optics Letters*, 2008, 33(4): 324–326.

[27] Y. Song, L. Zhan, S. Hu, Q. Ye, and Y. Xia, "Tunable multiwavelengthBrillouin-erbium fiber laser with a polarization-maintaining fiber sagnac loop filter," *IEEE Photonics Technology Letters*, 2004, 16(9): 2015–2017.

[28] H. Dong, G. Zhu, Q. Wang, H. Sun, N. K. Dutta, J. Jaques, *et al*., "Multiwavelength fiber ring laser source based on a delayed interferometer," *IEEE Photonics Technology Letters*, 2005, 17(2): 303–305.

[29] Y. Han, X. Dong, C. Kim, M. Jeong, and J. Lee, "Flexible all fiber Fabry-Pérot filters based on superimposed chirped fiber Bragg gratings with continuous FSR tunability and its application to a multiwavelength fiber laser," *Optics Express*, 2007, 15(6): 2921–2926.

[30] Y. G. Han, C. S. Kim, J. U. Kang, and U. C. Peak, "Multiwavelength Raman fiber-ring laser based on tunable cascaded long-period fiber gratings," *IEEE Photonics Technology Letters*, 2003, 15(3): 383–385.

[31] J. Yang, S. C. Tjin, and N. Q. Ngo, "Multiwavelength tunable fiber ring laser based on sampled chirp fiber Bragg grating," *IEEE Photonics Technology Letters*, 2014, 16(4): 1026–1028.

[32] X. Jiang, Q. Yang, G. Vienne, Y. Li, L. Tong, J. Zhang, *et al*., "Demonstration of microfiber knot laser," *Applied Physics Lett*ers, 2006, 89(14): 143513.

[33] W. Fan, J. Gan, Z. Zhang, X. Wei, S. Xu, and Z. Yang, "Narrow linewidth single frequency microfiber laser," *Optics Letters*, 2012, 37(20): 4323–4325.

[34] A. Sulaiman, S. W. Harun, H. Arof, and H. Ahmad, "Compact and tunable erbium-doped fiber laser with microfiber Mach-Zehnder interferometer," *IEEE Journal of Quantum Electronics*, 2012, 48(9): 1165–1168.

[35] W. Jia, Q. Sun, X. Sun, J. Wo, Z. Xu, D. Liu, *et al*., "Wideband microfiber Fabry-Perot filter and its application to multiwavelength fiber ring laser," *IEEE Photonics Technology Letters*, 2014, 26(10), 961–964.

[36] W. Jia, Q. Sun, Z. Xu, X. Sun, and D. Liu, "Channel-spacing tunable multiwavelength erbium-doped fiber laser based on a microfiber Fabry-Perot filter," in *2014 Conference on Lasers and Electro-Optics*, San Jose, pp. 1–2, 2014.

[37] J. Geng and S. Jiang, "Fiber lasers: the 2µm market heats up," *Optics and Photonics News*, 2014, 25(7): 4–41.

[38] F. McALeavey, B. D. MacCraith, J. O'Gorman, and J. Hegarty, "Tunable and efficient diode-pumped Tm^{3+}-doped fluoride fiber laser for hydrocarbon gas sensing," *Fiber & Integrated Optics*, 1997, 16(4): 355–368.

[39] Q. Mao and J. W. Y. Lit, "Switchable multiwavelength erbium-doped fiber laser with cascaded fiber grating cavities," *IEEE Photonics Technology Letters*, 2002, 14(5): 612–614.

[40] S. Liu, F. Yan, T. Feng, B. Wu, Z. Dong,k and G. K. Chang, "Switchable and spacing-tunable dual-wavelength thulium-doped silica fiber laser based on a nonlinear amplifier loop mirror," *Applied Optics*, 2014, 53(24): 5522–5526.

[41] X. Ma, S. Luo, and D. Chen, "Switchable and tunable thulium-doped fiber laser incorporating a Sagnac loop mirror," *Applied Optics*, 2014, 53(20): 4382–4385.

[42] S. Liu, F. Yan, W. Peng, T. Feng Z. Dong, and G. Chang, "Tunable dual-wavelength thulium-doped fiber laser by employing a HB-FBG," *IEEE Photonics Technology Letters*, 2014, 26(18): 1809–1812.

[43] W. Peng, F. Yan, Q. Li, S. Liu, T. Feng, S. Y. Tan, *et al*., "1.94 µm switchable dual-wavelength Tm^{3+} fiber laser employing high-birefringence fiber Bragg grating," *Applied Optics*, 2013, 52(19): 4601–4607.

[44] M. Delgado-Pinar, J. Mora, A. Díez, J. L. Cruz, and M. V. Andrés, "Wavelength-switchable fiber laser using acoustic waves," *IEEE Photonics Technology Letters*, 2005, 17(3): 552–554.

[45] S. Zhao, P. Lu, D. Liu, and J. Zhang, "Switchable multiwavelength thulium-doped fiber ring lasers," *Optical Engineering*, 2013, 52(8): 086105–086111.

[46] W. Yang, P. Lu, S. Wang, D. Liu, and J. Zhang, "2-µm switchable, tunable and power-controllable dual-wavelength fiber laser based on parallel cavities using 3×3 coupler," *Applied Physics B*, 2015, 120(2): 349–354.

[47] M. A. Putnam, M. L. Dennis, I. N. Duling, C. G. Askins, and E. J. Friebele, "Broadband square-pulse operation of a passively mode-locked fiber laser for fiber Bragg grating interrogation," *Optics Letters*. 1998, 23(2): 138–140.

[48] M. A. Solodyankin, E. D. Obraztsova, A. S. Lobach, A. I. Chernov, A. V. Tausenev, V. I. Konov, *et al*.,

"Mode-locked 1.93 mm thulium fiber laser with a carbon nanotube absorber," *Optics Letters*, 2008, 33(12): 1336–1338.

[49] F. Bonaccorso, Z. Sun, T. Hasan, and A. C. Ferrari, "Graphene photonics and optoelectronics," *Nature Photonics*, 2010, 4(9): 611–622.

[50] X. Liu, "Interaction and motion of solitons in passively-mode-locked fiber lasers," *Physical Review A*, 2011, 84(5): 1688–1690.

[51] Z. Sun, T. Hasan, F. Wang, A. G. Rozhin, I. H. White, and A. C. Ferrari, "Ultrafast stretched-pulse fiber laser mode-locked by carbon nanotubes," *Nano Research*, 2010, 3(6): 404–411.

[52] K. Kieu and F. W. Wise, "All-fiber normal-dispersion femtosecond laser," *Optics Express*, 2008, 16(15): 11453–11458.

[53] L. R. Chen, G. E. Town, P. Y. Cortès, S. LaRochelle, and P. W. E. Smith, "Dual-wavelength, actively mode-locked fibre laser with 0.7 nm wavelength spacing," *Electronics Letters*, 2000, 36(23): 1921–1923.

[54] S. Pan and C. Lou, "Stable multiwavelength dispersion-tuned actively mode-locked erbium-doped fiber ring laser using nonlinear polarization rotation," *IEEE Photonics Technology Letters*, 2006, 18(13): 1451–1453.

[55] D. Pudo and L. R. Chen, "Actively mode locked, quadruple-wavelength fibre laser with pump-controlled wavelength switching," *Electronics Letters*, 2003, 39(3): 272–274.

[56] Z. Yan, X. Li, Y. Tang, P. P. Shum, X. Yu, Y. Zhang, et al., "Tunable and switchable dual-wavelength Tm-doped modelocked fiber laser by nonlinear polarization evolution," *Optics Express*, 2015, 23(4): 4369–4376.

[57] J. Sotor, G. Sobon, I. Pasternak, A. Krajewska, W. Strupinski, and K. M. Abramski, "Simultaneous mode-locking at 1565 and 1944 nm in fiber laser based on common graphene saturable absorber," *Optics Express*, 2013, 21(16): 18994–19002.

[58] Y. Luo, Q. Sun, Z. Wu, Z. Xu, S. Fu, L. Zhao, et al., "258-MHz group velocity locked vector dissipative solitons in a dispersion-managed short-cavity fiber laser," in *Optoelectronic Global Conference*, China, pp. 29–31, 2015.

[59] Y. Wang, B. Wang, and A. Wang, "Chaotic correlation optical time domain reflectometer utilizing laser diode," *IEEE Photonics Technology Letters*, 2008, 20(19): 1636–1638.

[60] A. Wang, N. Wang, Y. Yang, B. Wang, M. Zhang, and Y. Wang, "Precise fault location in WDM-PON by utilizing wavelength tunable chaotic laser," *Journal of Lightwave Technology*, 2012, 30(21): 3420–3426.

[61] L. Xia, D. Huang, J. Xu, and D. Liu, "Simultaneous and precise fault locating in WDM-PON by the generation of optical wideband chaos," *Optics Letters*, 2013, 38(19): 3762–3764.

[62] C. Jáuregui, J. M. López-Higuera, A. Cobo, O. M. Conde, and J. Zubía, "Multiparameter sensor based on a chaotic fiber-ring resonator," *Journal of the Optical Society of America B*, 2006, 23(10): 2024–2031.

[63] X. Zhang and L. Yang, "A fiber Bragg grating quasi-distributed sensing network with a wavelength-tunable chaotic fiber laser," *Systems Science and Control Engineering*, 2014, 2(1): 268–274.

[64] Z. Ma, M. Zhang, Y. Liu, X. Bao, H. Liu, Y. Zhang, et al., "Incoherent Brillouin optical time-domain reflectometry with random state correlated Brillouin spectrum," *IEEE Photonics Journal*, 2015, 7(4): 1–7.

[65] L. Xia, C. Yu, Y. Ran, J. Xu, and W. Li, "Static/dynamic strain sensing applications by monitoring the correlation peak from optical wideband chaos," *Optics Express*, 2015, 23(20): 26113–26123.

[66] Y. Luo, L. Xia, Z. Xu, C. Yu, Q. Sun, W. Li, et al., "Optical chaos and hybrid WDM/TDM based large capacity quasi-distributed sensing network with real-time fiber fault monitoring," *Optics Express*, 2015, 23(3): 2416–2423.

[67] C. G. Askins, M. A. Putnam, and E. J. Friebele, "Instrumentation for interrogating many-element fiber Bragg grating arrays," in *Proc. SPIE*, vol. 2444, pp. 257–266, 1995.

[68] A. D. Kersey, M. A. Davis, and H. J. Patrick, "Fiber grating sensors," *Journal of Lightwave Technology*, 1997, 15(8): 1442–1463.

[69] A. D. Kersey, T. A. Berkoff, and W. W. Morey, "Multiplexed fiber Bragg grating strain-sensor system with a fiber Fabry-Perot wavelength filter," *Optics Letters*, 1993, 18(16): 1370–1372.

[70] G. Yang, J. H. Guo, G. L. Xu, L. D. Lv, G. J. Tu, and L. Xia, "A novel fiber Bragg grating wavelength demodulation system based on F-P etalon," in *Proc. SPIE*, vol. 9270, pp. 92700V-1–92700V-7, 2014.

[71] D. M. Baney, B. Szafraniec, and A. Motamedi, "Coherent optical spectrum analyzer," *IEEE Photonics Technology Letters*, 2002, 14(3): 355–357.

[72] Q. Sun, J. Cheng, F. Ai, X. Li, D. Liu, and L. Zhang, "High speed and high resolution demodulation system for hybrid WDM/FDM based fiber microstructure sensor network by using Fabry-Perot filter," in *2015 Conference on Lasers and Electro-Optics*, San Jose, pp. 1–2, 2015.

[73] F. Ai, Q. Sun, J. Cheng, and D. Liu, "High resolution demodulation platform for large capacity hybrid WDM/FDM microstructures sensing system assisted by tunable FP filter," in *Progress in Electromagnetics Research Symposium*, Cambridge, pp. 1200–1203, 2015.

[74] M. Legre, R. Thew, and H. Zbinden, "High resolution Optical Time Domain Reflectometer based on 1.55 μm up-conversion photon-counting module," *Optics Express*, 2007, 15(13): 8237–8242.

[75] J. H. Park, J. S. Baik, and C. H. Lee, "Fault-detection technique in a WDM-PON," *Optics Express,* 2007, 15(4): 1461–1466.

[76] Y. Luo, L. Xia, Z. Xu, C. Yu, Q. Sun, W. Li, *et al.*, "Optical chaos and hybrid WDM/TDM based large capacity quasi-distributed sensing network with real-time fiber fault monitoring," *Optics Express*, 2015, 23(3): 2416–2423.

[77] J. Hodgkinson and R. P. Tatam, "Optical gas sensing: a review," *Measurement Science & Technology*, 2012, 24(1): 111–123.

[78] H. Waechter, J. Litman, A. H. Cheung, J. A. Barnes, and H. P. Loock, "Chemical sensing using fiber cavity ring-down spectroscopy," *Sensors*, 2010, 10(3): 1716–1742.

[79] X. Sun, Q. Sun, S. Zhu, Y. Yuan, Z. Huang, X. Liu, *et al.*, "High sensitive ammonia gas sensor based on graphene coated microfiber," in *PIERS Proceedings*, Prague, pp. 1196–1199, 2015.

[80] Z. Gu, Y. Xu, and K. Gao, "Optical fiber long-period grating with solgel coating for gas sensor," *Optics Letters*, 2006, 31(16): 2405–2407, 2006.

[81] W. Jin, H. Ho, Y. Cao, J. Ju, and L. Qi, "Gas detection with micro- and nano-engineered optical fibers," *Optical Fiber Technology*, 2013, 19(6): 741–759.

[82] J. Henningsen and J. Hald, "Dynamics of gas flow in hollow core photonic bandgap fibers," *Applied Optics*, 2008, 47(15): 2790–2797.

[83] J. P. Parry, B. C. Griffiths, N. Gayraud, E. D. McNaghten, A. M. Parkes, W. N. MacPherson, *et al.*, "Towards practical gas sensing with micro-structured fiber," *Measurement Science and Technology*, 2009, 20(7): 190–190.

[84] G. Stewart, C. Tandy, D. Moodie, M. A. Morante, and F. Dong, "Design of a fiber optic multi-point sensor for gas detection," *Sensors and Actuators B: Chemical*, 1998, 51(1): 227–232.

[85] W. Jin, "Performance analysis of a time-division-multiplexed fiber-optic gas-sensor array by wavelength modulation of a distributed-feedback laser," *Applied optics*, 1999, 38(25): 5290–5297.

[86] G. Whitenett, G. Stewart, H. B. Yu, and B. Culshaw, "Investigation of a tuneable mode-locked fiber laser for application to multipoint gas spectroscopy," *Journal of Lightwave Technology*, 2004, 22(3): 813–819.

[87] M. Završnik and G. Stewart, "Theoretical analysis of a quasi-distributed optical sensor system using FMCW for application to trace gas measurement," *Sensors and Actuators B: Chemical*, 2000, 71(1): 31–35.

[88] F. Ye, L. Qian, and B. Qi, "Multipoint chemical gas sensing system based on frequency-shifted interferometry," *Conference on Optical Fiber Communication/National Fiber Optic Engineers Conference, 2008*, San Diego, United States, pp. 1–3, 2008.

[89] W. Zhang, Y. Lu, L. Duan, Z. Zhao, W. Shi, and J. Yao, "Intracavity absorption multiplexed sensor network based on dense wavelength division multiplexing filter," *Optics Express*, 2014, 22(20): 24545–24550.

[90] OptoSniff/Optosci, http://www.optosniff.com/.

[91] C. Caliendo, "Latest trends in acoustic sensing," *Sensors*, 2014, 14(4): 5781–5784.

[92] C. K. Kirkendall and A. Dandridge, "Overview of high performance fibre-optic sensing," *Journal of Physics D: Applied Physics*, 2004, 37(18): R197–R216.

[93] J. G. Teixeira, I. T. Leite, S. Silva, and O. Frazão, "Advanced fiber-optic acoustic sensors," *Photonic Sensors*, 2014, 4(3): 198–208.

[94] S. Foster, A. Tikhomirov, M. Milnes, J. van Velzen, and G. Hardy, "A fiber laser hydrophone," in *Proc. SPIE*, vol. 5855, pp. 627–630, 2005.

[95] R. Chen, G. F. Fernando, T. Butler, and R. A. Badcock, "A novel ultrasound fibre optic sensor based on a fused-tapered optical fibre coupler," *Measurement Science and Technology*, 2004, 15(8): 1490–1495.

[96] B. Xu, Y. Li, M. Sun, Z. Zhang, X. Dong, Z. Zhang, *et al.*, "Acoustic vibration sensor based on

nonadiabatic tapered fibers," *Optics Letters*, 2012, 37(22): 4768–4770.

[97] J. P. Mutschlecner and R. W. Whitaker, "Infrasound from earthquakes," *Journal of Geophysical Research: Atmospheres*, 2005, 110(110): 372–384.

[98] S. Wang, P. Lu, L. Zhang, D. Liu, and J. Zhang, "Optical fiber acoustic sensor based on nonstandard fused coupler and aluminum foil," *IEEE Sensors Journal*, 2014, 14(7): 2293–2298.

[99] S. Wang, P. Lu, L. Zhang, D. Liu, and J. Zhang, "Intensity demodulation-based acoustic sensor using dual fiber Bragg gratings and a titanium film," *Journal of Modern Optics*, 2014, 61(12): 1033–1038.

[100] S. Wang, P. Lu, H. Liao, L. Zhang, D. Liu, and J. Zhang, "Passively mode-locked fiber laser sensor for acoustic pressure sensing," *Journal of Modern Optics*, 2013, 60(21): 1892–1897.

[101] Y. Ran, L. Xia, Y. Han, W. Li, J. Rohollahnejad, Y. Wen, *et al.*, "Vibration fiber sensors based on SM-NC-SM fiber structure," *IEEE Photonics Journal*, 2015, 7(2): 1–7.

Interferometric Distributed Sensing System With Phase Optical Time-Domain Reflectometry

Chen WANG[1*], Ying SHANG[1], Xiaohui LIU[1], Chang WANG[1], Hongzhong WANG[2], and Gangding PENG[3]

[1]*Shandong Provincial Key Laboratory of Optical Fiber Sensing Technologies, Laser Institute of Shandong Academy of Sciences, Jinan, 250014, China*

[2]*Shengli Oilfield Xinsheng Geophysical Technology Co. Ltd., No. 23 Xingfu Road, Dongying, 257086, China*

[3]*School of Electrical Engineering & Telecommunications, The University of New South Wales, NSW, 2052, Australia*

*Corresponding author: Chen WANG   E-mail: jgwangchen@163.com

Abstract: We demonstrate a distributed optical fiber sensing system based on the Michelson interferometer of the phase sensitive optical time domain reflectometer (φ-OTDR) for acoustic measurement. Phase, amplitude, frequency response, and location information can be directly obtained at the same time by using the passive 3×3 coupler demodulation. We also set an experiment and successfully restore the acoustic information. Meanwhile, our system has preliminary realized acoustic-phase sensitivity around −150 dB (re rad/μPa) in the experiment.

Keywords: Fiber optics sensors; Rayleigh scattering; optical time domain reflectometry; interferometry

1. Introduction

The distributed optical fiber acoustic sensors (DAS) offer the capability of measurement at thousands of points simultaneously, using a simple and unmodified optical fiber as the sensing element. It has been extensively studied and adopted for industrial applications during the past decades. Up to now, the distributed optical fiber measurements mainly include optical fiber interferometer sensors and optical backscattering based sensors. Interferometer sensors acquire distributed information by integration of the phase modulation signals, and usually two interferometers are used to determine the position, including combining the Sagnac to a Michelson interferometer [1], modified

Sagnac/Mach-Zehnder interferometer [2], twin Sagnac[3]/Michelson [4]/Mach-Zehnder [5] interferometers, and adopting a variable loop Sagnac [6]. Another distinguished technique is the use of optical backscattering based sensors. A promising technique is phase sensitive optical time domain reflectometer (φ-OTDR) using a narrow line-width laser [7, 8]. Brillouin-based dynamic strain sensors have been researched recently [9]. Recently, a hybrid interferometer-backscattering system is demonstrated [10], but the interferometer and the backscattering parts are working separately.

A major limitation of those distributed sensors above is that they are incapable of determining the full vector acoustic field, namely the amplitude, frequency, and phase, of the incident signal, which is a necessity for seismic imaging. Measuring the full acoustic field is a much harder technical challenge to overcome, but in doing so, it is possible to achieve high resolution seismic imaging and also make other novel systems, for example a massive acoustic antenna.

In this paper, we demonstrate the design and characterization of a distributed optical fiber sensing system based on the Michelson interferometer of the φ-OTDR for acoustic measurement. Phase, amplitude, frequency response, and location information can be directly obtained at the same time. Experiments show that our system successfully restores the acoustic information and has preliminarily realized the acoustic-phase sensitivity around −150 dB (re rad/μPa). Our system offers a versatile new tool for acoustic sensing and imaging, such as through the formation of a massive acoustic camera/telescope. The new technology can be used for surface, seabed, and downhole measurements all by using the same optical fiber cable.

2. Experimental setup and signal processing

The experimental setup of the Michelson interferometer of the φ-OTDR is shown in Fig. 1. The light source is a narrow linewidth laser with the maximum output power of 30 mW and linewidth of 5 kHz. The continuous wave (CW) light with a wavelength of 1550.12 nm is injected into an acoustic-optic modulator (AOM) to generate the pulses, whose width is 200 ns and the repetition rate is fixed at 20 kHz. The maximum detection length is related to the repetition rate of the pulse. The time interval among the pulses should be larger than the round trip time that the pulses travel in the detection fiber to keep only one pulse inside the detection fiber. For the 20 kHz repetition rate, the detection range is around 5 km which is determined by $L<c/2nf$. The detection frequency range is also related to the repetition rate. In our case, the highest detection frequency is no more than 20 kHz theoretically.

An erbium-doped fiber amplifier (A) is used to amplify the pulses, and the ASE noise is filtered by an optical fiber Bragg grating filter (F). Then the amplified pulses are launched into a single mode detection fiber (Corning SMF-28e) by a circulator. The Rayleigh back-scattering is amplified (A) and filtered (F) again to obtain better signal-to-noise-ratio (SNR) improvement and then injected into a Michelson interferometer which consists of a circulator, a 3×3 coupler, and two Faraday rotation mirrors (FRMs) [11]. The half arm length of the Michelson interferometer s is set to 5 m. The final interference signals outputting from the 3×3 coupler are collected by three photodetectors (PD1–3), and then the signal processing scheme is accomplished by a software program. Theoretically, there is a 120° phase shift between two adjacent PDs. Accordingly, the outputs of the three PDs can be expressed as

$$I_k = D + I_0 \cos[\phi(t) - (k-1) \times (2\pi/3)], \quad k=1,2,3 \quad (1)$$

where $\phi(t)=\phi_s+\phi_n+\phi_0$. ϕ_s, ϕ_n, and ϕ_0 are respectively the signal to be detected, the noise, and the intrinsic phase of the system. For each point on the detection fiber, ϕ_s is obtained after the demodulation process shown in Fig. 2. It can directly demodulate all the information from the signal detected at the same time without any Fourier transforms.

In our experiment, 2000 periods for detection fiber scanning are recorded by a high-speed oscilloscope with 100 MHz sampling rate, and the total data acquisition time is 0.1 s. Here, we choose a 200 m detection fiber and several individual acoustic frequencies within the detection length and frequency range as a test example. Two piezoelectric transducer (PZT) cylinders with 10 m single mode fiber wound are put at 100 m and 160 m over 200 m detection fiber in our system as the acoustic sources. Both PZTs are driven by two function generators. To eliminate the different frequency responses of the detection fiber, we set the two function generators to output the same sine-wave with the same frequency of 200 Hz but the amplitudes are 1 V and 2 V.

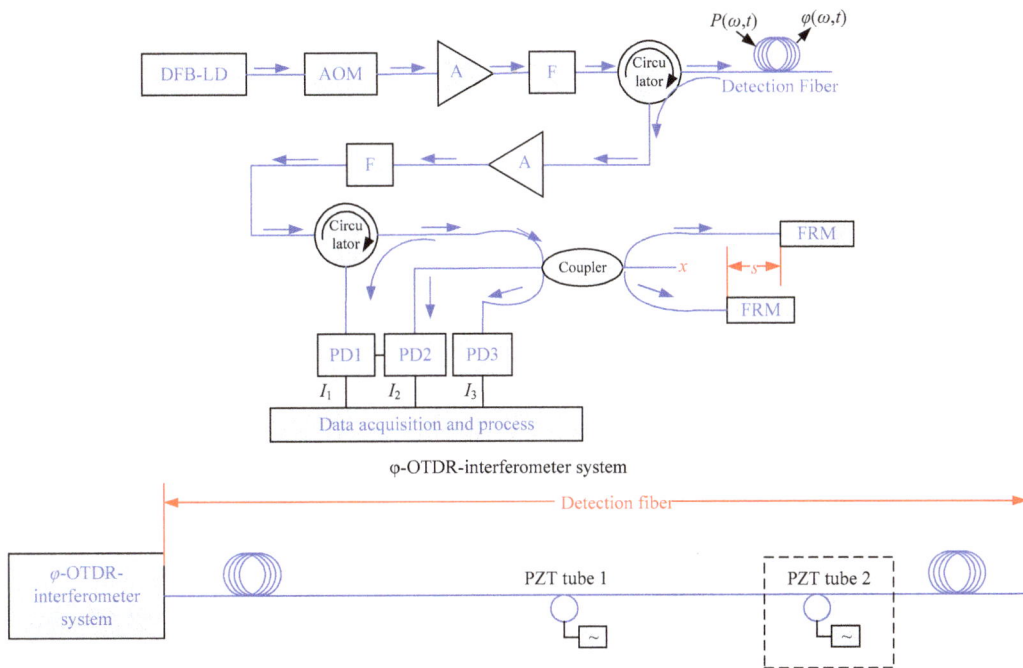

Fig. 1 Experimental setup for the Michelson interferometer of the φ-OTDR (DFB-FL: distributed feedback fiber laser; AOM: acoustic-optic modulator; A: erbium-doped fiber amplifier; F: optical fiber grating filter; FRM: Faraday rotation mirror; PD1–3: photodetectors; PZT: piezoelectric transducer).

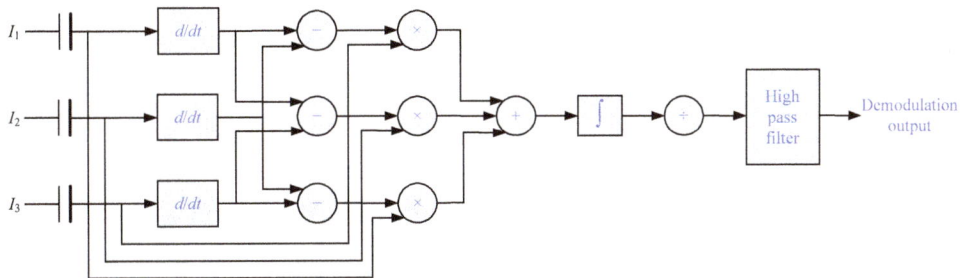

Fig. 2 Demodulation system based on the 3×3 coupler.

Figure 3(a) shows the global demodulation result. The system demodulates the whole acoustic situations along the detection fiber when both two function generators are working. We can see that besides the 1 V demodulated signal between 85 m and 105 m, our system does demodulate another signal between 155 m and 175 m. Between these two sine-like signals, there is about 50 m fiber without any vibrations in the global demodulation result, indicating no influence of the Rayleigh scatterings from different points of the detection fiber during the demodulation process. The difference of the demodulated PZT fiber position is probably caused by the change in the acquisition starting point in the oscilloscope. Figure 3(b) shows the instantaneous frequency extracted from Fig. 3(a) at 95 m and 165 m along with their spectral analyses via fast Fourier transform (FFT) of the two demodulated signals. The amplitude of the 1 V demodulated signal at 95 m is −0.791 dB [=20 lg(A_{signal}), 0.913 rad], and the amplitude of the 2 V demodulated signal at 165 m is about 5.460 dB (1.875 rad). The amplitude rate A_{2V}/A_{1V}=1.875/0.913 ≈ 2.054, nearly twice between the 2 V and 1 V signal amplitudes. Also the background noise of the two demodulated signals are all around −60 dB [=10 lg(A_{signal}/A_{noise}), 1×10^{-3} rad)], so that the SNR is 29.6 dB. This result indicates that our system can well recreate the signals by their own proportions.

(a)

(b)

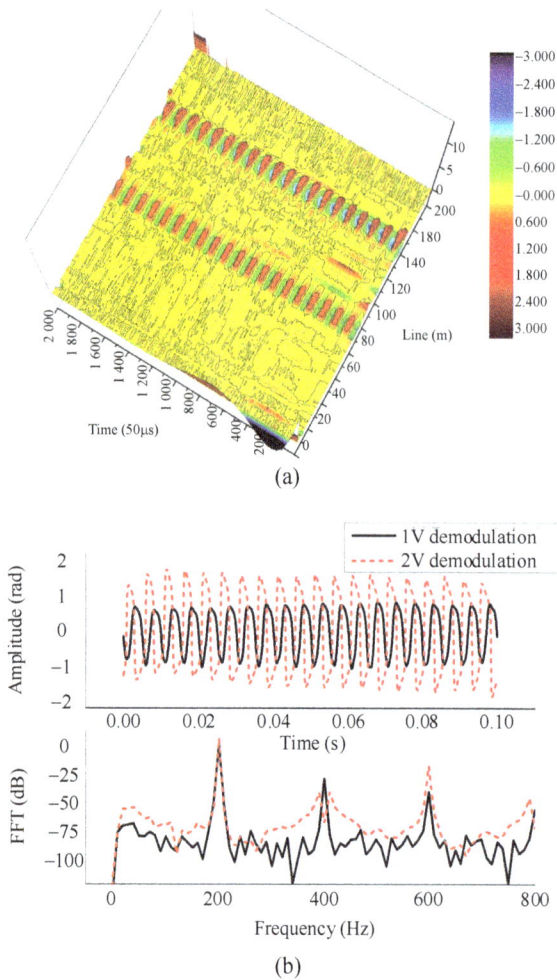

Fig. 3 Demodulated scalograms of (a) two 200 Hz acoustic events, with an amplitude of 1 V at the 100 m position and an amplitude of 2 V at the 160 m position, respectively and (b) time and frequency responses of the two demodulation results.

Moreover, we use a water tank system to test the demodulation capability of our system (Fig. 4). An underwater speaker is fixed in the tank and driven by a function generator. The function generator is used to drive the underwater speaker with a 200 Hz separate sinusoidal signal. We wrap the sensing fiber into a 10 m length of fiber ring from the DAS instrument. A commercial piezoelectric hydrophone is also placed close to the fiber ring to measure the acoustic pressure amplitude. The fiber ring and the piezoelectric hydrophone are placed 5 cm away from the underwater speaker so that the sound wave produced by the speakers can be directly transmitted to the fiber.

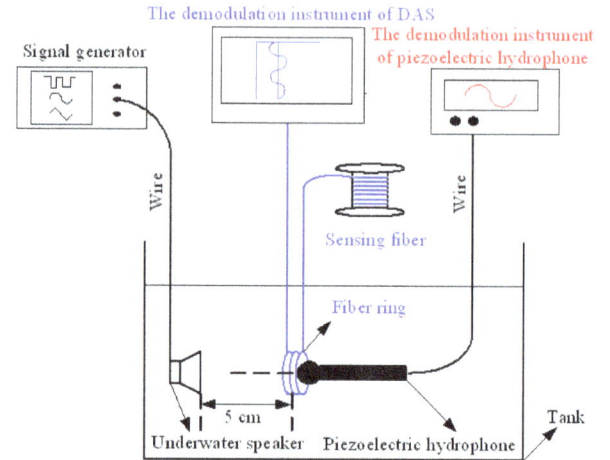

Fig. 4 Schematic diagram of underwater distributed acoustic testing experiment.

The hydrophone signal is to relate the phase measurement with acoustic pressure. So the acoustic pressure amplitudes in different acoustic intensities at 200 Hz are measured using the piezoelectric hydrophone, and the phase-pressure sensitivity of our φ-OTDR-interferometry system is calculated. Table 1 shows the phase-pressure data and results of our system. With a decrease in the acoustic pressure amplitudes, the demodulated phase changes also decrease. But the phase-pressure sensitivities are almost the same around 0.026 rad/Pa (−150 dB, $=20 \lg(A_{signal})$, re rad/μPa), using the changed phase amplitude our system demodulated divided by the actual acoustic pressure amplitude the piezoelectric hydrophone detected, indicating that our system can well demodulate the amplitude, frequency, and phase of the acoustic signals with a sensitivity of −150 dB.

Table 1 Experimental data of sensitivity.

Piezoelectric hydrophone		DAS	
Amplitude (mV)	Acoustic pressure (Pa)	Phase changes (rad)	Acoustic phase sensitivity dB (re rad/μPa)
30	15	0.654	−147
24	12	0.312	−152
18	9	0.243	−151
12	6	0.154	−151

3. Further discussion

Another point that should be explained is that due to the participation of the interferometer the width of the source demodulated by our φ-OTDR-interferometer system is broadened than its origin. The extended length between the original source and the demodulated one equals just to the length of the arm length of the interferometer. On the contrary, the introduction of the Michelson interferometer has its advantage to the signal demodulation and could improve the sensitivity of our system because the effective detecting fiber is expanded. It could increase the dynamic sensitivity of our system significantly. These parameters should be chosen wisely in real applications. The polarization of the Rayleigh backscattering is also important to our system. The advantage of using Michelson interferometer rather than Mach-Zehnder one is that the FRMs keep the polarization states of the input and output lights independent from the fiber birefringence. And also the interferometer is not used as the sensing fiber, so the polarization is not a critical issue. Further experiment will be done to eliminate the polarization influence with certain polarizer at the beginning of the detection fiber.

In application, the mapping of acoustic events is very important, which directly gives what is happening around the detecting area. Classically, point sensors have been used as a serial of arrays to determine when, what, and where the acoustic event is, thus making a high cost of the monitor. Distributed sensors are much cheaper but have a major limitation that they are incapable of determining the full vector acoustic field, namely the amplitude, frequency, and phase, of the incident signal, which is a necessity for seismic imaging. By using the method of our φ-OTDR interferometry, it offers a versatile new tool for acoustic mapping and imaging in one single optical fiber, such as through the formation of a massive acoustic camera/telescope. For example, it is possible to incorporate our system as the optical hydrophone or directional accelerometer arrays and even to measure on existing arrays directly with the appropriate wavelength choice. It also can be used in many seismic acquisitions to date, encompassing vertical seismic profiling, in both flowing and non-flowing wells, and surface seismic surveys.

4. Conclusions

In this paper, we demonstrate the design and characterization of a distributed optical fiber sensing system based on the Michelson interferometer of the φ-OTDR for acoustic measurement. The phase, amplitude, frequency response, and location information can be directly obtained at the same time by using the passive 3×3 coupler demodulation. Experiments show that our system successfully restores the acoustic information with the acoustic-phase sensitivity around $-150\,dB$ (re rad/μPa). Our system offers a versatile new tool for acoustic sensing and imaging, such as through the formation of a massive acoustic camera/telescope. The new technology can be used for surface, seabed, and downhole measurements. The use of the system in downhole applications allows a continuum of benefits extending to flow profiling and condition monitoring, all using the same optical fiber cable.

Acknowledgment

This work was supported by the Shandong Natural Science Foundation (No. ZR2013FL028), Science and Technology Development Project of Shandong Province (2014GGX103019), and Innovation and Achievement Transformation Projects of Shandong Province (2014ZZCX04206).

References

[1] S. J. Spammer, P. L. Swart, and A. A. Chtcherbakov, "Merged Sagnac-Michelson interferometer for distributed disturbance detection," *Journal of Lightwave Technology*, 1997, 15(6): 972−976.
[2] A. A. Chtcherbakov, P. L. Swart, S. J. Spammer, and B. M. Lacquet, "Modified Sagnac/Mach-Zehnder

interferometer for distributed disturbance sensing," *Microwave and Optical Technology Letters*, 1999, 20(1): 34−36.

[3] S. J. Russell, K. R. C. Brady, and J. P. Dakin, "Real-time location of multiple time-varying strain disturbances acting over a 40 km fiber section using a novel dual-Sagnac interferometer," *Journal of Lightwave Technology*, 2001, 19(2): 205−213.

[4] X. Hong, J. Wu, C. Zuo, F. Liu, H. Guo, and K. Xu, "Dual Michelson interferometers for distributed vibration detection," *Applied Optics*, 2011, 50(22): 4333−4338.

[5] Q. Sun, D. Liu, J. Wang, and H. Liu, "Distributed fiber-optic vibration sensor using a ring Mach-Zehnder interferometer," *Optics Communications*, 2008, 281(6): 1538−1544.

[6] X. J. Fang, "Fiber-optic distributed sensing by a two-loop Sagnac interferometer," *Optics Letters*, 1996, 21(6): 444−446.

[7] J. C. Juarez, E. W. Maier, K. N. Choi, and H. F. Taylor, "Distributed fiber-optic intrusion sensor system," *Journal of Lightwave Technology*, 2005, 23(6): 2081−2087.

[8] J. C. Juarez and H. F. Taylor, "Field test of a distributed fiber-optic intrusion sensor system for long perimeters," *Applied Optics*, 2007, 46(11): 1968−1971.

[9] Y. Dong, L. Chen, and X. Bao, "Time-division multiplexing-based BOTDA over 100 km sensing length," *Optics Letters*, 2011, 36(2): 277−279.

[10] T. Zhu, Q. He, X. Xiao, and X. Bao, "Modulated pulses based distributed vibration sensing with high frequency response and spatial resolution," *Optics Express*, 2013, 21(3): 2953−2963.

[11] M. D. Todd, M. Seaver, and F. Bucholtz, "Improved, operationally-passive interferometric demodulation method using 3×3 coupler," *Electronics Letters*, 2002, 38(15): 784−786.

Distributed Fiber Optic Interferometric Geophone System Based on Draw Tower Gratings

Ruquan XU[*], Huiyong GUO, and Lei LIANG

National Engineering Laboratory for Fiber Optic Sensing Technology, Wuhan University of Technology, Wuhan, 430070, China

[*]Corresponding author: Ruquan XU   E-mail: xuruquan@whut.edu.cn

Abstract: A distributed fiber optic interferometric geophone array based on draw tower grating (DTG) array is proposed. The DTG geophone array is made by the DTG array fabricated based on a near-contact exposure through a phase mask during the fiber drawing process. A distributed sensing system with 96 identical DTGs in an equal separation of 20 m and an unbalanced Michelson interferometer for vibration measurement has been experimentally validated compared with a moving-coil geophone. The experimental results indicate that the sensing system can linearly demodulate the phase shift. Compared with the moving coil geophone, the fiber optic sensing system based on DTG has higher signal-to-noise ratio at low frequency.

Keywords: Fiber optic sensing; vibrating sensing; draw tower grating; interferometric geophone

1. Introduction

Seismic exploration technology has been widely employed for the exploration of oil and other mineral resources, detection of geological hazards, hydrographic surveys, rail vibrations, debris flow, and engineering quality inspection [1–3]. Seismic exploration and monitoring for oil and gas reservoirs is a peculiar application that requires a large number of geophones deployed outdoors over large areas to detect backscattered waves from artificial sources. A storing and processing system collects the data from all the geophones to get an image of the sub-surface. The existed cabling system to connect sensors is known to cause inefficiencies, electromagnetic interference, and weight cost, as well as insufficient flexibility in survey design [4]. Therefore, the development of a new type geophone with higher multiplexing capacity is one of the keys to improve the detection capabilities of seismic surveys. Compared with traditional geophones, the fiber Bragg grating (FBG) geophone owns many advantages, such as immunity to electromagnetic interference, remote sensing, and multiplexing capacity, which makes FBG particularly suitable for multipoint sensing [5]. The FBG accelerometer can receive the acceleration with the help of wavelength demodulation schemes such as scanning Fabry-Perot filters [6], edge filter methods [5, 7], or interferometric techniques [8]. An eight-element fiber laser geophone array system was presented in [9]. However, these fiber optic sensors multiplexing capacity on a single fiber only reaches 20–30. In this paper, we propose a distributed fiber optic interferometric geophone system based on identical DTGs. The sensor system is constructed by an identical low reflectivity DTG array fabricated based on a near-contact exposure through a phase mask during the fiber drawing process. The adjacent DTGs construct a Fabry-Perot (F-P) vibration sensor, which is sensitive to vibration. For the ultralow reflectivity of the DTGs and no fusion points between DTGs, this system has larger capacity than traditional fiber optic geophone system.

2. DTG array fabrication

The fabrication technique presented here is based on a near-contact exposure through a phase mask during the fiber drawing process. The draw tower grating fabrication setup is shown in Fig. 1. For increasing the single pulse photosensitivity, we used a highly photosensitive Ge/B co-doped preform. The fiber preform was heated in a furnace up to a temperature of more than 2 000 ℃ and was then drawn with a velocity of approximately 12 m/min into a bare fiber with a final diameter of 125 μm. Before the fiber was coated, Bragg gratings were inscribed using a line-narrowed ArF excimer laser (OptoSystems, CL5300), with a beam size of 4×12 (mm), pulse width of 10 ns, maximum pulse energy of 40 mJ, and maximum laser repetition rate of 300 Hz based on phase mask method.

The UV (ultraviolet) laser writing setup was placed on the draw tower to synchronize with the grating inscription to reduce the vibration between the bare fiber and the laser beam. The laser beam (pulse energy of 25 mJ) was normally incident on the 1 550 nm phase mask and diffracted entirely. A slit was attached to the mask and transferred only 5 mm of the exposing beam to the bare fiber. The DTG was formed by the interference between the +1 and −1 diffracted orders of the phase mask. Thus, the refractive index of the fiber core was periodically changed. The interference fringes were determined by the structure of the phase mask and not influenced by light source coherence, which ensured the good wavelength consistency of the DTGs.

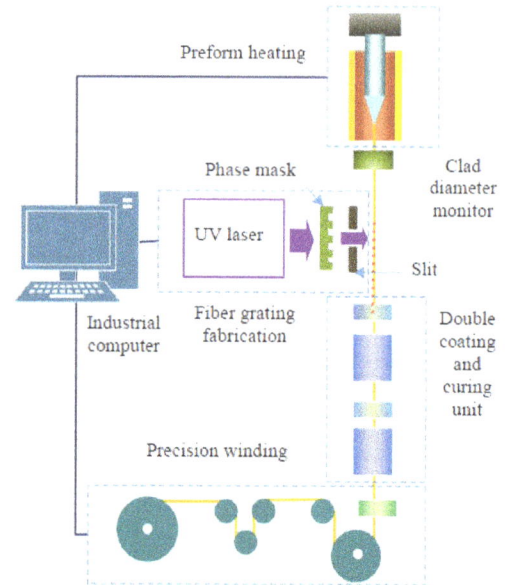

Fig. 1 Diagram for DTGs fabrication experimental setup.

The control computer fixed the laser pulse repetition at 0.01 Hz to synchronize the grating inscription, and the distance between adjacent DTGs was 20 m. Directly after the grating inscription, the DTG fiber was coated, and the coating was cured under UV light.

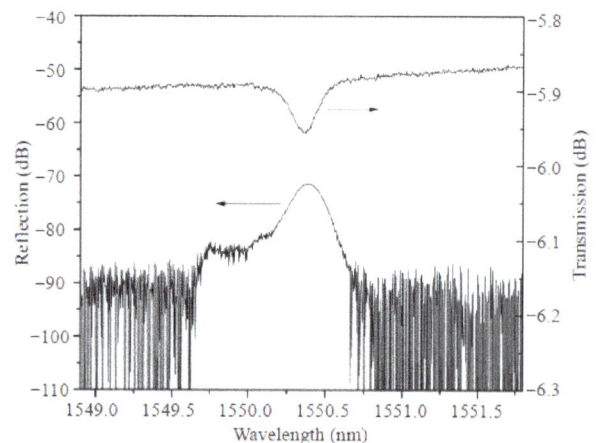

Fig. 2 Reflection and transmission spectra of the DTG array.

For measuring the reflection spectrum of DTGs, a standard setup with a superluminescent diode optical source (spectral range from 1 510 nm to 1 570 nm), an optical circulator, and an optical spectrum analyzer (OSA, Yokogawa, AQ6370C) was used. The reflection and transmission spectra of the DTGs are shown in Fig. 2. The central wavelength of the DTGs is 1 550.4 nm, the spectrum width of the DTGs is 0.2 nm (−3 dB). The results indicate that on-line writing DTGs using the phase mask method have a central wavelength with good uniformity. The DTG array contains 96 DTGs, and the distance between adjacent DTGs is 20 m. The DTG average reflectivity is 0.015%.

3. Experimental setup and operation principle

The experimental setup for the measurement of the ground vibration signals is shown in Fig. 3. The sensor system is constructed with 48 DTG F-P interferometers (containing two adjacent identical DTGs) and an unbalanced Michelson interferometer. To get distributed information along a DTG fiber, the reflection signals from each DTG F-P interferometer are detected. Narrow linewidth coherent light pulses are used as probe pulses with a wavelength close to the central wavelength of the DTG array. The reflection signals are coupled into an unbalanced Michelson interferometer comprised of a 2×2 and a 3×3 fiber coupler. The difference in the fiber length between the two arms of the Michelson interferometer is equal to two times of the difference in fiber length between two adjacent DTGs. The reflected signals of the first pulse from DTG1 and the reflected signals of the second pulse from DTG2 are coupled into an unbalanced Michelson interferometer, and then interference occurs at the 3×3 coupler. Therefore, with a probe pulse injected into the DTGs, the reflected waves at the two arms of the unbalanced Michelson interferometer can be expressed respectively as

$$E_a(t) = \frac{\sqrt{2}}{2} \sum_{i=1}^{N} E_0 \sqrt{(1-R)^{2(i-1)} R} \times$$
$$\exp\{j2\pi f(t-\tau_i)\} \mathrm{rect}\left(\frac{t-\tau_i}{W}\right) \quad (1)$$

$$E_b(t) = \frac{\sqrt{2}}{2} \sum_{i=1}^{N} E_0 \sqrt{(1-R)^{2(i-1)} R} \times$$
$$\exp\{j2\pi f(t-\tau_i-\tau_0)\} \mathrm{rect}\left(\frac{t-\tau_i-\tau_0}{W}\right) \quad (2)$$

where R is the reflectivity of the DTG$_i$

($i = 1, 2, \cdots, N$), N is the total number of the DTGs, L_0 is the fiber distance of the adjacent two DTGs, f is the frequency of the probe pulse, and τ_0 is the round trip time of the probe pulse between two adjacent DTGs. $\tau_0 = 2n_{\mathrm{eff}} L_0 / c$, τ_i is the time delay of DTG$_i$. For avoiding interfering with each other, the width of the probe pulse $W < 2n_{\mathrm{eff}} L_0 / c$. n_{eff} is the effective refractive index, and c is the vacuum light speed.

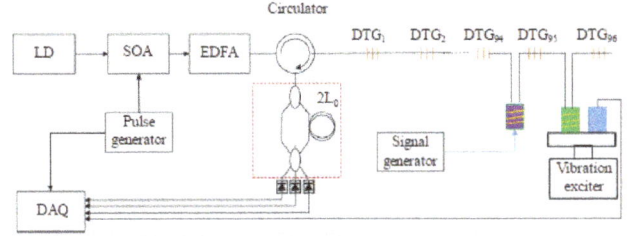

Fig. 3 Diagram of experimental setup.

Therefore, $E_a(t)$ interferes with $E_b(t)$ at the 3×3 coupler, and the three interference outputs of the symmetric 3×3 coupler in the Michelson interferometer can be expressed as

$$I_k(t) = \sum_{i=1}^{N-1} \left[D_i + E_i \cos\left(\varphi_{i,i+1} - (k-2)\frac{2\pi}{3} \right) \right] \times$$
$$\mathrm{rect}\left(\frac{t-\tau_{i+1}}{W} \right), \quad k = 1,2,3 \quad (3)$$

with

$$D_i = \frac{1}{3} E_0^2 R(1-R)^{2(i-1)} \quad (4)$$

$$E_i = \frac{1}{3} E_0^2 R(1-R)^{2(i-1)} \quad (5)$$

where $\varphi_{i,i+1}$ is the phase shift between DTG$_i$ and DTG$_{i+1}$. D_i is the direct component determined by the optical power and gain in the detection electronics, and E_i is the AC component determined by the 2×2 coupler splitting ratio and the polarization states of the two arms of the Michelson interferometer.

If the Michelson interferometer is stable, the phase shift $\varphi_{i,i+1}$ between DTG$_i$ and DTG$_{i+1}$ is given as

$$\varphi_{i,i+1} = 2\pi f n_{\mathrm{eff}} (L_{i+1} - L_i)/c - \varphi_0 \quad (6)$$

where L_i and L_{i+1} are the distances from the input to DTG$_i$ and DTG$_{i+1}$, respectively; φ_0 is the phase difference of the Michelson interferometer. The three interference outputs of the symmetric 3×3 coupler in the Michelson interferometer can be expressed as [10]

$$V_{\text{out}}(t) = \sum_{i=1}^{N-1} S_i \varphi_{i,i+1} \text{rect}\left(\frac{t - \tau_{i+1}}{W}\right) \qquad (7)$$

where S_i is dependent on the electronics amplifier and is set as $\sqrt{3}$ in the software program. Therefore, $V_{\text{out}}(t)$ at τ_2, τ_3, $\cdots$, τ_N can be expressed as $V_{i,\text{out}} = S_i \cdot \varphi_{i,i+1}$, and the phase shift is as follows:

$$\varphi_{i,i+1} = \frac{1}{S_i} V_{i,\text{out}} = \frac{V_{i,\text{out}}}{\sqrt{3}} . \qquad (8)$$

So we can obtain the phase shift by measuring the amplitude of the demodulated signal, $V_{i,\text{out}}$. Thus, for a symmetric 3×3 coupler, the demodulated result is free from the optical power and the reflectivity of the DTGs.

The intrinsic multiple reflections crosstalk phenomenon in the inline DTG-FFP sensor system is a serious problem that limits the multiplexing number of sensors, which comes from the light reflected two more times by DTGs other than the target DTG but reaching the 3×3 coupler at the same time with the real signal. For the DTG array with identical low reflectivity, only the first-order crosstalk (signals undergoing three reflections) needs to be considered. Then the worst crosstalk for the DTG sensor system with n DTGs is given as follows [11]:

$$C_{\text{DTG}_i,\text{DTG}_{i+1}} = 20\lg\left(\frac{iR}{1-R}\right). \qquad (9)$$

The worst crosstalk is dependent on the reflectivity of the DTGs and the multiplexing number of DTGs. Figure 4 shows the worst crosstalk in the DTG sensor systems with different multiplexing numbers of DTGs when the reflectivities are 1%, 0.1%, 0.01%, and 0.001%, which shows the worst crosstalk increases as an increase in the multiplexing number of DTGs with fixed reflectivity.

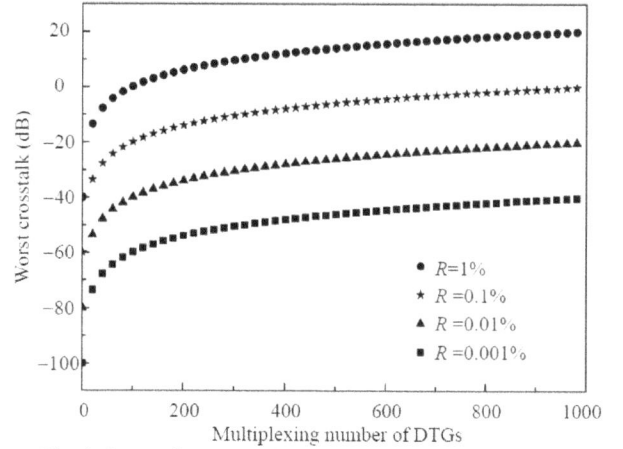

Fig. 4 Crosstalk verses the multiplexing number of DTG at different reflectivities.

4. Experiment and results

An 1550 nm coherent laser with 5 mW (Santec, TLS510) was pulsed by an optical semiconductor amplifier (SOA, INPHNEX 1502) to create 100 ns optical pulses with 100 kHz repetition rate. The SOA driving current was modulated by signal generator (RIGOL, DG1022) with pulsed signals. The optical pulses were amplified by an erbium-doped fiber amplifier (EDFA) and filtered by a narrow band optical filter and then coupled into DTG sensor array by a circulator. The reflections of the DTGs injected into an unbalanced Michelson interferometer consisting of a 2×2 coupler and a 3×3 coupler. The interference signals converted from the symmetric 3×3 coupler were coupled into three photodetectors, and then the signals were acquired by DAQ (NI, PXI-5122) and processed by a software program.

4.1 Optical phase demodulation

In order to further demonstrate the multiplexing system phase shift demodulation performance, the fiber between DTG_{94} and DTG_{95} was coiled on a commercial piezoelectric ceramic transducer (PZT), which was driven by a signal generator (RIGOL, DG1022). The PZT could change the phase shift linearly on the fiber between DTG_{94} and DTG_{95}. The PZT was driven by 10 Hz and 200 Hz sine signals, respectively. Figures 5(a) and 5(b) demonstrate the time domain figures of the phase shift of 10 Hz and 200 Hz sine signals, which shows the sensing system can demodulate the phase shift in real time. Figure 6 shows the frequency spectrum of phase shift of 10 Hz and 200 Hz sine signals provided in Fig. 5. The system dynamic scope reaches 27 dB at 10 Hz and 40 dB at 200 Hz, respectively. Therefore, the technique presented in this paper provides a practical method of interrogating as many as 96 sensors with a large dynamic range. Further gains may be realized by decreasing the reflection of the DTGs, as can be seen from Fig. 4. When the driving signal frequency was fixed at 10 Hz, we obtained the linearity of demodulated phase shift by tuning the amplitude of the driving signal. Figure 7 shows that the linearity of demodulated phase shift is 0.99. The result indicates that our fiber optic interferometric geophone system based on identical DTGs can properly demodulate the instantaneous time and frequency domain signals with good linearity.

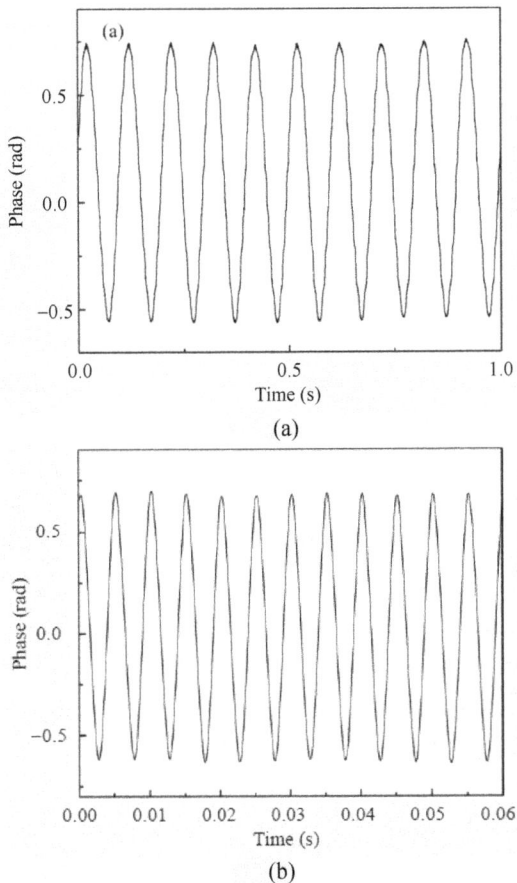

(a)

(b)

Fig. 5 Time domain signal of (a) 10 Hz and (b) 200 Hz.

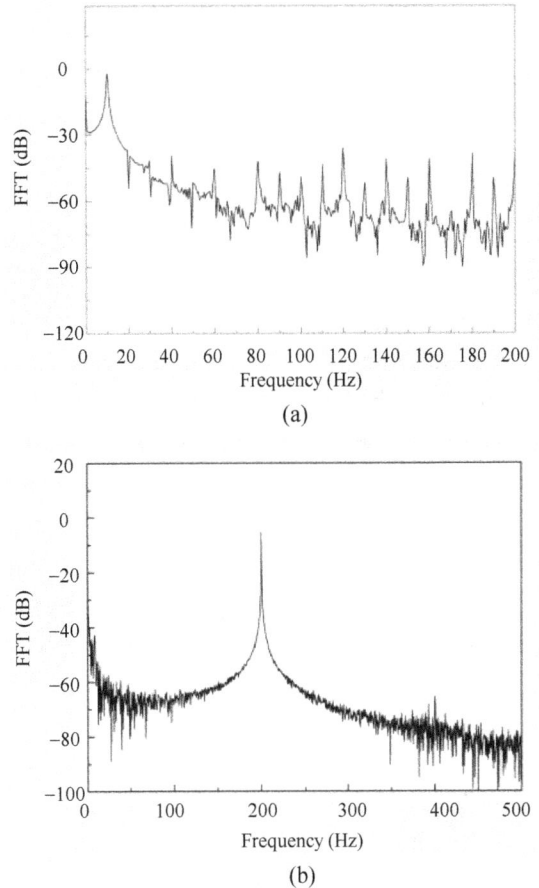

(a)

(b)

Fig. 6 Frequency domain signal of (a) 10 Hz and (b) 200 Hz.

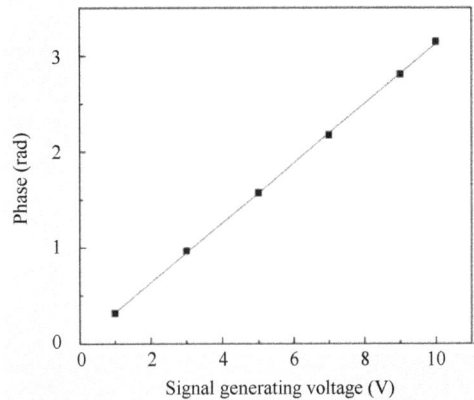

Fig. 7 Demodulated amplitude with different signal generating voltages at 10 Hz.

4.2 Vibration experiment

For validating the vibration demodulation performance, the fiber between DTG_{95} and DTG_{96} was coiled on a compliant cylinder to detect the vibration signals [12]. In order to compare with the conventional moving-coil geophone, we fixed the fiber sensor and a moving-coil geophone

(SGO-15HT) on the same vibrator (DH40500). The output signal of the SGO-15HT was amplified and filtered by an amplifier (SR810). When the vibrator outputs 80 Hz vibration signal to simulate seismic wave, the response signals from 0 s to 0.1 s demodulated by the fiber optic sensor and SGO-15HT are shown in Figs. 8(a) and 8(b), respectively. Figure 8 indicates that the demodulated 80 Hz signal by the fiber sensor system fits well with the signal detected by the conventional moving-coil geophone, which has a good sinusoidal shape. Figures 9(a) and 9(b) show the frequency spectrum of the Figs. 8(a) and 8(b), respectively. Compared with the moving-coil geophones, the signal demodulated by the fiber sensor system has higher signal-to-noise ratio at low frequency, but it contains more harmonic peaks, which is probably caused by the 3×3 coupler with asymmetry splitting ratio.

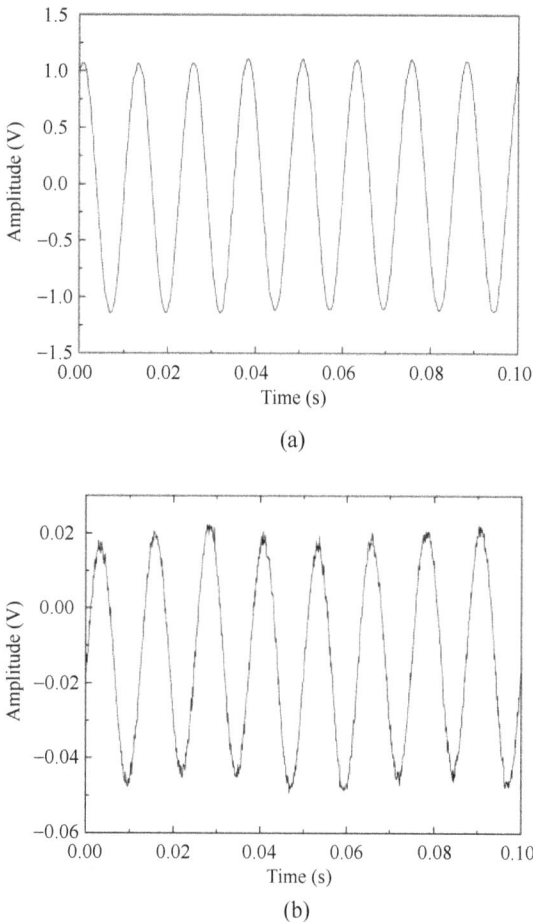

(a)

(b)

Fig. 8 Demodulated vibration signals by (a) fiber sensor and (b) SGO-15HT.

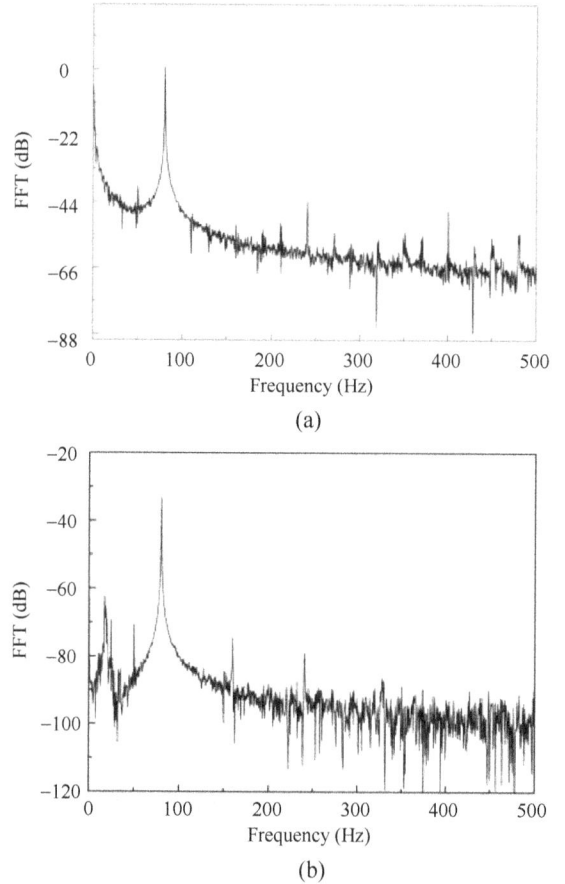

(a)

(b)

Fig. 9 Frequency responses of the demodulated vibration signals by (a) fiber sensor and (b) SGO-15HT.

5. Conclusions

A fiber optic interferometric multiplexing vibration sensing system based on DTG array is proposed and experimentally demonstrated in the paper. A DTG array with 96 DTGs was on line fabricated on the draw tower with 20 m distance between adjacent two DTGs. Phase modulation experimental results indicate that the sensing system phase demodulation dynamic scope reaches 27 dB at 10 Hz and 40 dB at 200 Hz, and the linearity of phase demodulation is 0.99 at 10 Hz when the DTGs multiplexing number reaches 96. Vibration experimental results show that the fiber optic sensing system based DTGs has larger multiplexing capacity and higher signal-to-noise ratio at low frequency than the moving coil geophone. The next step of the work is to optimize the sensor package to enhance the sensitivity and reduce the volume of the sensors.

Acknowledgment

This work was partly supported by the Major Program of the National Natural Science Foundation of China (Grant No. 61290311), the Fundamental Research Funds for the Central Universities (Grant No. 2015III056), and Key Consulting Project of Chinese Academy of Engineering (Grant No. 2016-XZ-13).

References

[1] G. Tao, X. L. Zhang, X. R. Liu, S. H. Chen, and T. Y. Liu, "A new type of fiber Bragg grating based seismic geophone," *Applied Geophysics*, 2009, 6(1): 84–92.

[2] I. Crespo-Chacon, J. L. Garcia-de-la-Oliva, and E. Santiago-Recuerda, "On the use of geophones in the low-frequency regime to study rail vibrations," *Procedia Engineering*, 2016, 143: 782–794.

[3] T. C. Liang and Y. L. Lin, "A fiber-optic sensor for the ground vibration detection," *Optics Communications*, 2013, 306(18): 190–197.

[4] S. Savazzi, U. Spagnolini, L. Goratti, D. Molteni, M. Latva-aho, and M. Nicoli, "Ultra-wide band sensor networks in oil and gas explorations," *IEEE Communications Magazine*, 2013, 51(4): 150–160.

[5] J. Y. Wang, B. X. Hu, W. Li, G. D. Song, L. Jiang, and T. Y. Liu, "Design and application of fiber Bragg grating (FBG) geophone for higher sensitivity and wider frequency range," *Measurement*, 2016, 79: 228–235.

[6] Z. G. Wang, W. T. Zhang, J. Han, W. Z. Huang, and F. Li, "Diaphragm-based fiber optic Fabry-Perot accelerometer with high consistency," *Journal of Lightwave Technology*, 2014, 32(24): 4208–4213.

[7] Q. Zhang, T. Zhu, J. D. Zhang, and K. S. Chiang, "Micro-fiber-based FBG sensor for simultaneous measurement of vibration and temperature," *IEEE Photonics Technology Letters*, 2013, 25(18): 1751–1753.

[8] A. D. Kersey, T. A. Berkoff, and W. W. Morey, "High resolution fiber grating based strain sensor with interferometric wavelength-shift detection," *Electronics Letters*, 1992, 28(3): 236–238.

[9] Z. H. Sun, X. H. Liu, F. X. Zhang, S. J. Li, X. L. Zhang, C. Wang, *et al.*, "High sensitivity fiber laser geophone array and field test analysis," *Measurement*, 2016, 79: 216–221.

[10] F. F. Chen, Y. Jiang, and L. Jiang, "3×3 coupler based interferometric magnetic field sensor using a TbDyFe rod," *Applied Optics*, 2015, 54(8): 2085–2090.

[11] H. Z. Lin, L. N. Ma, Z. L. Hu, Q. Yao, and Y. M. Hu, "Multiple reflections induced crosstalk in inline TDM fiber Fabry-Perot sensor system utilizing phase generated carrier scheme," *Journal of Lightwave Technology*, 2013, 31(16): 2651–2658.

[12] Y. S. Zhang, X. G. Qiao, Q. P. Liu, D. K. Yu, H. Gao, M. Shao, *et al.*, "Study on a fiber Bragg grating accelerometer based on compliant cylinder," *Optical Fiber Technology*, 2015, 26: 229–233.

15

An Arc Tangent Function Demodulation Method of Fiber-Optic Fabry-Perot High-Temperature Pressure Sensor

Qianyu REN[1,2], Junhong LI[3], Yingping HONG[1,2],
Pinggang JIA[1·], and Jijun XIONG[1,2*]

[1]*Science and Technology on Electronic Test & Measurement Laboratory, North University of China, Taiyuan, 030051, China*

[2]*Key Laboratory Instrumentation Science & Dynamic Measurement, Ministry of Education, North University of China, Taiyuan, 030051, China*

[3]*Department of Automation, Shanxi University, 030006, China*

[*]Corresponding author: Jijun XIONG   E-mail: xiongjijunnuc@126.com

Abstract: A new demodulation algorithm of the fiber-optic Fabry-Perot cavity length based on the phase generated carrier (PGC) is proposed in this paper, which can be applied in the high-temperature pressure sensor. This new algorithm based on arc tangent function outputs two orthogonal signals by utilizing an optical system, which is designed based on the field-programmable gate array (FPGA) to overcome the range limit of the original PGC arc tangent function demodulation algorithm. The simulation and analysis are also carried on. According to the analysis of demodulation speed and precision, the simulation of different numbers of sampling points, and measurement results of the pressure sensor, the arc tangent function demodulation method has good demodulation results: 1 MHz processing speed of single data and less than 1% error showing practical feasibility in the fiber-optic Fabry-Perot cavity length demodulation of the Fabry-Perot high-temperature pressure sensor.

Keywords: Arc tangent; Fabry-Perot; demodulation; pressure sensor

1. Introduction

The traditional pressure sensor can't be well applied in the field of high-temperature pressure measurement. With the rapid development of engineering and technology, the accurate measurement of pressure in the harsh environment is increasingly required [1]. In this case, the fiber-optic Fabry-Perot pressure sensor has been widely researched due to its excellent advantages, such as the small size, immunity to electromagnetic interference, wide range of dynamic measurement, and high resolution, which have been applied in high-temperature environments, such as the high-speed missile surface and combustion chamber of space engine [2].

In the fiber-optic Fabry-Perot sensor measurement system, Fabry-Perot cavity length demodulation is of great importance. The popular demodulation methods include intensity demodulation and phase demodulation [3–6]. The former method has merits of simple structure, fast speed, ease to implement, small size, and low cost. However, the demodulation accuracy is generally low. The relevant proposed compensation methods all have influence on the light source to some extent, which also limit the measurement precision. The phase demodulation method has the advantages of high accuracy and good stability. However, most of the current phase demodulation systems are composed of large and expensive optical interferometers, which are difficult to be used in the practical engineering. In order to realize the application of the fiber-optic Fabry-Perot high-temperature pressure sensor in practical engineering, developing a high-precision, fast-rate, good-stability, and small-volume demodulation system is largely imperative [7–10].

The phase demodulation methods mainly include the fringe counting method, Fourier-transform method, and correlation method. The fringe counting method is mostly based on the peak-peak tracing method, thus the demodulation accuracy is affected by the accuracy of the peak wavelength, leading to low accuracy. The principle of the Fourier-transform method is complex. The correlation demodulation system is expensive. With the widespread use of digital techniques in instrumentation and communication systems, full digital demodulation has become the current trend of fiber-optic Fabry-Perot sensor demodulation, which has the advantages of high speed, high precision, and good stability. However, the traditional full digital demodulation method still combines with large optical instruments [11–14]. This paper proposes a new demodulation algorithm based on the phase generated carrier (PGC) arc tangent function and the analysis of several mature demodulation methods. The MATLAB simulation is carried out, the new algorithm is simulated using field-programmable gate array (FPGA) hardware, and the results are analyzed. As the results shown,

the new algorithm expands the range of the original algorithm, verifying the feasibility of the arc tangent function demodulation method in the fiber-optic Fabry-Perot high-temperature pressure sensor demodulation.

2. Demodulation mechanism of the PGC arc tangent function method

The mechanism of the PGC arc tangent function method is to modulate the initial signal with the high-frequency and large-amplitude phase modulation signal [15–19]. The modulated optical signal is transmitted to the sensor, after which the phase variation of the optical signal is caused by the measured parameters. Then the signal propagates into the circulator, and the optical signal is converted into the electrical signal in the photodetector, outputting two signals, which are respectively mixed with the modulation signal and the double frequency modulation signal. Then the two modulation signals pass through the low-pass filter, and two quadrature signals are received. The optical circuit of modulation and demodulation system is shown in Fig. 1.

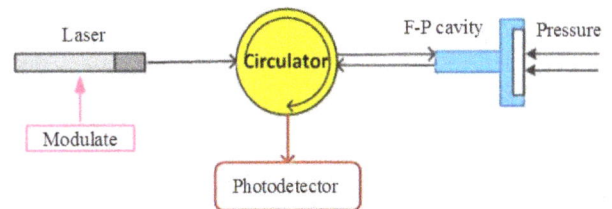

Fig. 1 PGC modulation and optical circuit schematic.

The light intensity signal is calculated as

$$I = A + B\cos\left[C\cos(\omega_0 t) + \varphi(t)\right] \tag{1}$$

where A and B are constants which are related to the power of the input light source and the response of the detector; $C\cos(\omega_0 t)$ is the carrier signal; $\varphi(t)$ is the signal to be measured, and this signal includes the low-frequency signal caused by the environmental drift.

Equation (1) is expanded into the first kind of Bessel function as

$$I = A + B\left\{\left[J_0(C) + 2\sum_{k=1}^{\infty}(-1)^k J_{2k}(C)\cos 2k\omega_0 t\right] \times \cos\varphi(t) - \left[2\sum_{k=0}^{\infty}(-1)^k J_{2k+1}(C)\cos(2k+1)\omega_0 t\right]\sin\varphi(t)\right\}$$

$$(2)$$

where $J_k(C)$ is the first kind of Bessel function of order k.

The signal I is mixed with the output signals $G\cos\omega_0 t$ and $H\cos 2\omega_0 t$, respectively. Then the high-frequency signal is filtered by a low-pass filter, and the two quadrature signals are received. The two signals are shown as follows:

$$-BGJ_1(C)\sin(\quad) \qquad (3)$$

$$-BHJ_2(C)\cos(\quad). \qquad (4)$$

And $(3)/(4)$ can be obtained as follows:

$$\frac{G \times J_1(C)}{H \times J_2(C)}\tan\varphi(\quad). \qquad (5)$$

In general, for the operation simple, we can take $G = H$, and the coefficient of (5) will be $J_1(C)/J_2(C)$. Firstly, the coefficient of (5) is normalized. And then the arc tangent operation is carried out. Finally, using the high-frequency filter to filter out the low-frequency signal caused by the environment, the signal $\varphi(t)$ is received. The traditional PGC arc tangent demodulation method is shown in Fig. 2.

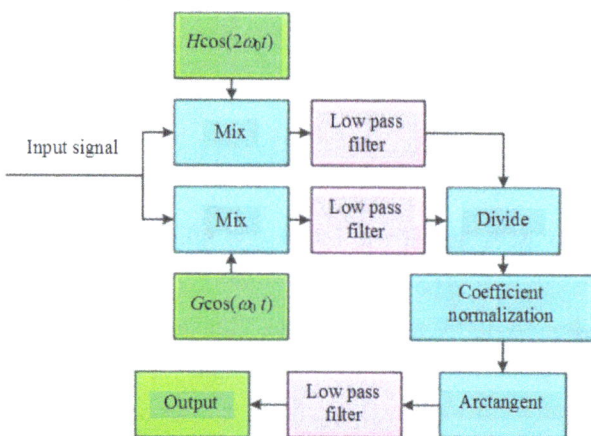

Fig. 2 Schematic of the PGC arc tangent demodulation method.

3. Algorithm simulation and analysis

The maximum detectable signal amplitude of the PGC demodulation algorithm is affected by the amplitude C and the frequency of the modulated signal. If the signal amplitude is beyond or near this range, the demodulation result will be distorted. It is enough to be applied in the interference hydrophones which have a little range of phase change. However, it can't meet the use requirement of the fiber-optic Fabry-Perot pressure sensors which have a large range of phase variation.

The relationship between the cavity length and phase is described as

$$\Delta L = \frac{\Delta\varphi(t)\lambda}{2\pi} \qquad (6)$$

where ΔL is the Fabry-Perot cavity length variation; $\Delta\varphi(t)$ is the phase variation; λ is the laser wavelength.

Using (6) to calculate the cavity length variation of the fiber-optic Fabry-Perot high-temperature pressure sensor developed by our laboratory, in the range of $0-5$bar, the phase change range is $0-4.6\pi$. It is not enough to use the traditional arc tangent function method. Therefore, it is necessary to enlarge the range of the demodulation algorithm.

The output signal of the PGC demodulation algorithm is simulated by MATLAB. As shown in Fig. 3, $\varphi(t) = 10\sin t$. As shown in Fig. 4, the phase jump will appear in the traditional algorithm of phase demodulation. At the jump position, the difference between the phase value of the latter point which is output by the phase demodulation and the original phase value is π. According to the result, a new algorithm which can compensate the phase is designed by using MATLAB. The result is shown in Fig. 5.

The first part of this algorithm is the setup of two thresholds: one is positive, and the other is negative. When the previous data are greater than the positive threshold and the latter data are less than the negative threshold, it is regarded as a jump point, and the FPGA will add all the following data to π. Conversely, if the previous data are less than the negative threshold and the latter data are greater than the positive threshold, the FPGA will subtract π from all the following data. Through the new arc tangent algorithm, the initial signal data are restored. Thus, the demodulation range of the arc tangent function method is expanded.

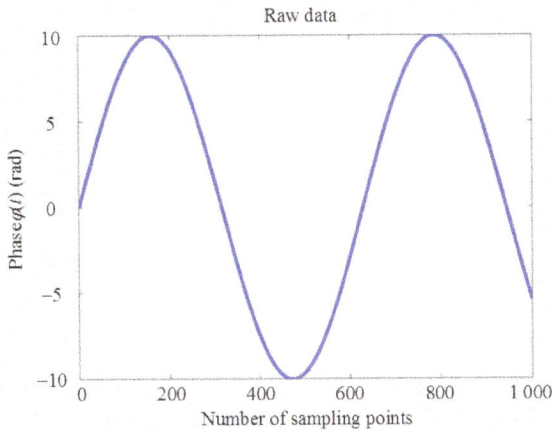

Fig. 3 Original signal $\Delta\varphi(t)$.

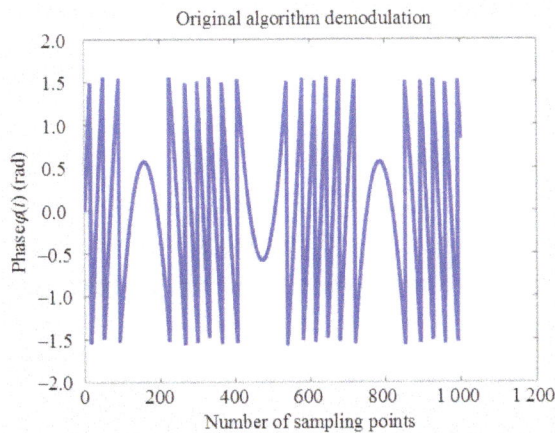

Fig. 4 Original algorithm demodulation signal.

It can be seen from Fig. 3 to Fig. 5 that the algorithm solves the shortcomings of the traditional arc tangent function demodulation method in theory. The requirement of the application of the fiber-optic Fabry-Perot high-temperature pressure sensor is satisfied. The phase change range can satisfy the change range of the fiber-optic Fabry-Perot cavity. Using this new algorithm, the demodulation of the fiber-optic Fabry-Perot sensor is easier to realize full digital demodulation, and it is easy to use the FPGA

to achieve the method. At the same time, the new demodulation method will also use the optical system for the optical signals orthogonal processing. After the orthogonal processing, optical signals are converted into electrical signals by the photoelectric converter which can reduce the interference caused by the circuit.

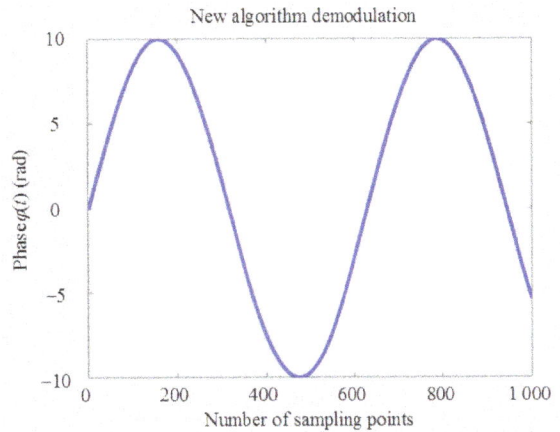

Fig. 5 New algorithm demodulated signal.

4. FPGA design and data simulation

This paper selects the Cyclone IV series EP4CE30F23C8N chip of Altera Company. Cyclone IV E has low power consumption, and it can achieve high functionality with very low cost. EP4CE30F23C8N chip which uses 1.2 V power supply, with 329 user I/O ports, the 28848 logic units, 594 KB embedded memory and the 4 general PLLs, which can satisfy the requirements of the hardware design.

The realization of the design process is shown in Fig. 6. Orthogonal signals 1 and 2 are generated by the MATLAB simulation of the two orthogonal signals. Using very-high-speed integrated circuit hardware description language (VHDL) to compile the FPGA module to control the IP core computing and data processing, the results are stored in the first input first output (FIFO) IP core. Finally, the serial port will output the data. The two data are simulated, and the initial signals of the two simulations are sine wave with an amplitude of 10 rad. The numbers of sampling points are 64 and 256, respectively. The simulation results are shown in Figs. 7 and 8.

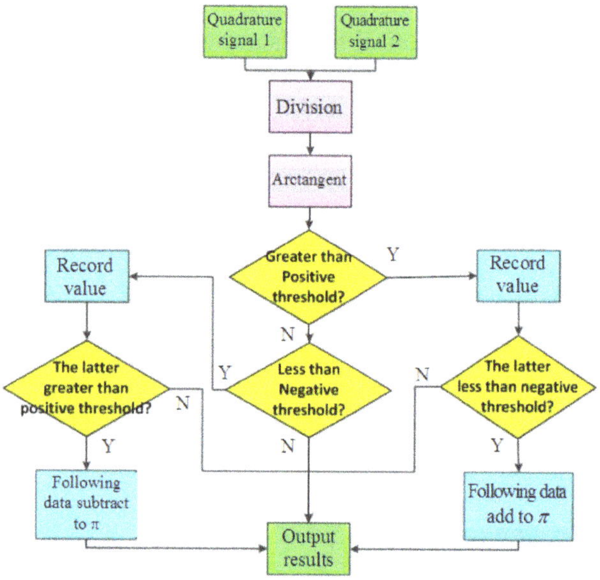

Fig. 6 Flow chart of new algorithm.

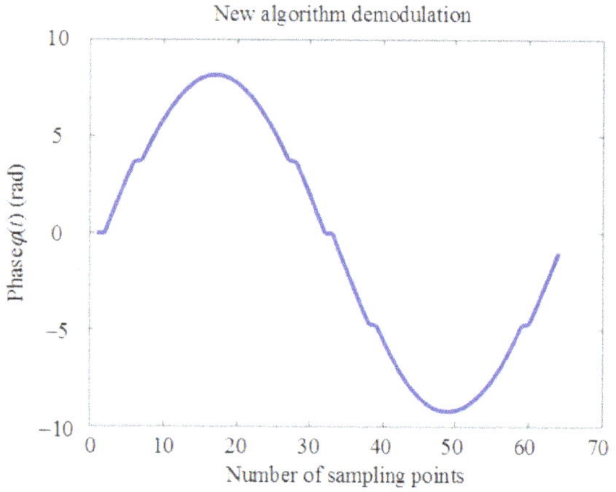

Fig. 7 FPGA simulation result of 64 sampling points.

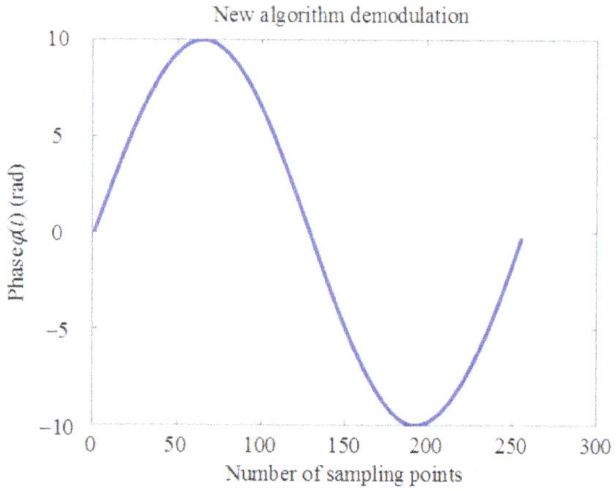

Fig. 8 FPGA simulation result of 256 sampling points.

It can be seen from Figs. 7 and 8 that when the signal amplitude is 10 rad, it is not enough to

demodulate the signal which samples 64 points in each cycle. When the number of sampling points is increased to 256, the waveform of the signal is demodulated completely, and the error is less than 1%. Because the time of a single data demodulation is about 0.7 µs, the demodulation time of the full waveform is about 179 µs. Theoretically, if the sampling rate is equal to the above, when the signal frequency of 10 rad is below 5 kHz, the signal can be demodulated completely. Because the output phase signal of the fiber-optic Fabry-Perot high temperature pressure sensor is dominated by the low- frequency signal, the current pressure tests can be satisfied. Using the fiber-optic Fabry-Perot pressure sensor developed in our laboratory, each pressure changes from 0 bar to 1 bar, and the phase variation is approximately 0.92π. That means, in theory, if the pressure variation is less than 1 bar within 12.5 µs, the new demodulation algorithm can satisfy the test requirements.

5. Conclusions

By the MATLAB simulation, FPGA simulation, and results analysis, the feasibility of the new algorithm is verified preliminarily. The new algorithm overcomes the existing shortcomings of the traditional PGC demodulation algorithm, and it expands the range to satisfy the requirements of fiber-optic Fabry-Perot high-temperature pressure sensor demodulation. Compared with the general fiber-optic Fabry-Perot pressure sensor demodulation algorithm, the new arc tangent algorithm has the advantages of fast speed, simple principle, high accuracy, and low cost, so it is expected to become a new mainstream demodulation method, providing a new idea for the future of the arc tangent function demodulation algorithm in the fiber-optic Fabry-Perot pressure sensor in high temperature application.

Acknowledgment

This work was supported by the National

Science Fund for Distinguished Young Scholars (No. 51425505) and the National Natural Science Foundation of China (No. 51405454).

References

[1] T. George, K. A. Son, R. A. Powers, L. Y. D. Castillo, and R. Okojie, "Harsh environment microtechnologies for NASA and terrestrial applications," in *Proceeding of IEEE 4th International Conference on Sensors*, Irvine, CA, USA, 2005, pp. 1253–1258.

[2] G. C. Fang, P. G. Jia, T. Liang, Q. L. Tan, Y. P. hong, W. Y. Liu, *et al.*, "Diaphragm-free fiber-optic Fabry-Perot interferometer based on tapered hollow silica tube," *Optics Communications*, 2016, 371: 201–205.

[3] Z. H. Yu and A. B. Wang, "Fast demodulation algorithm for multiplexed low-finesse Fabry-Perot interferometers," *Journal of Lightwave Technology*, 2016, 34(3): 1015–1019.

[4] J. F. Jiang, T. G. Liu, Y. M. Zhang, L. N. Liu, Y. Zha, F. Zhang, *et al.*, "Development of a parallel demodulation system used for extrinsic Fabry-Perot interferometer and fiber Bragg grating sensors," *Applied Optics*, 2006, 45(3): 528–535.

[5] B. Wu, Y. Yuan, J. Yang, A. Zhao, and L. Yuan, "Improved signal demodulation method in optical fiber seismometer," *Sensor Letters*, 2012, 10(7): 1402–1406.

[6] W. R. Allan, Z. W. Graham, J. R. Zayas, D. P. Roach, and D. A. Horsley, "Multiplexed fiber Bragg grating interrogation system using a microelectromechanical Fabry-Perot tunable filter," *IEEE Sensors Journal*, 2009, 9(8): 936–943.

[7] Q. X. Yu and X. L. Zhou, "Pressure sensor based on the fiber-optic extrinsic Fabry-Perot interferometer," *Photonic Sensors*, 2011, 1(1): 72–83.

[8] M. Li, B. Tong, N. Arsad, and J. J. Guo, "A double-fiber Fabry-Perot sensor based on modified fringe counting and direct phase demodulation," *Measurement Science & Technology*, 2013, 24(9): 094012.

[9] Y. F. Tao, M. Wang, D. M. Guo, X. Q. Ni, and H. Hao, "Nine-point phase demodulation for interferometric measurement," *Optik*, 2016, 127(14): 5654–5662.

[10] E. Lu, Z. L. Ran, F. Peng, Z. W. Liu, and F. G. Xu, "Demodulation of micro fiber-optic Fabry-Perot interferometer using subcarrier and dual-wavelength method," *Optics Communications*, 2012, 285(6): 1087–1090.

[11] K. Toge and F. Ito, "Recent research and development of optical fiber monitoring in communication systems," *Photonic Sensors*, 2013, 3(4): 304–313.

[12] J. H. Xie, F. Y. Wang, Y. Pan, J. J. Wang, Z. L. Hu, Y. M. Hu, "High resolution signal-processing method for extrinsic Fabry-Perot interferometric sensors," *Optical Fiber Technology*, 2015, 22: 1–6.

[13] N. Wang, Y. Zhu, T. C. Gong, L. H. Li, and W. M. Chen, "Multichannel fiber optic Fabry-Perot nonscanning correlation demodulator," *Chinese Optics Letters*, 2013, 11(7): 10–12.

[14] H. T. Chen and Y. C. Liang, "Analysis of the tunable asymmetric fiber F-P cavity for fiber sensor edge-filter demodulation," in *Proceeding of IEEE 2014 International Conference on Consumer Electronics*, Shenzhen, 2014, pp. 338–343.

[15] "Diaphragm based long cavity Fabry-Perot fiber acoustic sensor using phase generated carrier," *Optics Communications*, 2017, 382: 514–518.

[16] Y. Liu, L. W. Wang, C. D. Tian, M. Zhang, and Y. Liao, "Analysis and optimization of the PGC method in all digital demodulation systems," *Journal of Lighawave Technology*, 2008, 26(17–20): 3225–3233.

[17] Y. J. Wei and Z. H. Zhai, "Error analysis of dual wavelength quadrature phase demodulation for low-finesse Fabry-Perot cavity based fibre optic sensor," *Optik*, 2011, 122(14): 1309–1311.

[18] A. L. Zhang and S. Zhang, "High stability fiber-optics sensors with an improved PGC demodulation algorithm," *IEEE Sensors Journal*, 2016, 16(21): 7681–7684.

[19] S. C. Huang, Y. F. Huang, and F. H. Hwang, "An improved sensitivity normalization technique of PGC demodulation with low minimum phase detection sensitivity using laser modulation to generate carrier signal," *Sensors and Actuators A: Physical*, 2013, 191: 1–10.

16

Study of All-Optical Sampling Using a Semiconductor Optical Amplifier

Chen WU, Yongjun WANG*, Lina WANG, and Fu WANG

School of Electronic Engineering, Beijing University of Posts and Telecommunications, Beijing, 100876, China

*Corresponding author: Yongjun WANG   E-mail: wangyj@bupt.edu.cn

Abstract: All-optical sampling is an important research content of all-optical signal processing. In recent years, the application of the semiconductor optical amplifier (SOA) in optical sampling has attracted lots of attention because of its small volume and large nonlinear coefficient. We propose an optical sampling model based on nonlinear polarization rotation effect of the SOA. The proposed scheme has the advantages of high sampling speed and small input pump power, and a transfer curve with good linearity was obtained through simulation. To evaluate the performance of sampling, we analyze the linearity and efficiency of sampling pulse considering the impact of pulse width and analog signal frequency. We achieve the sampling of analog signal to high frequency pulse and exchange the positions of probe light and pump light to study another sampling.

Keywords: All-optical sampling; semiconductor optical amplifier; nonlinear polarization rotation

1. Introduction

In recent years, the all-optical signal processing system has become a hot research field of optical communication, and it is generally achieved by nonlinear effects of nonlinear elements, in which the semiconductor optical amplifier (SOA) is an important device. On the one hand, the SOA can achieve online amplification in the optical communication link to make up line loss; on the other hand, it has advantages of high nonlinear coefficient, compact structure, low power consumption, and ease to integrate. In particular, the SOA can produce many optical nonlinear effects, such as cross phase modulation (XPM) [1, 2], cross gain modulation (XGM) [3, 4], four-wave mixing (FWM) [5, 6], and nonlinear polarization potation

(NPR) [7, 8], which can achieve wavelength conversion, optical switch, optical buffer, and all-optical sampling in the signal processing [9–12]. Compared with other nonlinear effects of the SOA, NPR is applied to more various aspects due to that it is high-speed and easy to implement, and can also achieve a positive and inverted beam at the same time. Since 1990s, the SOA has attracted extensive attention in the study of all-optical signal processing, and its application methods and scope are growing steadily.

2. Working principle

The NPR effect of the SOA is an important aspect of its nonlinear characteristics. The polarization-related characteristics of the SOA include polarization dependent gain (PDG) and polarization dependent phase, resulting in that SOA's output polarization state changes as the input signal. PDG characteristic means that the gains of the transverse electric (TE) mode and transverse magnetic (TM) mode are different, polarization dependent phase characteristic refers to that there is birefringence between the TE and TM modes in the SOA, which leads to that phase change is not the same in two modes after the input light passes through the SOA.

Ignoring the influence of surface remaining reflectance and spontaneous emission, the transmission equations of SOA's photon density and phase in the TE and TM directions are

$$\frac{\partial S^i(z,t)}{\partial z} = \left[\Gamma^i g^i(z,t) - \alpha_{int}^i\right]S^i, \quad i = \text{TE or TM} \quad (1)$$

$$\frac{\partial \phi^i(z,t)}{\partial z} = -\frac{1}{2}\alpha^i \Gamma^i g^i(z,t), \quad i = \text{TE or TM} \quad (2)$$

where Γ is the mode field limiting factor, α is the linewidth enhancement factor (LEF), $g(z,t)$ is the gain, and α_{int} the is mode loss.

The relationship between the photon density S and input light power P is

$$S^i = \frac{P^i}{\hbar \omega}\frac{L}{\upsilon_g}, \quad i = \text{TE or TM}. \quad (3)$$

The gains $g^{\text{TE}}(z,t)$ and $g^{\text{TM}}(z,t)$ are

$$g^{\text{TE}}(z,t) = \xi^{\text{TE}}\left[n_c(z,t) + n_x(z,t) - N_0\right]/V_a \quad (4)$$

$$g^{\text{TM}}(z,t) = \xi^{\text{TM}}\left[n_c(z,t) + n_y(z,t) - N_0\right]/V_a. \quad (5)$$

We introduce the non-equilibrium factor f to measure the anisotropy of the SOA, so the carrier rate equations of TE and TM modes are

$$\frac{\partial n_x(z,t)}{\partial t} = -\frac{n_x(z,t) - \overline{n_x}}{T_1} - \frac{n_x(z,t) - fn_y(z,t)}{T_2} \\ -g^{\text{TE}}(z,t)S^{\text{TE}}(z,t) \quad (6)$$

$$\frac{\partial n_y(z,t)}{\partial t} = -\frac{n_y(z,t) - \overline{n_y}}{T_1} - \frac{fn_y(z,t) - n_x(z,t)}{T_2} \\ -g^{\text{TM}}(z,t)S^{\text{TM}}(z,t) \quad (7)$$

where at equilibrium,

$$\overline{n_x} = \frac{\overline{n}f}{1+f}, \quad \overline{n_y} = \frac{\overline{n}}{1+f}, \quad \overline{n} = \frac{I}{e}T_1. \quad (8)$$

As shown in Fig. 1, the proposed optical sampling scheme is based on the NPR effect of the SOA. The optical sampling switch is constituted by the SOA, polarization controller (PC), isolator, circulator, polarizing beam splitter (PBS), and band pass filter (BPF). A narrow Gaussian pulse sequence is used as a probe light, and a sinusoidal optical signal of different wavelengths from probe light is used as a pump light. When PC1 is adjusted to make the input light polarization state be at 45° with SOA's horizontal direction, then PC2 is adjusted to make that there is no effect of the pump light, so that the probe light can have its minimum output through the PBS.

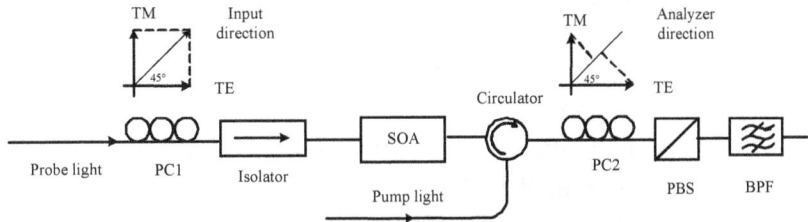

Fig. 1 All-optical sampling scheme based on the NPR effect of the SOA.

3. Simulation results and analysis

3.1 Optical switch performance

For all-optical sampling utilizing the NPR effect, its transfer curve linearity of the PBS port is the key factor that influences the efficiency of the sampling system. To obtain a transfer curve with a good linearity, we analyze the static output curve of 0.1 mW, 0.3 mW, 0.5 mW, 0.7 mW, and 0.9 mW probe lights, the bias current is 350 mA, as shown in

Fig. 2(a). It is observed that the linearity and sampling scopes of 0.1 mW and 0.9 mW are both bad, the sampling scopes of 0.5 mW and 0.7 mW are large, but their linearity is not as good as 0.3 mW, therefore, in the subsequent analysis of the optical sampling application, the probe light power is set to 0.3 mW.

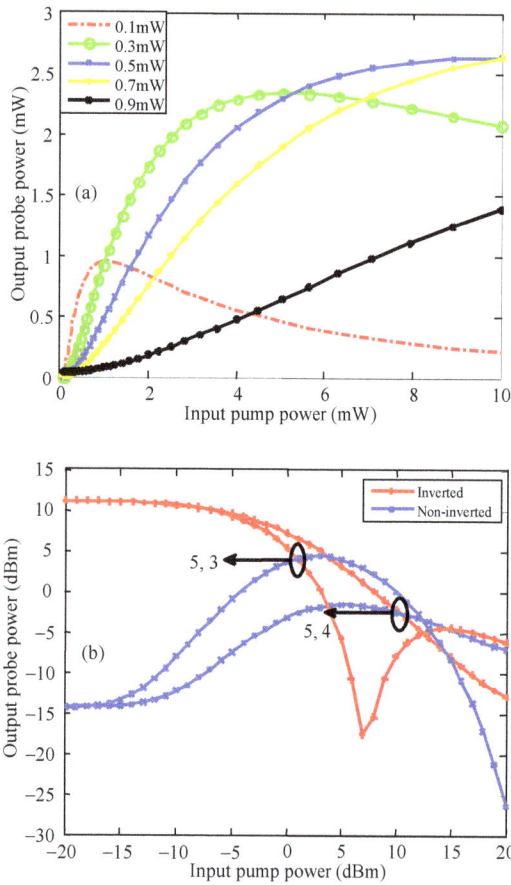

Fig. 2 Output power with respect to the pump power: (a) input probe light power and (b) linewidth enhancement factor.

The probe light passes through the SOA and outputs from the PBS. Static output results of the PBS port are analyzed in Fig 2(b). It can be seen from two non-inverted output curves of the PBS that there is a transfer curve with a good linearity which can be applied to the research on optical sampling. For the curve whose LEFs are 5 and 3, the polarization state is rotated when the pump power is 7 dBm, so it can achieve the polarization optical switch. While another group of curves with small linewidth gain difference, the phase difference is

hard to reach π, so it cannot achieve the polarization optical switch even though the pump power is 15 dBm.

3.2 Phase difference factors

Variable curves of phase difference between the TE and TM modes with respect to the bias current, pump power, and probe power are acquired. Figure 3(a) shows the phase difference change with the SOA bias current in the NPR process. With an increase in the bias current, the phase difference increases and becomes saturated, and at the same bias current, when the probe light power is greater, the phase difference is smaller. Figure 3(b) analyzes the phase difference change with the pump light power at different bias currents of different probe light powers. It can be seen from the three groups of curves that when the probe light power is bigger, the phase difference at the same pump power is smaller and easier to saturate. Electrically and optically controlled optical switches can be realized by making the phase difference to π through the appropriate choice of the probe power, pump power, and bias current.

In addition, the impacts of the PDG and LEF on the phase difference are discussed, respectively. The results show that the PDG can affect the phase difference change, but its role is limited. LEF difference is the major influencing factor, and the selection of the SOA with the larger LEF difference can obtain the smaller pump light power to achieve optical sampling. As shown in Fig. 3(c), its maximum phase difference change with the pump power is less than 1 rad by comparing several curves of the PDG from 0.6 dB to 4 dB. Figure 3(d) analyzes effects of different PDGs and LEFs on the phase difference change. It is obvious that when α^{TE} and α^{TM} are 5 and 3, it is still able to reach π phase difference of the 0.6-dB PDG, but it is difficult for other two curves no matter how much the PDG is.

(a)

(b)

(c)

(d)

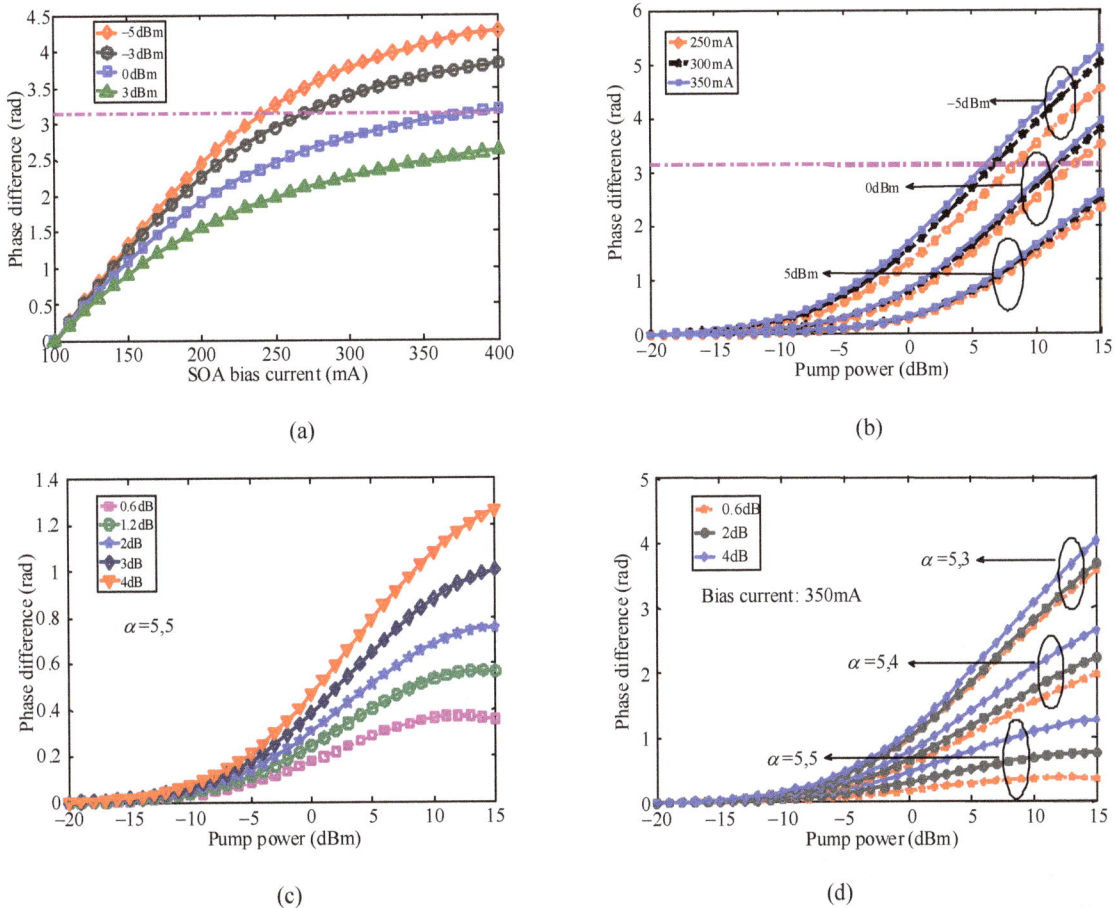

Fig. 3 Curves of phase difference: (a) with bias current and probe power, (b) with pump power, probe power, and bias current, (c) with PDG and pump power, and (d) with PDG, LEF, and pump power.

3.3 Sampling results

We establish an ultrafast all-optical sampling scheme by analyzing SOA's NPR effect, in which the input probe light is a narrow Gaussian pulse sequence, and the input pump light is an analog cosine signal with different wavelengths from the probe light. In the process of optical sampling, the incident pump light modulates the carrier density and gain of the SOA to rotate the polarization state of sampling pulse light, and the polarization state modulation is converted to the intensity modulation through the PBS. The BPF filters analog signal, so that only the ultrafast pulse-modulated light can pass through, and the all-optical sampling is achieved.

Figures 4(a) and 4(b) are the normalized powers of the input probe light and pump light, and the corresponding power curves of the output probe

light and their pulse envelopes are shown in Figs. 4(c) and 4(d), respectively. It can be seen from the figures that the information of the analog signal is modulated to the output pulse envelope well, so the sampling of the analog signal to high frequency

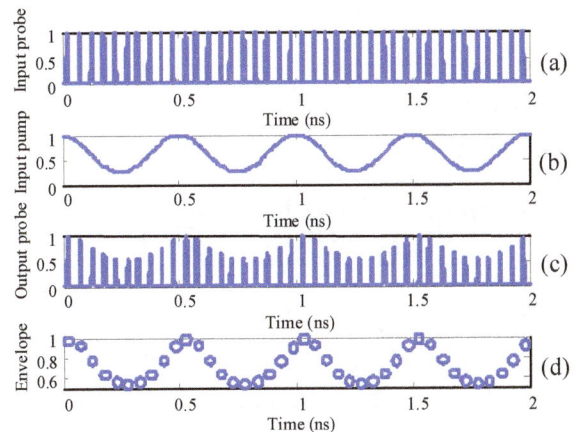

Fig. 4 Normalized powers of the (a) input probe light and (b) input pump light; normalized powers of the (c) output probe light and (d) output probe envelope.

pulse is achieved successfully. For calculation, the bias current of the SOA is set to 350 mW. The peak power of the probe light is 0.3 mW, and its repetition rate is 20 GHz. The pump light has the 4-mW peak power and 1-mW valley power, and its frequency is different from the probe light. The incident polarization angles are both at the degree of 45.

After that, the positions of the probe light and pump light are exchanged to achieve the sampling of the high frequency pulse to the analog signal. Figure 5(a) shows the sampling result, and it is obvious that there is a big distortion in the output probe light. When there is no time-varying input optical signal, the carrier concentration and photon density will reach a steady state of dynamic equilibrium in SOA's active region. But when there is a pulse light injection, this balance is broken, and the carrier will experience a slow recovery process to reach the stable state.

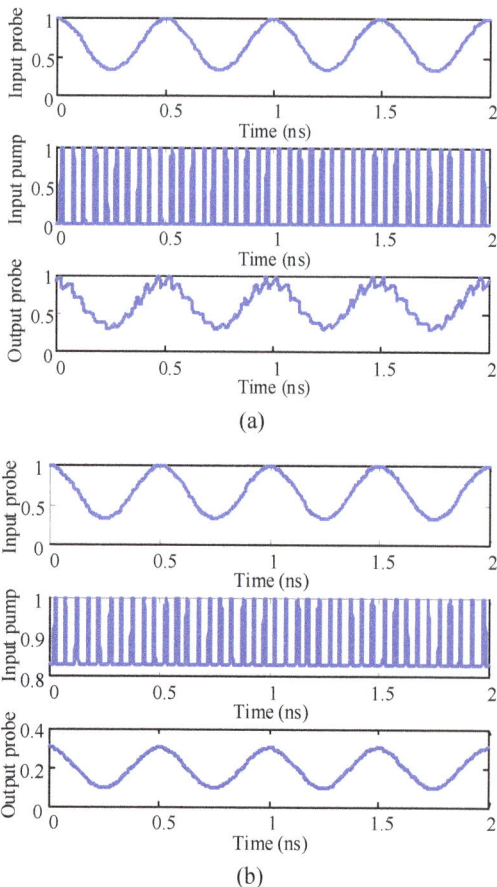

(a)

(b)

Fig. 5 Sampling results of exchanged probe and pump lights: (a) without holding light and (b) with holding light.

In order to reduce that distortion, we add a holding light to the pump light, and the improved sampling result is shown in Fig. 5(b). It is observed that holding light injection can improve the linearity of the all-optical sampling effectively, which is because suitable holding light power accelerates carrier recovery greatly, so it can speed up the response to the ultrafast signals and suppress signal distortion introduced by slow gain recovery to eliminate signal frontier overshoot. However, the disadvantage of this approach is that the gain will decline leading to negative impacts.

3.4 Linearity and efficiency of sampling

Taking into account the influence of the pulse width on the sampling results, we apply the Fourier curve to fit the pulse envelope of the 20-GHz sampling pulse of different pulse widths, and calculate the fundamental efficiency E_1 and k-order harmonic distortion D_k, and the correlation coefficient is near 0.99999. Due to that the harmonic distortion is mainly from the second harmonic, this paper mainly compares the values of E_1 and D_2 in the analysis. The band effect needs to be considered for the narrow pulse of less than 2-ps width, so we only analyze the case of a pulse width of several ps. The E_1 and D_2 values of different pulse widths of 1-G, 2-G, and 2.5-G analog signals are shown in Figs. 6(a) and 6(b), respectively. As can be seen from the figures, the sampling pulse width affects both the fundamental efficiency and second harmonic distortion, and has a nearly linear relationship. This is because of the impact of SOA carrier recovery time that when the sampling pulse width is narrower, the carrier recovery concentration is greater, and the obtained pulse value is greater after the SOA, so the fundamental efficiency is higher, and the second harmonic distortion is smaller.

We analyze the sampling performance of the analog signals of different frequencies. Each sampling pulse width is 6.5 ps, and the fitting

(a)

(b)

(c)

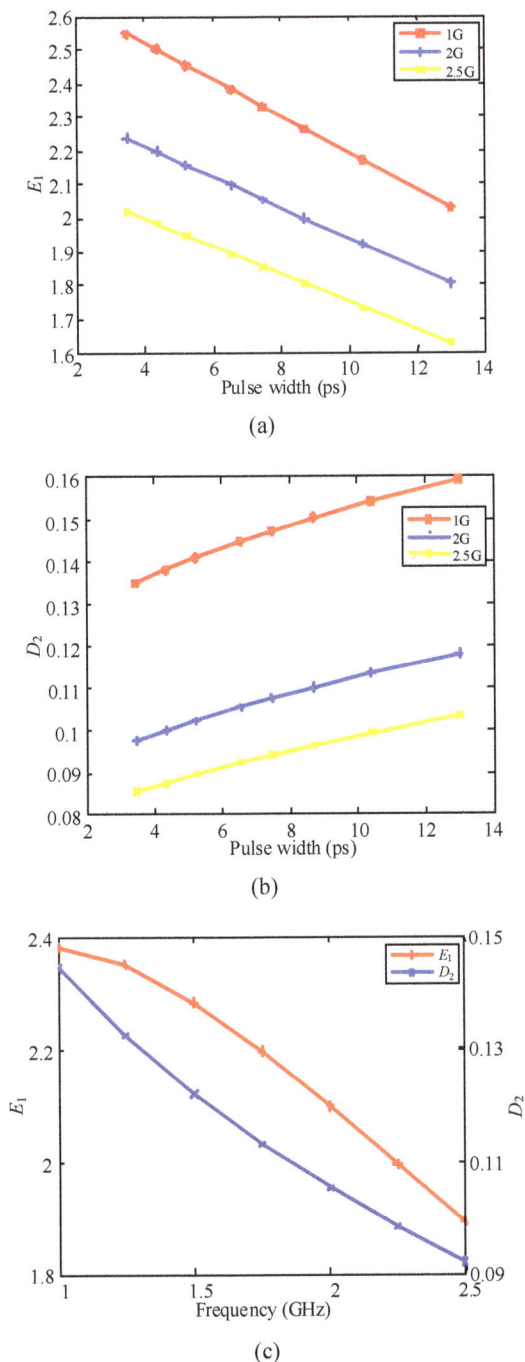

Fig. 6 Sampling linearity and efficiency: (a) E_1 of different pulse widths, (b) D_2 of different pulse widths, and (c) E_1 and D_2 of different frequencies.

correlation coefficient is near 0.99999. Figure 6(c) shows the corresponding value curves of E_1 and D_2, as can be seen from the figure, with an increase in the frequency, the second harmonic distortion decreases, but its fundamental efficiency also decreases. For the analog signals of the same power

range, when the frequency is higher, the power difference between adjacent pulse sampling points is greater, and the carrier concentration change magnitude is bigger, so the second harmonic distortion is smaller. Affected by carrier recovery time, the signal with the lower frequency has greater carrier concentration difference between power peaks and troughs, and the obtained gain and its fundamental efficiency are higher. Thus, the analog signal frequency is higher, at the same time of getting better second harmonic distortion, the fundamental efficiency is sacrificed.

4. Conclusions

We construct a theoretical all-optical sampling model based on the nonlinear polarization rotation effect of the semiconductor optical amplifier and discuss the impact of the input light power, bias current, PDG, LEF, pulse width, and frequency on the sampling effect. The feasibility of the proposed scheme has been proved, and the process of all-optical sampling is achieved by simulation. We study the evaluation methods of all-optical sampling efficiency and linearity, and obtain the fundamental efficiency and second harmonic distortion to evaluate the sampling performance. The presented optical sampling method has an important role in promoting the application of optical sampling in the all-optical signal processing.

Acknowledgment

This work was supported by the National Natural Science Foundations of China (Grant No. 61378061 and 61575026).

References

[1] H. Lee, H. Yoon, Y. Kim, and J. Jeong, "Theoretical study of frequency chirping and extinction ratio of wavelength-converted optical signals by XGM and XPM using SOA's," *IEEE Journal of Quantum Electronics*, 1999, 35(8): 1213−1219.

[2] M. Jabbari, M. K. Moravvej-Farshi, R. Ghayour, and A. Zarifkar, "XPM response of a chirped DFB-SOA all-optical flip-flop injected with an assist light at transparency," *Journal of Lightwave Technology*, 2009, 27(13): 2199‒2207.

[3] J. Kim, M. Laemmlin, C. Meuer, D. Bimberg, and G. Eisenstein, "Theoretical and experimental study of high-speed small-signal cross-gain modulation of quantum-dot semiconductor optical amplifiers," *IEEE Journal of Quantum Electronics*, 2009, 45(3): 240‒248.

[4] C. Vagionas, D. Fitsios, G. T. Kanellos, N. Pleros, and A. Miliou, "All optical flip flop with two coupled travelling waveguide SOA-XGM switches," in *Conference on Lasers and Electro-optics*, San Jose, CA, pp. 1‒2, 2012.

[5] G. Contestabile, M. Presi, and E. Ciaramell, "Multiple wavelength conversion for WDM multicasting by FWM in an SOA," *IEEE Photonics Technology Letters*, 2004, 16(7): 1775‒1777.

[6] L. Ma, P. Ghelfi, M. Yao, F. Berizzi, and A. Bogoni, "Demonstration of optical sample parallelisation for high-speed photonic assisted ADCs," *Electronics Letters*, 2011, 47(5): 333‒335.

[7] S. Fu, W. Zhong, P. Shum, C. Wu, and J. Q. Zhou, "Nonlinear polarization rotation in semiconductor optical amplifiers with linear polarization maintenance," *IEEE Photonics Technology Letters*, 2007, 19(23): 1931‒1933.

[8] Y. Liu, Q. S. Zhang, S. J. Zhang, L. Huang, X. G. Tang, H. P. Li, *et al.*, "High-speed all-optical sampling based on nonlinear polarization rotation in a semiconductor optical amplifier," *Photonics and Optoelectronic*, 2010, 6838: 1‒3.

[9] T. Durhuus, B. Mikkelsen, C. Joergensen, S. L. Danielsen, and K. E. Stubkjaer, "All-optical wavelength conversion by semiconductor optical amplifiers," *Journal of Lightwave Technology*, 1996, 14(6): 942‒954.

[10] K. E. Stubkjaer, "Semiconductor optical amplifier-based all-optical gates for high-speed optical processing," *Selected Topics in Quantum Electronics*, 2000, 6(6): 1428‒1435.

[11] S. Fu, P. Shum, L. Zhang, C. Wu, and A. M. Liu, "Design of SOA-based dual-loop optical buffer with a 3×3 collinear coupler: guideline and optimizations," *Journal of Lightwave Technology*, 2006, 24(7): 2768‒2778.

[12] P. Li, L. Jiang, J. G. Zhang, J. Z. Zhang, and Y. C. Wang, "Low-complexity TOAD-based all-optical sampling gate with ultralow switching energy and high linearity," *IEEE Photonics Journal*, 2015, 7(4): 1‒8.

17

Powerful Narrow Linewidth Random Fiber Laser

Jun YE[1,2], Jiangming XU[1,2], Hanwei ZHANG[1,2], and Pu ZHOU[1,2*]

[1]*College of Optoelectronic Science and Engineering, National University of Defense Technology, Changsha, 410073, China*

[2]*Hunan Provincial Collaborative Innovation Center of High Power Fiber Laser, Changsha, 410073, China*

[*]Corresponding author: Pu ZHOU   E-mail: zhoupu203@163.com

Abstract: In this paper, we demonstrate a narrow linewidth random fiber laser, which employs a tunable pump laser to select the operating wavelength for efficiency optimization, a narrow-band fiber Bragg grating (FBG) and a section of single mode fiber to construct a half-open cavity, and a circulator to separate pump light input and random lasing output. Spectral linewidth down to 42.31 GHz is achieved through filtering by the FBG. When 8.97 W pump light centered at the optimized wavelength 1036.5 nm is launched into the half-open cavity, 1081.4 nm random lasing with the maximum output power of 2.15 W is achieved, which is more powerful than the previous reported results.

Keywords: Random fiber laser; distributed feedback; Rayleigh scattering; Raman scattering

1. Introduction

Since the concept of random laser was first demonstrated by Letokhov *et al.* [1] in 1966, random lasers have drawn more and more attention. As is well-known, two key elements are necessary in the traditional laser scheme: a cavity which provides positive feedback and a gain medium which creates amplification. But in random lasers, there is no traditional resonator, which provides random lasers with special features compared with conventional cavity lasers. In earlier decades, researchers obtained random lasing in powder of active crystals [2], nanowires [3], and polymers [4]. The simplicity of realization of random lasers gives them an upper hand over conventional lasers. But such systems require high peak powers pump, and the laser efficiency is relatively low due to small active areas and low directionality, and more importantly, which are cumbersome or almost no control over the spectral properties of the emission [5].

Low-dimensional random systems can be used to solve these problems [6] by using optical fibers providing a lot of possibilities because of its waveguide property. Random lasing was demonstrated in the photonic crystal fiber with the hollow core filled with a suspension of TiO_2 nanoparticles [7], polymer optical fibers [8], and rare-earth doped fiber with randomly spaced gratings [9].

Recently, Turitsyn *et al.* [10] proposed a radically different branch named random fiber laser (RDFL), which operated via extremely weak random scattering in a single mode fiber (SMF). The random distributed feedback (DFB) was provided by backward random Rayleigh scattering, and the amplification was produced by the stimulated Raman scattering (SRS) effect. Following this research, a number of works were published, in which the RDFLs have been designed to have characters of high efficiency and high power output [11, 12], tunable [13, 14], multi-wavelength [15, 16], and polarized output [17, 18].

Due to its unique features, RDFL has found a large variety of application potentials, such as telecom and remote sensing [19–21]. However, most of RDFLs have relatively broad spectra (several nanometers). For some practical application fields such as coherent sensing and detection, suppressing the linewidth is required to increase the system performance. Ultra-narrow lasing has been reported in a coherent Brillouin random fiber laser [22, 23]. As for RDFLs based on Rayleigh scattering and Raman amplification, laser generation with a narrow linewidth down to 0.05 nm through filtering by the FBG was demonstrated in [5], and the corresponding highest output power approached 100 mW level. In this paper, we demonstrate a more powerful narrow linewidth (~40 GHz) RDFL by employing a higher power tunable pump laser, of which the output power reaches more than 2 W, which is about one order higher in magnitude than that of the previously reported results.

2. Experimental setup

Figure 1 shows a schematic diagram of the experimental setup. The pump source we employ is a tunable fiber laser, which consists of an Yb-doped fiber laser (YDFL) with a ring cavity geometry, in which we utilize a 1030 nm – 1090 nm tunable bandpass filter (T-BPF) to select the operating wavelength. To achieve higher output power, a standard master oscillator power amplifier (MOPA) configuration is used to amplify the seed laser, and the maximal power of the tunable fiber laser is about 9 W. To avoid unwanted feedback that influences the

former system, an isolator (ISO) is positioned after the tunable pump source. The pump light input and the random lasing output are separated by a circulator. A 1% coupler is used to monitor the pump laser operation. A section of 3 km SMF-28e fiber functions as the Raman gain medium and the distributed feedback mirrors. To achieve narrow linewidth generation and decrease the lasing threshold, a narrow-band of 0.07 nm fiber Bragg grating (FBG) centered at 1081.29 nm is spliced to the SMF, which constructs a half-open cavity. So the feedback is provided both by FBG reflection and distributed Rayleigh scattering. The narrow-band lasing output is at the 3rd port of the circulator. To eliminate Fresnel reflection, all the end facets are cleaved at an angle of 8°. For the sake of simplicity, we define lasing output from circulator side and FBG side as Output A and Output B, respectively.

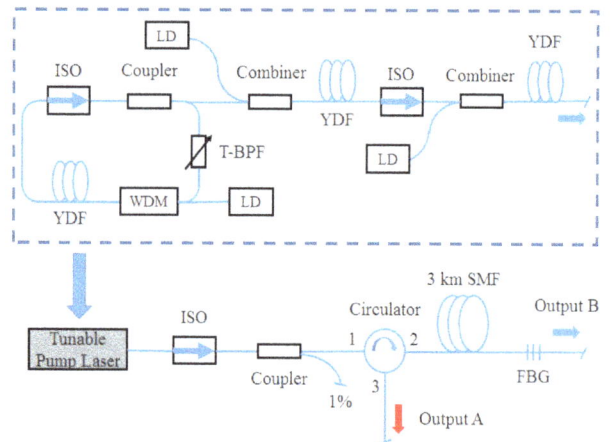

Fig. 1 Schematic of the narrow linewidth random fiber laser (ISO: isolator; YDF: Ytterbium-doped fiber; WDM: wavelength division multiplexing; T-BPF: tunable bandpass filter; LD: laser diode; SMF: single mode fiber; FBG: fiber Bragg grating).

3. Results and discussion

The wavelength tuning range of the pump laser is 1030 nm – 1090 nm. The spectra of lasing at 1030 nm – 1045 nm are shown in Fig. 2(a), and we can see that the amplified spontaneous emission (ASE) decreases with an increase in the wavelength. The ASE is about 25 dB lower than the signal laser at 1030 nm, and this value increases to about 45 dB at 1045 nm. Figure 2(b) shows the output power of the tunable fiber laser operating at different wavelengths, in which the maximal power of the pump laser reaches about 9 W when the wavelength is longer than 1033.5 nm, and the power of 1030 nm is relatively low because of stronger ASE.

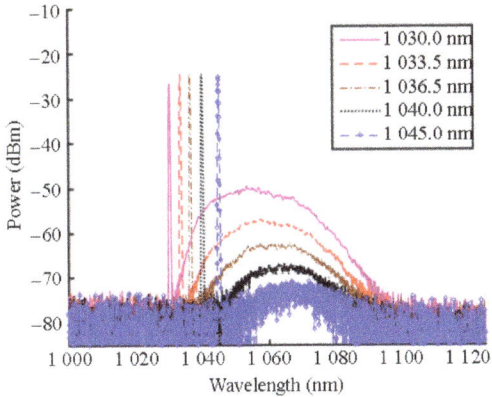

(a)

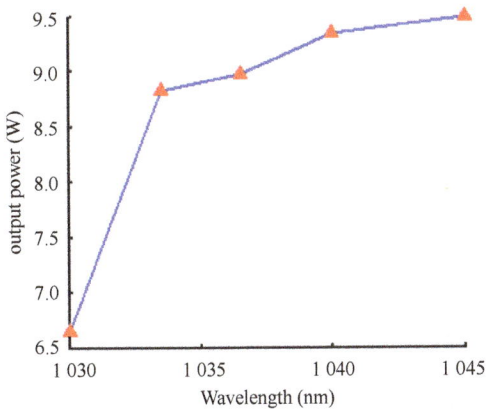

(b)

Fig. 2 Experimental results of the tunable pump laser: (a) spectra from 1030 nm to 1045 nm and (b) output power as a function of the wavelength.

The spectra of the laser emitting from Output A at different pump wavelengths are shown in Fig. 3. At 1030 nm, the 2nd order Raman Stokes light is measured, which is due to high peak power caused by the instability of temporal domain. However, with an increase in the pump wavelength, the spontaneous Raman scattering noise gets stronger. When the pump wavelengths are 1035 nm, 1040 nm, and 1045 nm, the powers of spontaneous Raman scattering noise account for 10%, 52%, and 96% of the total output power, respectively. To achieve a

high enough power of the 1st order narrow linewidth emission, it is important to make sure that the spontaneous Raman scattering noise is relatively low, and no 2nd order Raman Stokes light exists, so we experimentally choose 1036.5 nm as the optimized operating pump wavelength.

Figure 4 depicts the output powers dependence on the pump power. The threshold pump value is about 2.3 W in this case, and the output powers grow linearly in both Output A and Output B while the pump power is higher than the threshold. When 8.97 W pump light centered at 1036.5 nm is launched into the half-open cavity, we obtain the maximum output power of 2.15 W from Output A. At the same time, by doing integration based on the spectrum data and measuring the total power from Output B, residual 1036.5 nm pump power is down to 22 mW, and the power of spontaneous Raman scattering from Output B is 3.26 W. The power leakage is relatively high due to the narrow linewidth of the FBG, and the other reason is that the higher the generation power is, the more pronounced the nonlinear spectral broadening is [24]. Also, it cannot achieve narrow linewidth emission from Output B.

The spectrum of Output A at the maximal power in Fig. 5(a) shows neither 2nd order Raman Stokes light nor pump light, the central wavelength is around 1081.4 nm, and the full width at half maximum (FWHM) is about 0.16 nm (see the inset), which is nearly equal to 40 GHz in this wavelength range. The spontaneous Raman scattering noise is 25 dB lower than the laser line in the output spectrum. As shown in Fig. 5(b), the transmitted pump light and spontaneous Raman scattering noise can be observed in the spectrum of Output B.

Table 1 shows the 3 dB linewidth and the equivalent bandwidth of Output A with different output powers, and we can see that the bandwidth increases at lower output power while approaches stable at higher output power.

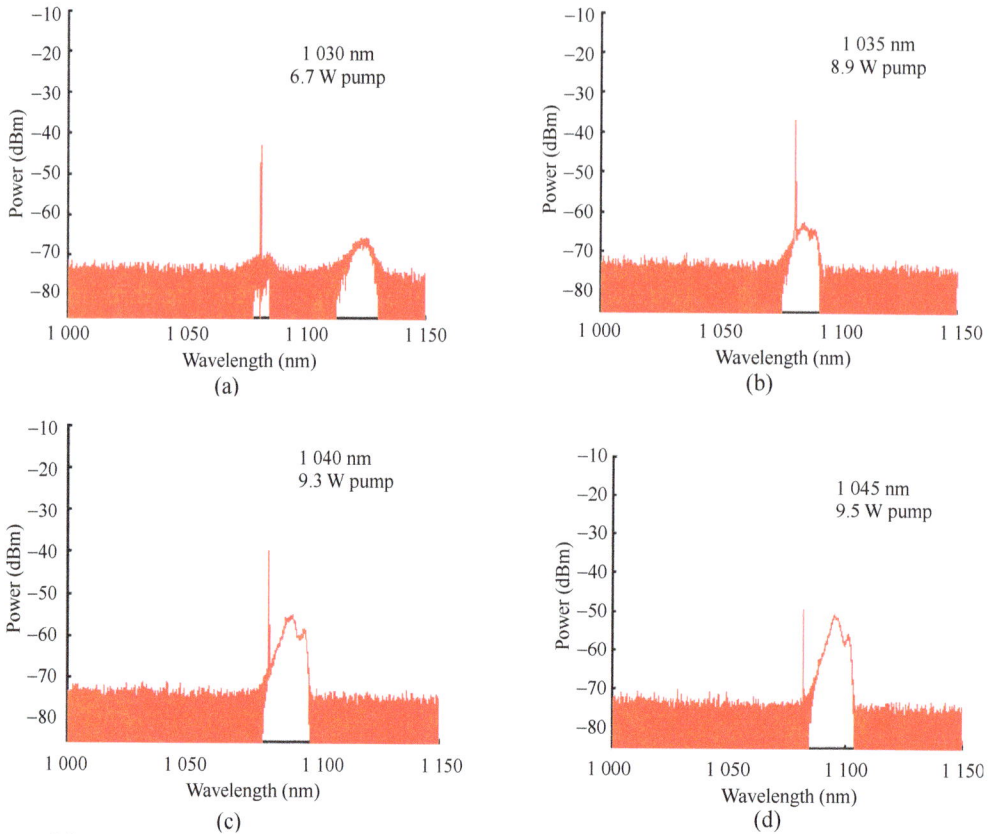

Fig. 3 Spectra of Output A with the maximal output power at different wavelengths of pump light: (a) 1030 nm, (b) 1035 nm, (c) 1040 nm, and (d) 1045 nm.

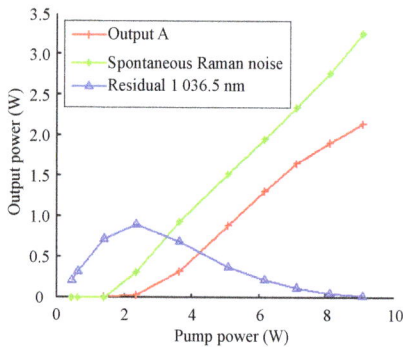

Fig. 4 Powers of Output A, spontaneous Raman scattering noise from Output B, and residual 1036.5 nm versus pump power.

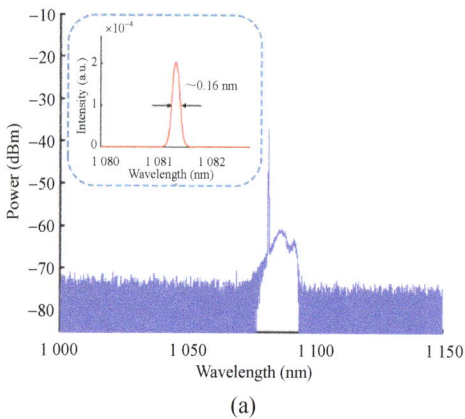

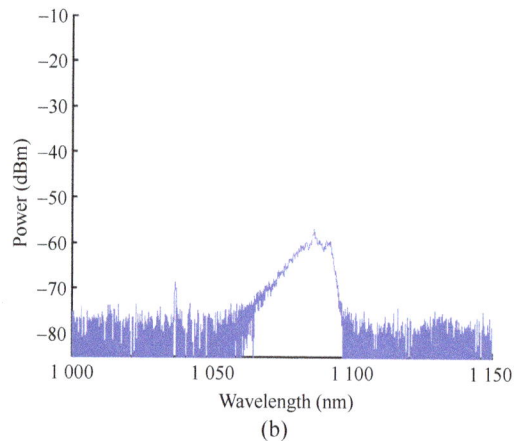

Fig. 5 Spectrum of (a) Output A and (b) Output B with the maximal power.

Table 1 Output linewidth and equivalent bandwidth.

Output power (W)	Output linewidth (nm)	Equivalent bandwidth (GHz)
0.026	0.1267	32.51
0.320	0.1424	36.53
0.881	0.1529	39.22
1.310	0.1710	43.87
1.650	0.1663	42.66
1.910	0.1671	42.87
2.150	0.1649	42.31

The developed laser system is formed by a section of passive fiber with a point reflector (FBG) at one end, and it can achieve a narrow spectrum following the FBG reflection bandwidth. However, the distributed Rayleigh backscattering is extremely weak. To achieve laser generation, a sufficient feedback is required to satisfy the condition with the integral gain overcoming accumulated fiber losses. We also find that the narrow linewidth output power grows linearly and reaches more than 2 W, having approximately 32% slope efficiency.

4. Conclusions

In this paper, we investigate the half-open cavity random fiber laser to obtain powerful narrow linewidth output. We utilize a tunable fiber laser to select the operating wavelength for efficiency optimization. The feedback in our case is provided both by FBG reflection and distributed Rayleigh scattering, and the gain is provided by the Raman scattering effect along 3 km SMF-28e fiber. As a result, we obtain an RDFL with the maximum output power of 2.15 W centered at 1081.4 nm, and the corresponding bandwidth is as narrow as about 40 GHz. The demonstrated powerful narrow linewidth RDFL has great potential for being used in practical applications such as coherent sensing and detection.

References

[1] R. V. Ambartsumyan, N. G. Basov, P. G. Kryukov, and V. S. Letokov, "Laser with nonresonant feedback," *Soviet Journal of Experimental & Theoretical Physics*, 1966, 3(2): 167.

[2] V. Markushev, V. Zolin, and C. M. Briskina, "Luminescence and stimulated emission of neodymium in sodium lanthanum molybdate powders," *Soviet Journal of Quantum Electronics*, 1986, 16(2): 427–430.

[3] H. C. Hsu, C. Y. Wu, and W. F. Hsieh, "Stimulated emission and lasing of random-growth oriented ZnO nanowires," *Journal of Applied Physics*, 2005, 97(6): 064315-1–064315-4.

[4] S. V. Frolov, M. Shkunov, A. Fujii, K. Yoshino, and Z. V. Vardeny, "Lasing and stimulated emission in π-conjugated polymers," *IEEE Journal of Quantum Electronics*, 2000, 36(1): 2–11.

[5] S. Sugavanam, N. Tarasov, X. Shu, and D. V. Churkin, "Narrow-band generation in random distributed feedback fiber laser," *Optics Express*, 2013, 21(14): 16466–16472.

[6] S. K. Turitsyn, S. A. Babin, D. V. Churkin, I. D. Vatnik, M. Nikulin, and E. V. Podivilov, "Random distributed feedback fibre lasers," *Physics Reports*, 2014, 542(2): 133–193.

[7] C. J. Matos, S. M. L. De, A. M. Brito-Silva, M. A. M. Gámez, A. S. Gomes, and C. B. Araújo, "Random fiber laser," *Physical Review Letters*, 2007, 99(15): 153903-1–15903-4.

[8] Z. Hu, B. Miao, T. Wang, Q. Fu, D. Zhang, H. Ming, et al., "Disordered microstructure polymer optical fiber for stabilized coherent random fiber laser," *Optics Letters*, 2013, 38(22): 4644–4647.

[9] N. Lizárraga, N. P. Puente, E. I. Chaikina, T. A. Leskova, and E. R. Méndez, "Single-mode Er-doped fiber random laser with distributed Bragg grating feedback," *Optics Express*, 2009, 17(2): 395–404.

[10] S. K. Turitsyn, S. A. Babin, A. E. El-Taher, P. Harper, D. V. Churkin, S. I. Kablukov, et al., "Random distributed feedback fiber laser," *Nature Photonics*, 2010, 4(4): 231–235.

[11] H. Zhang, P. Zhou, H. Xiao, and X. Xu, "Efficient Raman fiber laser based on random Rayleigh distributed feedback with record high power," *Laser Physics Letters*, 2014, 11(7): 075104.

[12] X. Du, H. Zhang, X. Wang, P. Zhou, and Z. Liu, "Short cavity-length random fiber laser with record power and ultrahigh efficiency," *Optics Letters*, 2016, 41(3): 571–574.

[13] S. A. Babin, A. E. El-Taher, P. Harper, E. V. Podivilov, and S. K. Turitsyn, "Tunable random fiber laser," *Physical Review A*, 2011, 84(2): 4903–4911.

[14] X. Du, H. Zhang, X. Wang, and P. Zhou, "Tunable random distributed feedback fiber laser operating at 1 μm," *Applied Optics*, 2015, 54(4): 908–911.

[15] A. E. El-Taher, P. Harper, S. A. Babin, D. V. Churkin, E. V. Podivilov, J. D. Ania-Castanon, et al.,

"Effect of Rayleigh-scattering distributed feedback on multiwavelength Raman fiber laser generation," *Optics Letters*, 2011, 36(2): 130–132.

[16] X. Du, H. Zhang, X. Wang, X. Wang, P. Zhou, and Z. Liu, "Multiwavelength Raman fiber laser based on polarization maintaining fiber loop mirror and random distributed feedback," *Laser Physics Letters*, 2015, 12(4): 045106.

[17] E. A. Zlobina, S. I. Kablukov, and S. A. Babin, "Linearly polarized random fiber laser with ultimate efficiency," *Optics Letters*, 2015, 40(17): 4074–4077.

[18] X. Du, H. Zhang, X. Wang, P. Zhou, and Z. Liu, "Investigation on random distributed feedback Raman fiber laser with linear polarized output," *Photonics Research*, 2015, 3(2): 28–31.

[19] X. H. Jia, Y. J. Rao, F. Peng, Z. N. Wang, W. L. Zhang, H. J. Wu, *et al.*, "Random-lasing-based distributed fiber-optic amplification," *Optics Express*, 2013, 21(5): 6572–6577.

[20] Z. N. Wang, Y. J. Rao, H. Wu, P. Y. Li, Y. Jiang, X. H. Jia, *et al.*, "Long-distance fiber-optic point-sensing systems based on random fiber lasers," *Optics Express*, 2012, 20(16): 17695–17700.

[21] W. L. Zhang, Y. J. Rao, J. M. Zhu, Z. X. Yang, Z. N. Wang, and X. H. Jia, "Low threshold 2nd-order random lasing of a fiber laser with a half-opened cavity," *Optics Express*, 2012, 20(13): 14400–14405.

[22] M. Pang, X. Bao, and L. Chen, "Observation of narrow linewidth spikes in the coherent Brillouin random fiber laser," *Optics Letters*, 2013, 38(11): 1866–1868.

[23] Y. Li, P. Lu, X. Bao, and Z. Ou, "Random spaced index modulation for a narrow linewidth tunable fiber laser with low intensity noise," *Optics Letters*, 2014, 39(8): 2294–2297.

[24] S. V. Smirnov and D. V. Churkin, "Modeling of spectral and statistical properties of a random distributed feedback fiber laser," *Optics Express*, 2013, 21(18): 21236–21241.

Huge Capacity Fiber-Optic Sensing Network Based on Ultra-Weak Draw Tower Gratings

Minghong YANG[*], Wei BAI, Huiyong GUO, Hongqiao WEN, Haihu YU, and Desheng JIANG

National Engineering Laboratory for Fiber Optic Sensing Technology, Wuhan University of Technology, Wuhan, 430070, China

[*]Corresponding author: Minghong YANG   E-mail: minghong.yang@whut.edu.cn

Abstract: This paper reviews the work on huge capacity fiber-optic sensing network based on ultra-weak draw tower gratings developed at the National Engineering Laboratory for Fiber Optic Sensing Technology (NEL-FOST), Wuhan University of Technology, China. A versatile drawing tower grating sensor network based on ultra-weak fiber Bragg gratings (FBGs) is firstly proposed and demonstrated. The sensing network is interrogated with time- and wavelength-division multiplexing method, which is very promising for the large-scale sensing network.

Keywords: Ultra-weak FBG; optical fiber sensors; sensing network

1. Introduction

Fiber Bragg grating (FBG) sensors network [1, 2] has attracted considerable interests for quasi-distributed sensing. Wavelength-division multiplexing (WDM) is a popular scheme because of its intuitionistic wavelength demodulation method. However, limited by the bandwidth of broadband light sources, often less than 100 nm, only tens of sensors can be multiplexed in one fiber. Childers *et al.* experimentally multiplexed 800 FBG sensors in a single array by using an optical frequency domain reflectometry (OFDR) scheme [3], but the maximum fiber span is limited by the coherence length of the tunable laser and polarization fading in the light interference. Time-division multiplexing (TDM), distinguishing the FBG sensors with an identical wavelength by detecting the different time delay between reflected pulses along the fiber, has substantial potential for increasing the number of the multiplexing sensors [4−7]. Several TDM schemes with a resonant cavity were reported [8, 9]. Notably, the required interrogated sensors of these schemes are traditional FBGs with reflectivity ≥ -20 dB, where the multiplexing capacity is seriously limited by crosstalks. Although the multiplexing capacity can be improved to some extent by using the hybrid scheme, such as WDM+TDM, these technologies are limited by the bandwidth and transmission loss. To obtain lower crosstalks and larger multiplexing capability, it is a great choice to use ultra-weak FBGs. Wang *et al.* experimentally demonstrated 1000 ultra-weak FBG (peak reflectivity ≤ -37 dB) by a serial TDM sensor network [10]. However, it was a relative long interval to obtain the result of one special FBG sensor, over 10 s, by reconstructing the reflection spectrum of each FBG during the entire scanning period in time domain.

In this paper, a new approach of on-line growing draw tower grating (DTG) and their large-scale sensing network is proposed and demonstrated, and a large-scale sensor network with wavelength scanning time division multiplexed (WSTDM) ultra-weak FBGs (reflectivity of about −40 dB) is realized. The novel ultra-weak DTG technology platform will open new possibility for further application of the large-scale FBG network with thousands of sensing elements. Furthermore, since the ultra-weak reflectance signal is still higher than the Brilloum signal, it can be very promising as a long distance sensing network with a faster response and a more precise positioning capability.

2. Ultra-weak draw tower grating fabrication

An on-line writing weak FBG array is the process that FBGs are directly inscribed into the fiber during drawing. In this case, FBG writing must be done before the fiber coating application, since the fiber coating is usually not transparent to ultro video (UV) light. In the traditional FBG preparation, the fiber has to be decoated before writing FBGs and recoated after the grating exposure. The conventional process is not only difficult to handle, but also it degrades fiber strength, which does damage to its engineering application. However, draw-tower FBGs can overcome these disadvantages, due to the simple operation and high mechanical stability [11, 12]. Furthermore, such a method can produce FBG sensor arrays without fiber fusion, which cannot be avoided in the traditional FBG array construction. No fiber fusion is importance because fusion loss greatly depresses the multiplexing capacity of a weak FBG array. These advantages therefore prompt researchers to begin the research of draw-tower FBGs early [13, 14].

The multiplexing capacity of the traditional FBG sensor arrays generally limits to dozens of FBGs, even though various multiplexing methods are used, e.g., wavelength division multiplexing and time division multiplexing [15−17]. Recently, some investigations reported that identical weak FBG array could greatly improve the multiplexing capacity and sensing distance due to its narrow bandwidth and weak reflection characteristics [18−20]. However, these bright characteristics require that weak FBG array itself has a good uniformity, especially good wavelength uniformity. The draw-tower grating technique is a good choice for preparing identical and weak FBG arrays.

Various factors of central wavelength separation including fluctuations of drawing tension and core temperature were mentioned [13]. However, by detecting the surface temperature of the bare fiber, we found that the core temperature of the bare fiber at the writing spot was stable in the stable drawing state. Nowadays, some draw-tower FBGs are already commercially available [21], however their wavelength accuracy of ≤0.4 nm did not meet the requirements for high quality FBG arrays. In their FBG writing platform, the interference fringes used for writing FBGs were easily influenced by air flow due to a long optical path of the Talbot interferometer [22]. Moreover, tiny changes in the angle and position of two reflecting mirrors would decrease the grating wavelength accuracy. Therefore, even if the phase mask technique was used in their FBG writing platform, the grating wavelength accuracy was not enough. In our experiment, the triangle interference area of the phase mask was directly used for on-line writing identical FBG arrays to depress the effect of the interferometer.

For on-line writing weak FBG arrays, we employed one production-type draw tower as shown in Fig. 1, which could operate at speeds ranging from 3 m/min to 200 m/min for a bare fiber diameter of 125 μm. The drawing speed and drawing tension could be automatically controlled and displayed. A line-narrowed ArF excimer laser (OptoSystems CL5300) with a beam size of 4 mm×12 mm, pulse width of 10 ns, maximum pulse energy of 40 mJ, and maximum laser repetition rate of 300 Hz was used in the FBG writing platform. The reflective spectra of FBGs were gained by one optical analyzer (YOKOGAWA-AQ6370B) with one homemade light emitting diode (LED) source at 1.3 μm wavelength band. The OTDR spectra of weak FBG array were acquired by an optical time domain reflectometer (OTDR, YOKOGAWA-AQ7260).

Fig. 1 Production-type draw-tower grating system.

FBG was written by the phase mask method which used periodic interference fringes of the ±1st diffraction light to irradiate photosensitive fiber and hence periodically change the refractive index of fiber core. In this method, the interference fringes were determined by the structure of phase mask, and not influenced by light source coherence, which ensured a good wavelength consistency of FBG arrays. The laser beam was as possible as large to ensure a uniform energy density in a wider writing spot. Additionally, FBG writing platform was mounted on the draw tower near the first coating to weaken fiber vibrating, as shown in Fig. 2. In the experiment, draw-tower FBG arrays were generally carried out under the relatively stable pulse energy of 25 mJ. The laser beam was focused from 4 mm× 12 mm to a line 0.7 mm×10 mm using three cylindrical lenses. The distance between the phase mask and bare fiber was controlled at 0.50 mm.

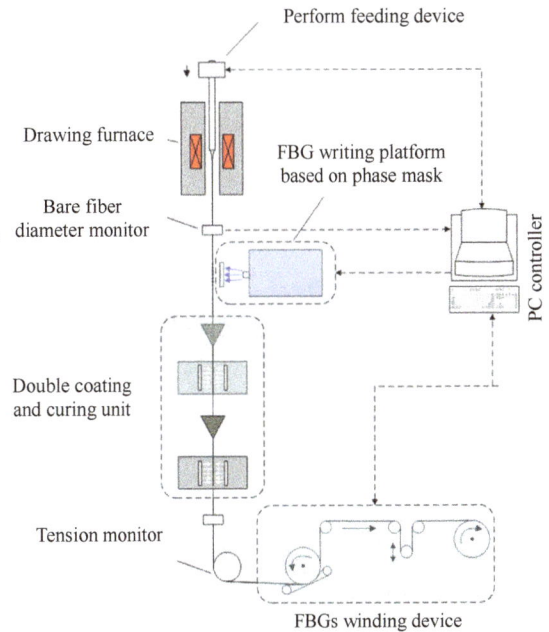

Fig. 2 Diagram of on-line writing FBG system.

The draw-tower FBG array required the fiber with a high single-pulse photosensitivity, since a single laser pulse used here had limited energy and irradiating time. Previous work reported that co-doping Ge/B could greatly improve the photosensitivity of optical fiber [23]. Here a Ge/B co-doped perform was carried out in the Optical Fiber Research and Development Department, FiberHome Communication Science and Technology Co., Ltd. The obtained fiber with a B-doped concentration of around 10 mol% showed an excellent single-pulse photosensitivity and an acceptable attenuation of 2.8 dB/km at 1300 nm.

Compared with the traditional hydrogen-loaded fiber (immersed in 9-Mpa hydrogen for 1 month), the Ge/B co-doped fiber showed the 15-folds grating reflectivity, also 2.1%, in the same condition: the laser pulse energy of 29 mJ and the beam size of 0.4 mm×10 mm. The Ge/B co-doped fiber meets on-line writing FBG arrays.

Figure 3(a) shows the OTDR trace of the continuous 200 FBGs in Array A. Each FBG was written successfully. Figure 3(b) displays the local enlargement between 0.8 km and 0.9 km. Therefore, the spatial separation among Array A was fixed at 5 m. At present, the system could fulfill a spatial separation at a millimeter level with the accuracy of ±1 mm.

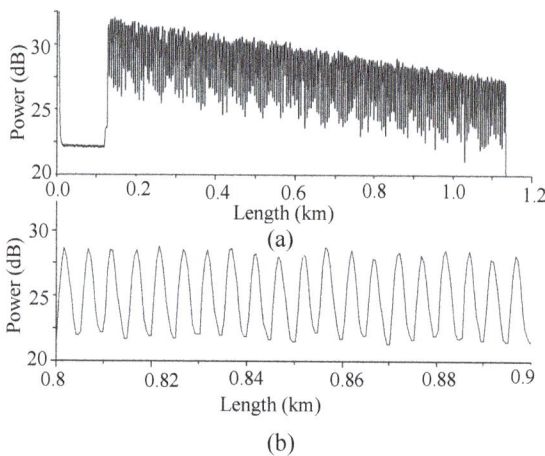

Fig. 3 OTDR fragment from 200 FBGs of Array A: (a) power evolution of 200 FBG in series and (b) enlargement of 20 FBGs.

Figures 4(a) and 4(b) shows the reflection spectra of Array A (b) and its single FBG (a) without tress. The overlapping spectrum of Array A was almost as same as that of its single FBG. Hence, results indicated that the online-writing FBG array by the phase mask method had a good uniformity of central wavelength.

Effect of the drawing speed on the quality of FBG was firstly studied. Figure 5 shows the reflection spectra of FBGs obtained at different drawing speeds. It can be seen that the shape of

reflection spectra was not affected by the drawing speed. In all the reflection spectra, the full width at half maximum (FWHM) and reflection power of the main peak were 0.09 nm and ~7 pW, respectively. This indicated that FBGs could be written perfectly at the drawing speeds of interest. Generally, the drawing tension was large at the high drawing speeds. This would deteriorate the central wavelength uniformity of FBG arrays (Fig. 6). Additionally, stabilizing the drawing process often took more than one hour at the high drawing speed. Therefore, in our experiment, on-line writing identical FBG arrays were done at the lower drawing speed of ~15 m/min.

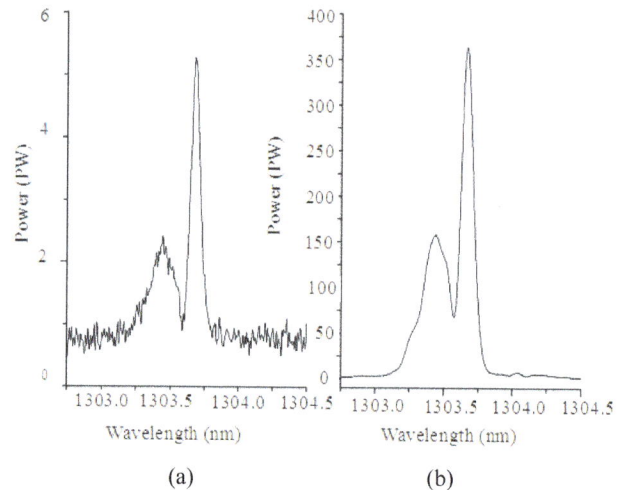

Fig. 4 Reflection spectra of the weak FBG investigated with the light source power of 2.5 nW at 1303.638 nm: (a) array of 200 FBGs and (b) single FBG.

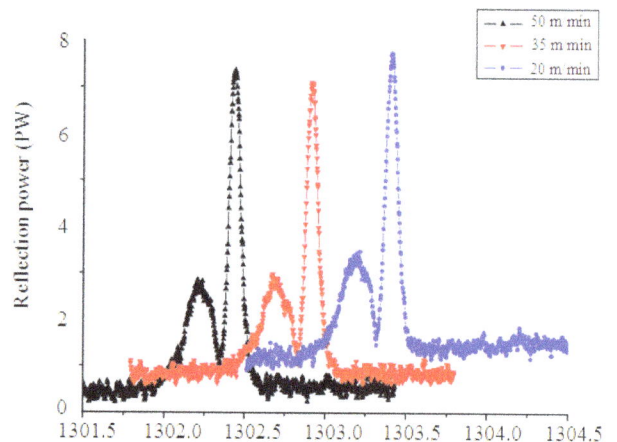

Fig. 5 Reflection spectra of three FBGs written at different speeds.

The wavelength uniformity of FBG arrays was important, because it was closely related to the data volume of terminal signal demodulation. Although the drawing process reached the stable work state, a small fluctuation of the drawing speed was always present to control the fiber diameter in a qualified range. Hence, the fluctuation of the drawing tension was recorded to follow the change in the drawing speed. We found that the central wavelength drifted reversely with an increase in drawing tension (Fig. 7). This suggested that the quality of FBG arrays was very sensitive to the tiny fluctuation of the drawing speed.

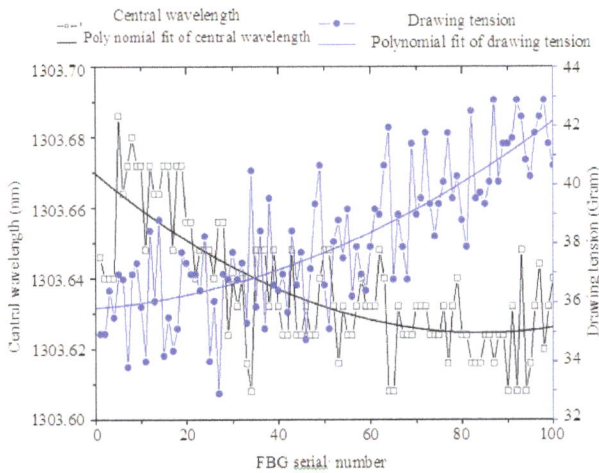

Fig. 6 Central wavelength and drawing tension traces of successive 100 FBGs in Array A.

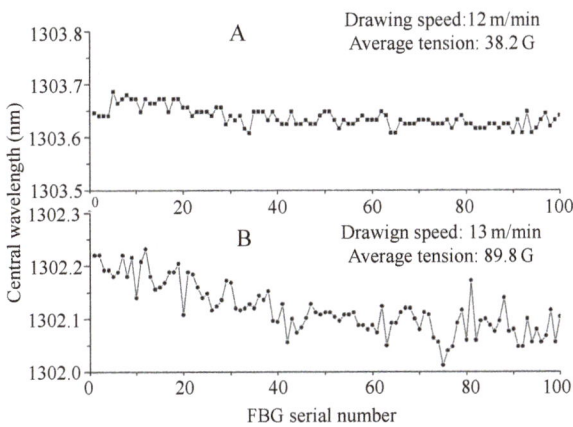

Fig. 7 Central wavelength distributions of successive 100 FBGs in Array A and Array B.

Here the central wavelength uniformity of on-line writing FBG array was investigated under different drawing tension (the drawing tension increased through decreasing the furnace temperature), as shown in Fig. 6. From Fig. 7, we can see that in Array A, the central wavelengths of 100 FBGs vary by ~0.1 nm from 1303.608 nm to 1303.686 nm. In Array B, fluctuation of the central wavelength reaches up to 0.25 nm, which covers from 1302.000 nm to 1302.250 nm. Additionally, the fluctuation of adjacent wavelengths became larger under the large drawing tension. Notably, the change in the drawing speed was small for both the FBG arrays, and hence results suggested that the large drawing tension was not a good condition. Our studies indicated that a stable and proper drawing tension was a critical factor that ensured online-writing FBG array with good wavelength uniformity.

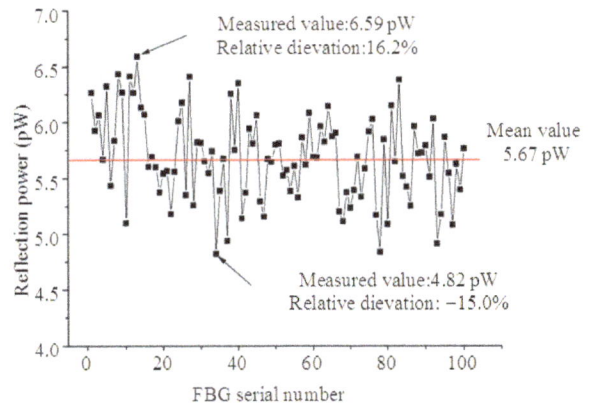

Fig. 8 Single FBG reflection power of 100 FBGs in Array A, investigated with the light source power of 2.5 nW at 1303.638 nm.

To investigate the uniformity of the reflection spectra, single FBG reflection power in Array A was measured successively. Figure 8 shows the reflection power of 100 FBGs in Array A. It could be seen that the relative deviation was about ±16%. During the draw-tower grating preparation, the output energy of laser pulse presented a fluctuation of ±8%. This suggested that effect of fiber vibration was not ignored on the fluctuation of FBG reflection power. Figure 9 exhibited the vibration of bare optical fiber in the exposure area. Since the laser energy density presented Guass distribution in the horizontal direction, it became more uneven from the middle to both sides after spot focus. Therefore, the fiber vibration in the beam spot would markedly affect the grating reflectivity.

To decrease the fiber vibration, effects of both the drawing speed and the drawing tension were investigated. When the drawing tension was smaller than 10 grams where the drawing speed was low, we observed that the lateral vibration of optical fiber was obvious. The lateral vibration was greatly improved with increasing the drawing tension up to 30 grams. However, it was difficult to improve greatly the reflectance consistency of FBG array by further decreasing fiber vibration. Other methods for improving the reflectivity uniformity of draw-tower FBG array should be found, e.g., by homogenizing the energy density and by further widening the horizontal dimension of writing spot.

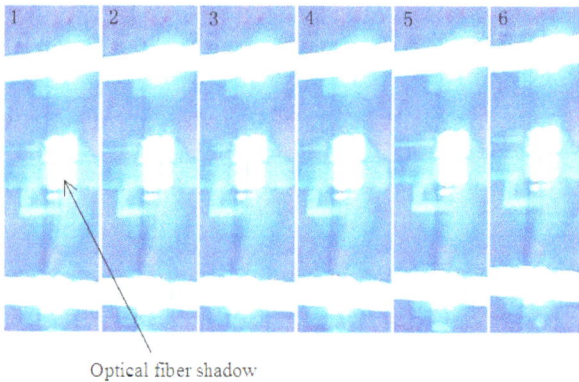

Fig. 9 Six shooting photos of the diffraction spot, taken under the condition: the drawing speed of 12 m/min, the average drawing tension of 38.2 grams, and laser frequency of 50 Hz.

3. Interrogation method of the ultra-weak draw tower gratings

The large capacity sensing network interrogation system with identical ultra-weak FBG is illustrated in Fig. 10. The light from ASE source was modulated to nanosecond pulses. A pulse generator, realized by field programmable gate array (FPGA) logic, driven the first semiconductor optical amplifier [SOA(1)] functioning as a modulator as well as the first stage optical amplifier. Then the pulses were amplified by the second stage optical

amplifier, Erbium-doped fiber amplifier, and launched into a FBG array. The reflected pulses from the FBGs returned to SOA(2) which acted as a gating device as well as an amplifier. Field programmable gate array (FPGA) generated a time-shifted pulse sequence to activate SOA(2) to filter and amplify one specific FBG signal while absorbing signals reflected from other FBGs. The pulses reflected by FBGs arrived at SOA(2) with time delays of:

$$\tau_i = \frac{2nL_i}{c} \qquad (1)$$

where τ_i is the time delay of pulses reflected by FBG_i, c is the speed of light in vacuum, n is the effective refractive index of optical fiber, and L_i is the distance of FBG_i from the circulator. When the time-shifted pulse sequence from FPGA have a time shift τ_i the pulse reflected by FBG_i will be amplified each time when they pass through SOA(2), and the pulses reflected by other FBGs are blocked. By changing delay time through FPGA, each FBG can be addressed separately.

A charge-coupled device (CCD) based demodulator (I-MON 256, Ibson Photonics) was used in the interrogation system to obtain the spectral information of FBGs. The interrogation speed of the system which is shown in (2) was limited by both the response time of CCD τ_{res} and the time interval of pulse sequence.

$$f \le \min(1/\tau_d, 1/\tau_{res}) \qquad (2)$$

where $\tau_d \ge 2nL/c$, L is the length of sensing fiber. The response time of CCD could be as low as tens of microseconds. For a sensing fiber less than 10 km in which 1000 FBGs are serial multiplexed, it takes less than 0.1 second to obtain the reflection wavelengths of all sensors. Once SOA(2) is turned on, a trigger pulse generated by FPGA will be sent to CCD to active an spectra sampling. Both CCD detector and SOAs are controlled by FPGA, so that the spectral sampled by CCD in time delay τ_i can be respond to FBG_i.

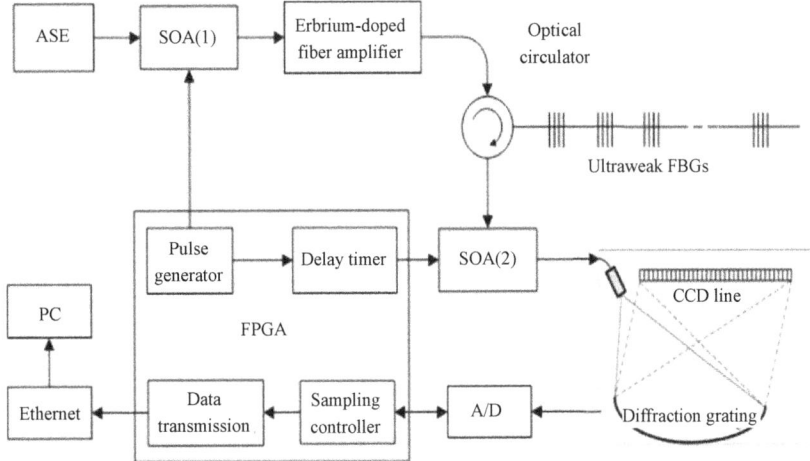

Fig. 10 Illustration of interrogation system.

The reflected spectra of FBG_i need to be reconstructed at different delay time τ_i through peak fitting algorithm because of the limited pixel number of CCD. The average pixel spacing of IMON 256 in wavelength is about 170 pm. By using Gaussian peak fitting algorithm, a peak wavelength resolution of 0.5 pm can be achieved. The adjusted Gaussian function is shown in (3):

$$f(\lambda_i) = A \cdot \exp\left[-\frac{(\lambda_i - B)^2}{2C^2}\right] \qquad (3)$$

where A, B, and C are the adjusted parameters (amplitude, center and deviation), and $f(\lambda_i)$ is the calculated spectra of λ_i. Gaussian peak fitting is computational intensive, so faster algorithms, such as centroid algorithm, can be used in high-speed interrogation in cost of higher fit noise.

In order to obtain two pulse sequences with time delay precision as high as 1 ps, a two-stage time delay controller (coarse tuning and fine tuning) was implemented. Figure 11 shows the procedure. The reference pulse sequence and delayed pulse sequence were named as 1# and 2#, respectively. The pulse waveform was realized by mapping into block random-access memory (BRAM). The width of pulse was expressed by the continuous digital "1" stored in BRAM, and low level of pulse was corresponding to continuous digital "0". Then the pulse cycle was expressed by the read/write BRAM

depth. Output parallel-to-serial logic resources (OSERDESE) were a dedicated parallel-to-serial converter with specific clocking and logic resources in FPGA. If converting the parallel data to serial bit signal of I/O interface by OSERDESE after reading BRAM sequentially, the coarse tuning was completed. Coarse tuning was the first stage delay control unit with the level of accuracy up to 0.625 ns for Xilinx Virtex7 FPGA.

Programmable output delay (ODELAY) near FPGA ports allowed outgoing signals to be delayed on an individual basis. This was the second stage delay control unit, fine tuning. The tap delay resolution was varied by selecting different FPGA series and different reference clock, generally 78 ps or 52 ps. In all 32 taps, adjustable delay can reach 2.496 ns or 1.664 ns.

The I-MON 256 module has a 256-pixel-InGaAs detector running at a clock frequency of 5 MHz in maximum. Figure 12 shows the system design scheme for spectra sampling. The detector began exposure integral when detecting a high level at a falling edge of CLK, and end exposure integral when detecting a low level at a falling edge of CLK. Pixels data will output serially after 8 CLK.

A/D chip AD9826, 3-Channel 16-bit Operation up to 15 MSPS A/D Converter, works at 1-channel SHA mode timing and converts analog signal to

16-bit valid pixel data for FPGA. Before InGaAs detector works at normal status, FPGA configures AD9826 by SCLK, SLOAD, and SDATA. CDSCLK2 and ADCCLK output the control clock with 1 MHz and phase difference of 90 degree. The

string and conversion will be reached when the high and low byte data are combined. The time measurement data have been received and packaged to PC by Ethernet work after data are cumulated to 256.

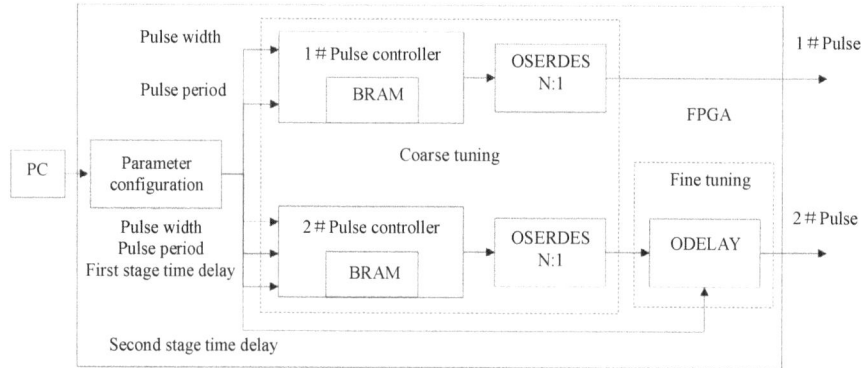

Fig. 11 Modulation pulses generated by FPGA method.

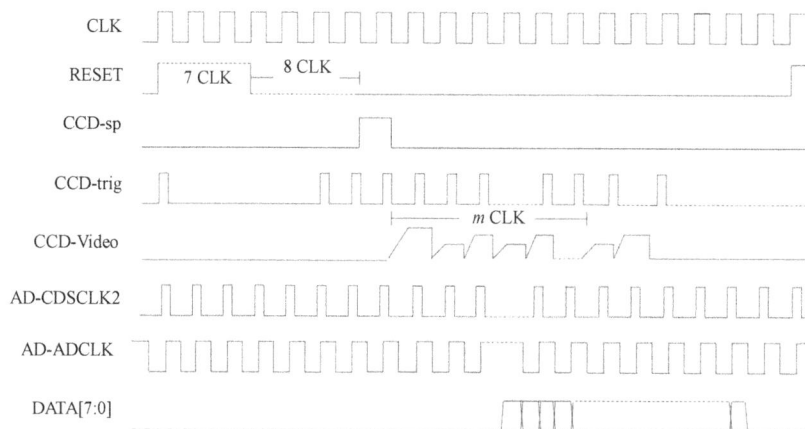

Fig. 12 FPGA timing of InGaAs detector.

The performance of the proposed interrogation system was further verified by demodulating an ultra-weak FBG array with 1009 FBGs and 2.5 meters equidistance between neighbor FBGs. The peak reflectivity of the FBGs was about −33 dB. The ultra-weak FBG array was fabricated by the on-line writing method [11]. All FBGs in the array were inscribed with a same phase mask so that they have almost the same peak wavelength. FPGA generated a pulse sequence with 20 ns width and 100 μs interval to drive SOA (1). By changing time delay with a step of 1 ns through delay timer, the reflectance along the sensing fiber was obtained by a

CCD detector. The peaks of the reflectance indicated the locations of FBGs. The normalized reflectance is shown in Fig. 13. The fluctuation of reflectivity was caused by fabrication errors induced during inscribing process.

Figure 14 shows peak wavelengths of all 1009 FBG sensors measured by the interrogation system. Peak wavelengths of the most of FBGs were distributed in the range from 1550.4 nm to 1550.65 nm. The CCD detector worked at 1000 Hz, and it took about 1 second to complete one interrogation process for a 1009-FBG-array with a distance of about 2.7 km.

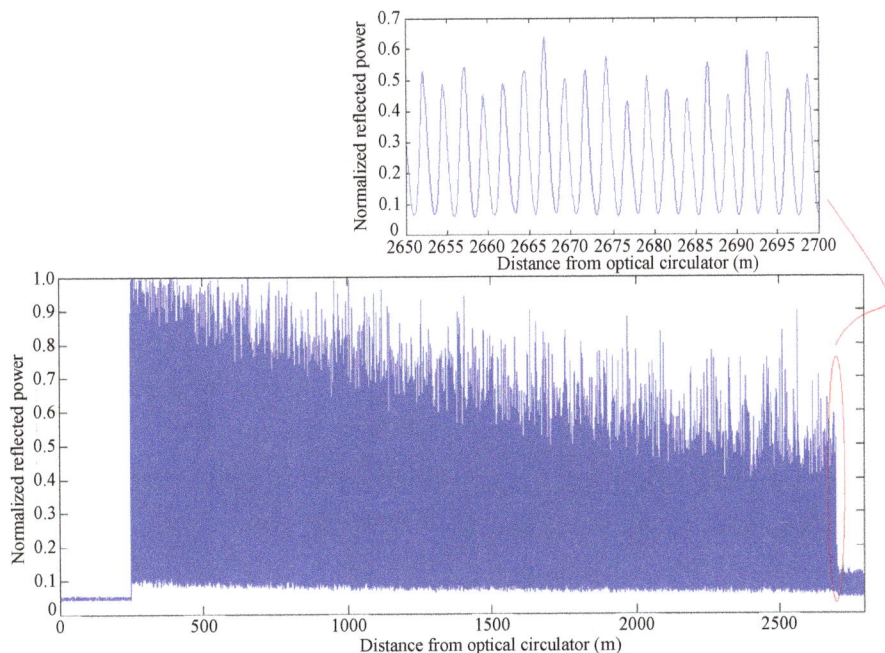

Fig. 13 Reflected power of FBG sensors (reflectivity = −33 dB).

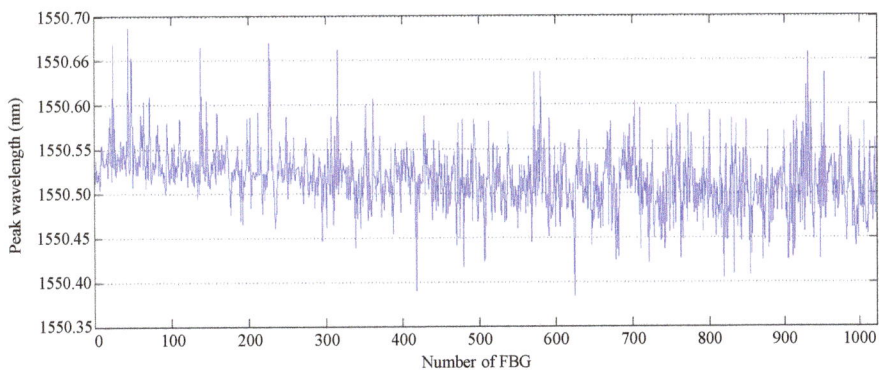

Fig. 14 Peak wavelength of FBG sensors.

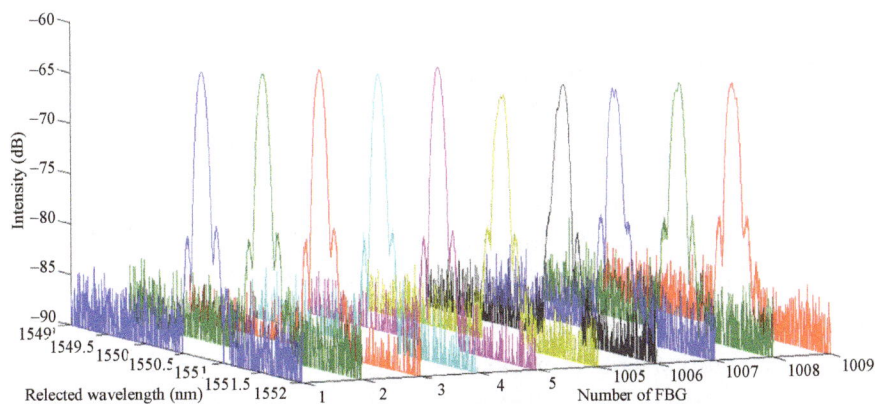

Fig. 15 Reflective spectrum of FBG sensors.

In order to investigate the spectral distortion induced by the crosstalks, the reflective spectrum of the first and last 5 FBGs of the array were measured by Spectrometer (YOKOGAWA AQ6370B), as shown in Fig. 15, by setting the delay timer to 2517 ns, 2543 ns, 2566 ns, 2592 ns, 2616 ns, 27716 ns, 27740 ns, 27765 ns, 27791 ns, and 27815 ns, respectively. The time delay of 2517 ns for FBG1 indicated there was a bare fiber of about 250 m between FBG1 and the circulator. The difference of delay time between two adjacent FBGs was about 25 ns. The peak reflectivity of the last 5 FBGs were −65.294 dBm, −64.619 dBm, −65.423 dBm, and −65.052 dBm, −65.522 dBm, respectively, which was 3 dB less than that of the first 5 FBGs because of multiple reflection crosstalks. There were tiny distortions in the spectrum of the last 5 FBGs which were also caused by multiple reflection crosstalks. The distortion in the end of sensors array will be further reduced by using FBG sensors with even lower reflectivity such as −40 dB.

To test the impact of the spectral distortion on measurement accuracy, the Bragg wavelength of the sensor array is measured 200 times in the save environment. Based on the data collected from CCD detector, the peak wavelength was calculated by a Gaussian peak fitting method. The measurement accuracy was less than 5 pm for all sensors in the array. Figure 16 shows the standard deviation of the first and last 5 FBGs. The wavelength measurement error (2σ) was limited in 2 pm. The impact of spectral distortion caused by crosstalks was negligible.

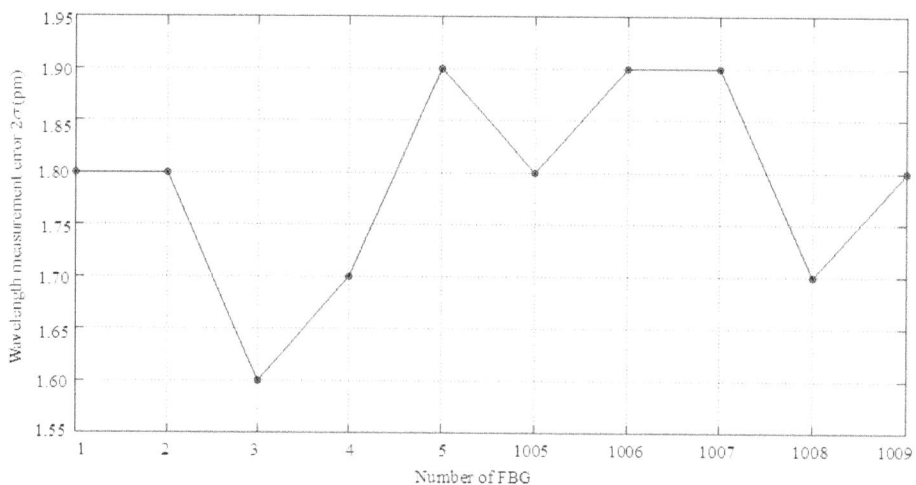

Fig. 16 Measurement error of peak wavelength for FBG1- FBG5 and FBG1005- FBG1009.

4. Experimental demonstration of fiber-optic sensing network based on ultra-weak FBG

The performance of the proposed interrogation system was demonstrated by demodulating an ultra-weak 2000-DTGs array with one meter equidistance between neighbored DTGs. The weak DTG array was fabricated by on-line writing method as mentioned above with one phase mask so that they have almost the same peak wavelength. The average reflectivity of the DTG array was about −31.5 dB, and all peak wavelengths of the ultra-weak DTG array fell within the range of 1552.075 nm ± 0.15 nm, in which those of 95% DTGs were within the smaller range of 1552.075 nm ± 0.05 nm. Figure 17(a) shows the reflective spectra of 38 representative DTGs in the array interrogated in the forward and reverse directions, respectively, of which serial numbers (in the forward direction) are 1, 2 , 3, 4, 5, 6, 7, 8, 9, 10, 20, 30, 40, 50, 60, 70, 80, 90, 100, 200, 300, 400, 500, 600, 700, 800, 900, 1000, 1100, 1200, 1300, 1400, 1500, 1600, 1700,

1800, 1900, and 2000. The results in the two interrogation indicated that the reflective spectra of the downstream DTGs in the array were distorted severely into the flat-top ones with some sharps and they have different top morphologies from each other. The flatting reason for the reflection peaks of the downstream DTGs is not difficult to understand, which should be attributed to the shadowing effect of upstream DTGs [24, 25]. According to the formula [24]:

$$C_i(\lambda) = \frac{(i-1)(i-2)}{2} R^3(\lambda)(1-R(\lambda))^{2i-4} I_0(\lambda), \quad i \geq 3.$$

The one-order crosstalks intensity of the ith DTG is not only related to its location in the array, but also is closely related to the reflectivity of wavelength λ of every upstream DTG. Due to the peak wavelength distribution in a certain range, the reflected spectra of the DTG array were dislocated mutually, and as for a certain wavelength λ, the reflectivity of every upstream DTG may be different; as for different wavelength dots, the one-order crosstalks of the ith will be different, which will give rise to some sharps in the top of the DTG reflection spectra. Due to an ideal

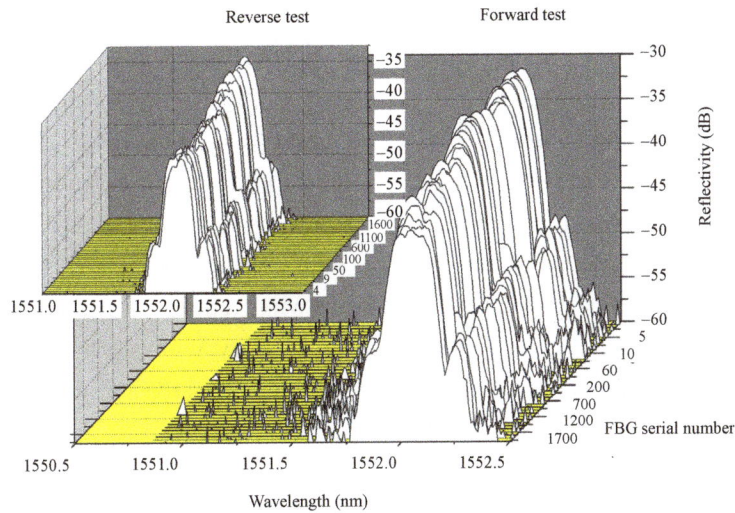

(a)

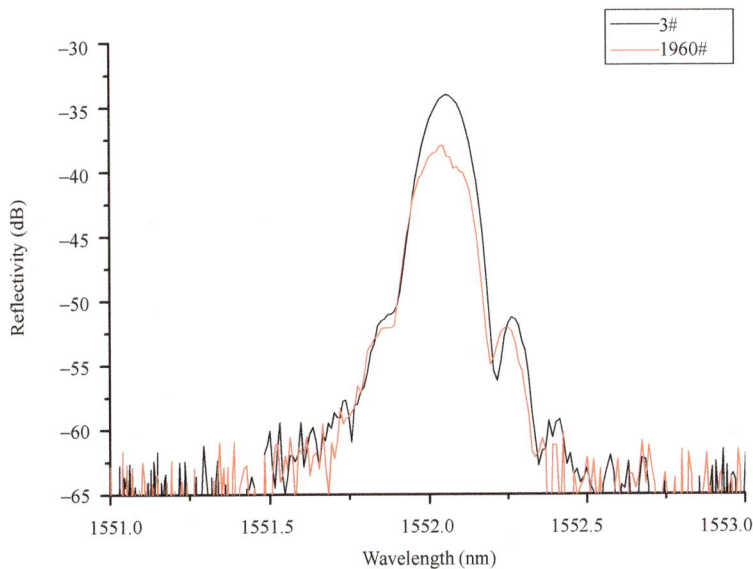

(b)

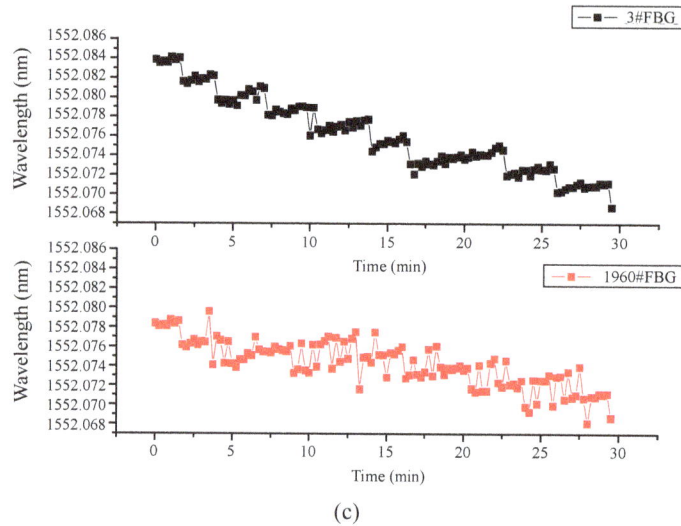

(c)

Fig. 17 Spectra of 1960 DTGs and their peak wavelength variation: (a) reflection spectra of 38 representative DTGs interrogated in the forward and reverse directions, (b) reflection spectra of the 3# and 1960# DTGs, and (c) their peak wavelengths interrogation trace over half an hour.

extinction ratio of fiber Bragg grating, the reflectivity of the spectral slope is extremely weak, and so the negative effect of the one-order crosstalks on the spectral slope is ignorable, which can be seen from the symmetrical slopes of all the spectra in Fig. 17(a). Based on the symmetrical slopes, the distorted spectra suffered from signal crosstalks are corrected by Gauss fitting to demodulate the peak wavelengths of the DTG array. The normal reflection spectrum (3rd DTG) and the distorted reflection spectrum (1960th DTG) are shown in Fig. 17(b), and their peak wavelength changes over half an hour were traced [Fig. 17(c)]. Compared with the 3rd DTG, the peak wavelength uncertainty of the 1960th DTG occurred. However, the peak wavelength uncertainty was limited within ±5 pm by the Guass-fitting method. When the peak wavelength was determined by the method of maximum value, its uncertainty can reach up to tens of picometers.

Obviously, if the distorted spectra are used as sensing media in engineer application, measurement accuracy will be worse. It will be crucial for an ideal measurement accuracy to guarantee gaining perfect reflection spectra from all DTGs in the sensing array. The simulation results told that the ultra-weak reflectivity of the DTG array can relieve greatly the

signal crosstalks [26]. On the other hand, a bad uniformity on peak wavelengths can also alleviate the crosstalks in the DTG array and guarantee to gain perfect reflection spectra. By adjusting the drawing tension in the DTG preparation process, a DTG array with a whole span of 4400 m, its DTG number of 1418, and a spacing of 3 m between two neighboring DTGs was prepared, and its peak wavelengths fluctuated mainly in the range of 1552.945 nm ± 0.2 nm [shown in Fig. 18(a)]. The DTG array had one average reflectivity of about −24 dB originally and had one degraded reflectivity of about −38 dB after one 80 °C isothermal aging time of 2 months. 16 representative DTGs were selected to show their reflection spectra in two cases, and their serial numbers were 1, 100, 200, 300, 400, 500, 600, 700, 800, 900, 1000, 1100, 1200, 1300, 1400, and 1418. Due to a high average reflectivity of −24 dB, the reflection spectra of the most DTGs in the wavelength-concentrated area had serverely distorted by the signal crosstalks, and this phenomenon occurred early on the reflection spectrum of the 100# DTG [Fig. 18(b)]. The peak wavelength reflectivities of 1#, 1400#, and 1418# DTGs are about −24 dB, −31 dB, and −24.5 dB, respectively. Compared with the 1# DTG, the reflectivity of the 1400# DTG dropped by 7 dB, while only 0.5 dB for the last 1418# DTG whose peak wavelength was drifted out from the wavelength-concentrated area. When the average reflectivity dropped to −38 dB, all the reflection spectra of the DTGs array appeared to be perfect and for some certain reasons, their wavelength uniformity became worse after the aging process [Fig. 18(c)]. The peak wavelength reflectivities of 1#, 1400#, and 1418# DTGs are about −38 dB, −40.5 dB, and −38.9 dB, respectively. Compared with the 1# DTG, the reflectivity of the 1400# DTG dropped by 2.5 dB, while only 0.9 dB for the last 1418# DTG. Experimental results indicated that the reflectivity decline and a worsen wavelength uniformity were beneficial to improve the grating spectra and gain a uniform power distribution in the DTG array, which were very helpful for a high precision detection of the sensor system.

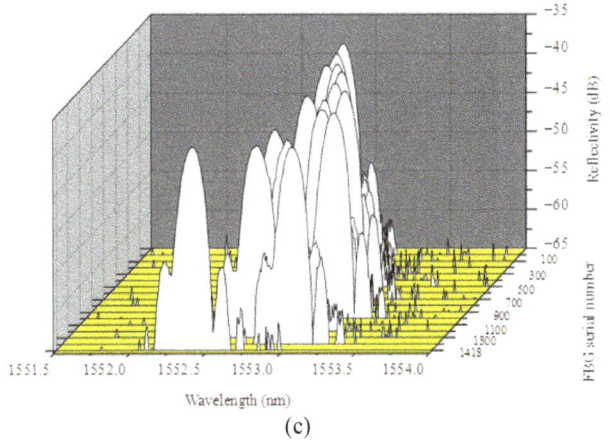

(a)

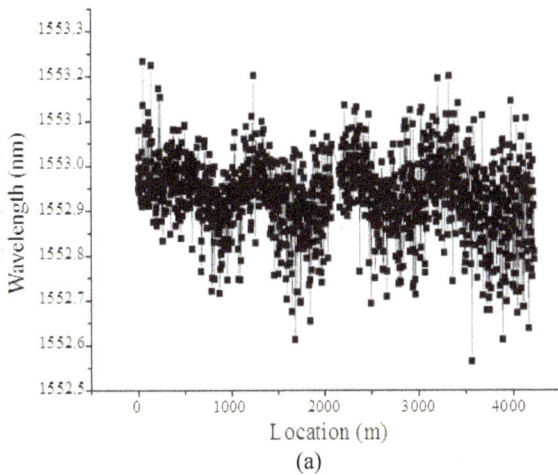

(b)

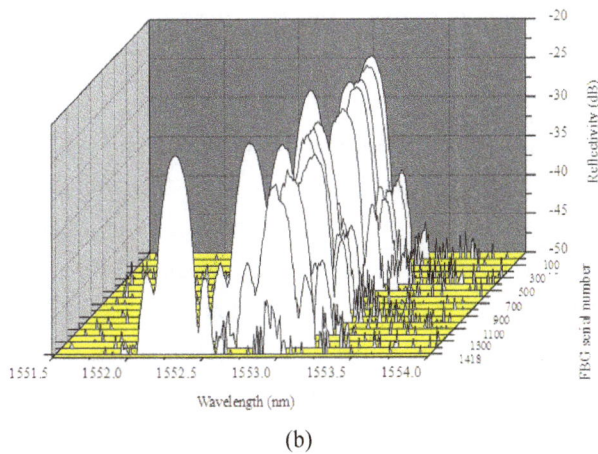

(c)

Fig. 18 Spectrum evolution of a 1418-DTG array: (a) fluctuant distribution of the peak wavelengths of a 1418-DTGs array, (b) reflection spectra of the 16 representative DTGs with a average reflectivity of −24 dB and (c) reflection spectra of the 16 representative DTGs with an average reflectivity of −38 dB

A further distributed temperature measurement is conducted to investigate the sensing performance of the network made up of 6108 DTGs with an average reflectivity of −38.5 dB in an about 10 km fiber. In order to depress notorious crosstalks in the large-capacity sensing array, the two wavelength bands of 1552.1 nm and 1552.6 nm were applied for it, and 3010 DTGs with a length of 6140 m and a neighboring DTG space of 2 m were in the 1552.6 nm band, and in the 1552.1 nm there were 3098 DTGs with a length of 3400 m and a neighboring DTG space of 1 m. The sensing fiber was segmented into six sections, and they were winded on five collection drums, in which 4# and 6# sections were winded on one same drum. The sections 1#, 3#, and 5# of the sensor array were heated in a temperature test chamber (SIDA TEMI300), while the rest of the DTG sensors were kept at room temperature of 25 ℃ (shown in Fig. 19(a)). The chamber temperature was stepped by 10 ℃ from 25 ℃ to 75 ℃ with an accuracy of 0.1 ℃. The result of temperature measurement is shown in Fig. 19(b). For the heated DTGs, their peak wavelengths showed red shift with an increase in temperature, and the two representative DTGs located at 242 m and 9000 m have a good linear response with a coefficient of 10.68 pm/℃ [Fig. 19(c)].

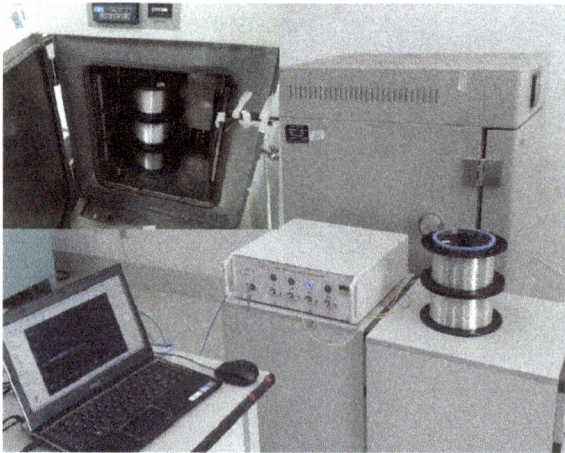

(a)

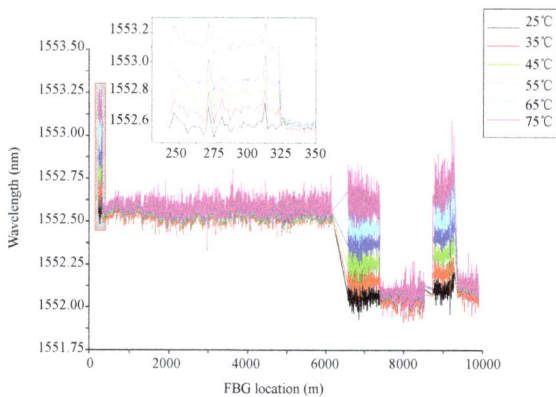

(b)

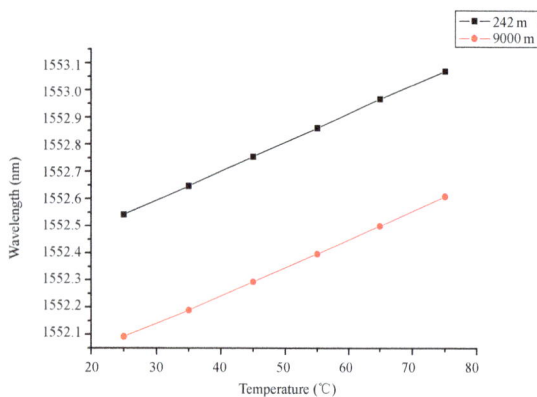

(c)

Fig. 19 Experimental demonstration: (a) images of the interrogation system of an ultra-weak DTG array, (b) temperature sensing performance of the 6018 ultra-weak DTG array, and (c) the static temperature response of the two representative DTGs located 242 m and 9000 m.

5. Conclusions

As conclusion, this paper reviews the work on fiber-optic sensing network with huge capacity

based on ultra-weak draw tower gratings developed at the National Engineering Laboratory for Fiber Optic Sensing Technology (NEL-FOST), Wuhan University of Technology, China. The draw-tower FBG arrays were prepared based on the phase mask technique, and the uniformity of the obtained FBG arrays was investigated. It was found that the uniformity of draw-tower FBG arrays greatly depended on the stability of the drawing process, especially the drawing tension. Under an optimized condition (the drawing speed: 12 m/min, the average drawing tension: 38.2 grams), one 300-FBG array with the average reflectivity of about 0.26% was prepared. They showed central wavelength bandwidth less than 0.1 nm, the mean reflection power with the relative deviation of ±16%, and excellent overlapping of the peak shape. These indicated that on-line writing FBG array by the phase mask technique had a good uniformity of central wavelength. However, due to the unstable laser pulse energy, the uneven distribution of the writing energy density and the optical fiber vibration, the uniformity of reflection power was not perfect.

A novel interrogation system for large scale sensing network with identical ultra-weak FBGs system based on serial TDM scheme was proposed and experimentally demonstrated in this paper. 1009 FBGs with reflectivity of about −33 dB and 2.5 meters equidistance between neighbor FBGs were multiplexed in series. The measurement accuracy of the Bragg wavelengths obtained by the interrogation system was less than 5 pm, and one round interrogation was completed within several seconds. The spectral distortion induced by multiple reflection crosstalks was measured and analyzed. The reflective spectra of FBGs in the end of the array showed a low transmission loss which can be further reduced by using even weaker FBGs, and this means the interrogation system has the ability to demodulate even larger number of sensors in serial.

Fiber-optic sensing network with huge capacity based on ultra-weak draw tower gratings is experimentally demonstrated as a new platform for optical fiber sensing technologies. The ultra-weak grating showed FWHM less than 0.1 nm with excellent overlapping of the peak fringe. Based on the developed ultra-weak FBG array, a versatile large-scale ultra-weak FBG sensor network with two SOAs and one high-speed CCD module has been proposed and demonstrated. Results show that the proposed sensor network can interrogate the large-scale array over 6000 FBGs by TDM+WDM. The pioneering design has low crosstalks and good sensing characteristic, and would be very promising for fiber optic sensing application.

Acknowledgment

This work was supported by the Major Program of the National Natural Science Foundation of China, NSFC (Grant No. 61290311) and the Natural Science Foundation of Hubei Province, China (No. 2014CFB269).

References

[1] Y. Rao, A. B. Ribeiro, D. A. Jackson, L. Zhang, and I. Bennion, "Simultaneous spatial, time and wavelength division multiplexed in-fibre grating sensing network," *Optics Communications*, 1996, 125(1−3): 53−58.

[2] W. Ecke, I. Latka, R. Willsch, A. Reutlinger, and R. Graue, "Fibre optic sensor network for spacecraft health monitoring," *Measurement Science and Technology*, 2001, 12(7): 974−980.

[3] B. A. Childers, M. E. Froggatt, S. G. Allison, T. C. Moore, D. A. Hare, C. F. Batten, *et al.*, "Use of 3000 Bragg grating strain sensors distributed on four 8-m optical fibers during static load tests of a composite structure," in *Proc. SPIE*, vol. 4332, pp. 133−142, 2001.

[4] D. J. F. Cooper, T. Coroy, and P. W. E. Smith, "Time-division-multiplexing of large serial fiber-optic Bragg grating sensor arrays," *Applied Optics*, 2001, 40(16): 2643−2654.

[5] Y. Wang, J. Gong, D. Wang, B. Dong, W. Bi, and Anbo Wang, "A quasi-distributed sensing network with time-division-multiplexed fiber Bragg gratings," *IEEE Photonics Technology Letters*, 2011, 2(23): 70−72.

[6] Y. J. Rao, A. B. Lobo Ribeiro, D. A. Jackson, L. Zhang, and I. Bennion, "Simulataneous spatial, time and wavelength division multiplexed in-fiber grating sensing network," *Optics Communications*, 2012, 125: 53−58

[7] M. Zhang, Q. Sun, Z. Wang, X. Li, D. Liu, and H. Liu, "A large capacity sensing network with identical weak fiber Bragg gratings multiplexing," *Optics Communications*, 2012, 285(13): 3082−3087.

[8] W. Chung, H.Y. Tam, P. K. A. Wai, and A. Khandelwal, "Time- and wavelength-division multiplexing of FBG sensors using a semiconductor optical amplifier in ring cavity configuration," *IEEE Photonics Technology letters*, 2005, 17(12): 2709−2711.

[9] G. D. Lloyd, L. A. Everall, K. Sugden, and I. Bennion, "Resonant cavity time-division-multiplexed fiber Bragg grating sensor interrogator," *IEEE Photonics Technology Letters*, 2004, 16(10): 2323−2325.

[10] Y. Wang, J. Gong, B. Dong, D. Y. Wang, T. J. Shilig, and A. Wang, "A large serial time-division multiplexed fiber Bragg grating sensor network," *Journal of Lightwave Technology*, 2012, 30(17): 2751−2756.

[11] C. G. Askins, M. A. Putnam, H. J. Patrick, and F. J. Friebele, "Fiber strength unaffected by on-line writing of single-pulse Bragg gratings," *Electronics Letters*, 1997, 33(15): 1333−1334.

[12] V. Hagemann, M. N. Trutzel, L. Staudigel, M. Rothhardt, H. R. Muller, and O. Krumpholz, "Mechanical resistance of draw-tower-Bragg-grating sensors," *Electronics Letters*, 1998, 34(2): 211−212.

[13] M. Xu, H. Geiger, J. L. Archambault, L. Reekie, and J. P. Dakin, "Novel interrogating system for fiber Bragg grating sensors using an acousto-optic tunable filter," *Electronics Letters*, 1993, 29(17): 1510−1511.

[14] C. G. Askins, M. A. Putnam, G. M. Williams, and E. J. Friebele, "Stepped-wavelength optical-fiber Bragg grating arrays fabricated in line on a draw tower," *Optics Letters*, 1994, 19(2): 147−149.

[15] S. Abad, F. M. Araujo, L. A. Ferreira, J. L. Santos, and M. Lopez-Amo, "Comparative analysis of wavelength-multiplexed photonic-sensor networks using fused biconical WDMS," *IEEE Sensors Journal*, 2003, 3(4): 475−483.

[16] Y. Dai, Y. Liu, J. Leng, G. Deng, and A. Asundi, "A novel time-division multiplexing fiber Bragg grating sensor interrogator for structural health monitoring,"

Optics and Lasers in Engineering, 2009, 47(10): 1028−1033.

[17] X. Wan and H. F. Taylor, "Multiplexing of FBG sensors using modelocked wavelength-swept fibre laser," *Electronics Letters*, 2003, 39(21): 1512−1514.

[18] M. Zhang, Q. Sun, Z. Wang, X. Li, D. Liu, and H. Liu, "A large capacity sensing network with identical weak fiber Bragg gratings multiplexing," *Optics Communications*, 2012, 285(13−14): 3082−3087.

[19] B. A. Childers, M. E. Froggatt, S. G. Allison, T. C. Moore, D. A. Hare, C. F. Batten, and D. C. Jegley, "Use of 3000 Bragg grating strain sensors distributed on four 8-m optical fibers during static load tests of a composite structure," in *Proc. SPIE*, 2001, 4332(133), doi: 10.1117/12.429650.

[20] D. L. Williams, B. J. Ainslie, J. R. Armitage, and R. Kashyap, "Enhanced UV photosensitivity in boron codoped germanosilicate fibers," *Electronics Letters*, 1993, 29(1): 45−47.

[21] http://www.fbgs-technologies.com/pagina.php?id=21 366

[22] B. Hartmut, S. Kay, U. Sonja, C. Christoph, R. Manfred, and L. Ines," Single-pulse fiber Bragg gratings and specific coatings for use at elevated temperatures," *Applied Optics*, 2007, 46(17): 3417−3424.

[23] D. L. Williams, B. J. Ainslie, R. Kashyap, G. D. Maxwell, J. R. Armitage, R. J. Campbell, *et al.*, "Photosensitive index changes in germania-doped silica glass fibers and waveguides," in *Proc. SPIE*, vol. 2044, pp. 56−68, 1993.

[24] Y. Wang, J. Gong, D. Wang, B. Dong, W. Bi, and A. Wang, "A quasi-distributed sensing network with time-division-multiplexed fiber Bragg gratings," *IEEE Photonics Technology Letters*, 2011, 2(23): 70−72.

[25] C. Chan, W. Jin, D. Wang, and M. S. Demokan, "Intrinsic crosstalk analysis of a serial TDM FBG sensor array by using a tunable laser," *Microwave and Optical Technology Letters*, 2003, 36(1): 2−4.

[26] C. G. Askins, M. A. Putnam, and G. M. Williams, "Stepped-wavelength optical-fiber Bragg grating arrays fabricated in line on a draw tower," *Optics Letters*, 1994, 19(2): 147−149.

19

Low-Frequency Vibration Measurement by a Dual-Frequency DBR Fiber Laser

Bing ZHANG, Linghao CHENG[*], Yizhi LIANG, Long JIN, Tuan GUO, and Bai-Ou GUAN

Guangdong Provincial Key Laboratory of Optical Fiber Sensing and Communications, Institute of Photonics Technology, Jinan University, Guangzhou, 510632, China

[*]Corresponding author: Linghao CHENG   E-mail: chenglh@ieee.org

Abstract: A dual-frequency distributed Bragg reflector (DBR) fiber laser based sensor is demonstrated for low-frequency vibration measurement through the Doppler effect. The response of the proposed sensor is quite linear and is much higher than that of a conventional accelerometer. The proposed sensor can work down to 1 Hz with high sensitivity. Therefore, the proposed sensor is very efficient in low-frequency vibration measurement.

Keywords: Fiber laser sensors; Doppler effect; vibration measurement

1. Introduction

Vibration measurement is widely employed in many applications, such as aerospace, architectural, automotive, security, and medical applications [1–4]. Frequently, vibration can be measured by measuring the acceleration of the target through attaching an accelerometer onto the target surface [5–6]. However, for many applications, the surface of the target may be difficult to be accessed, or may be too small or too hot to be attached a physical transducer. For these applications, non-contact measurement is highly demanded [7–8].

Doppler effect based vibration measurement is one method capable of measuring vibration without attaching transducer onto targets [9–12]. Moreover, it is quite efficient in measurement of vibration at low frequency at which conventional accelerometers

are highly inefficient. Lasers are frequently employed in this method by directing a laser beam onto the target surface and measuring the frequency variation of the reflected laser beam due to the Doppler effect resulted from the vibration. Among various lasers, fiber lasers are popular in vibration measurement for their light weight and compact size.

In this paper, we propose a low-frequency vibration measurement scheme based on the Doppler effect by employing a dual-frequency distributed Bragg reflector (DBR) fiber laser [13]. With dual-frequency output of the fiber laser, the scheme works in a heterodyne regime without employing a frequency shifter and hence results in a very compact design. The proposed sensor can work down to 1 Hz with high sensitivity and linear response.

2. Theory and experiment

The proposed scheme is shown in Fig. 1, which is based on a distributed Bragg reflector (DBR) fiber laser inscribed on an Er/Yb co-doped fiber by a 193-nm excimer [14]. The output power of the dual-frequency DBR laser is about −21 dBm, and the beat note power is about −10 dBm. The laser is composed of one 5-mm grating and one 5.5-mm grating, spaced in about 1 mm. Pumped at 980 nm, the laser outputs two wavelengths at 1531 nm with orthogonal polarizations and a frequency difference of about 3.256 GHz due to the intra-cavity birefringence. After being amplified by an erbium-doped fiber amplifier (EDFA), the two orthogonally polarized outputs are separated by a polarization beam splitter (PBS). One of the two polarizations is directed to a vibrating surface by a collimator, and the reflected light from the surface is collected by the collimator and combined after amplification with the other polarization through a circulator. The resulted light is then detected by a photo-detector, and the generated signal is sampled and processed to recover the vibration signal.

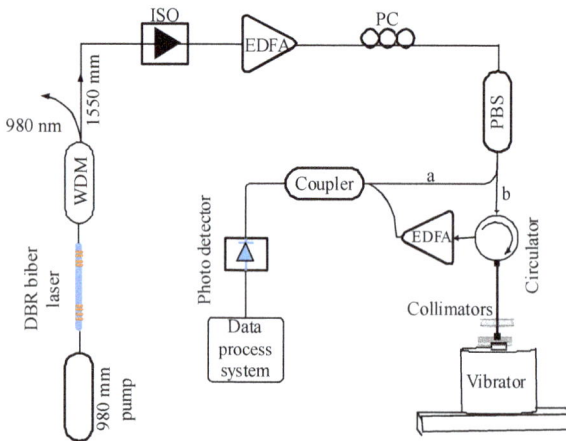

Fig. 1 Experimental setup: WDM: wavelength division multiplexer; ISO: isolator; EDFA: erbium-doped fiber amplifier; PC: polarization controller; PBS: polarization beam splitter.

The vibration is generated by a vibrator. Due to the Doppler effect, the vibrating surface induces a frequency shift to the light focused on the surface. This frequency shift is proportional to the vibration velocity. Therefore, the reflected light can be

expressed as

$$r_0(t) \propto \exp\left[j\omega_0 t + j\frac{\omega_0}{c}\int_0^t v(\tau)d\tau + j\theta_0(t)\right] \quad (1)$$

where ω_0 and θ_0 are the original angular frequency and the initial phase of the reflected light with phase noise included, respectively, and v is the vibration velocity. When the combined light of the reflected light and the other polarization light is photo-detected, the resulted photo-electrical current is written as

$$I(t) \propto \exp\left[j\Delta\omega t + j\frac{\omega_0}{c}\int_0^t v(\tau)d\tau + j\Delta\theta(t)\right] \quad (2)$$

where $\Delta\omega$ and $\Delta\theta$ are the frequency difference and the initial phase difference between the two polarizations. Through frequency demodulation, the instantaneous frequency of the photo-electrical current is obtained as

$$\omega(t) = \Delta\omega + \frac{\omega_0}{c}v(t) + \frac{d\Delta\theta(t)}{dt} \quad (3)$$

from which the vibration velocity can be obtained through filtering. Note that the last term in (3) results in some measurement error because the phase noise of the laser is also converted to frequency variation through frequency demodulation. However, majority of this noise can be filtered out by a properly designed filter. Frequently, vibration is also measured through acceleration, which can also be obtained through (3) by taking differential:

$$a(t) = \frac{d\omega(t)}{dt} = \frac{\omega_0}{c}\frac{dv(t)}{dt} + \frac{d^2\Delta\theta(t)}{dt^2}. \quad (4)$$

Figure 2 shows a measurement for a vibration at 20 Hz. The vibrator is driven at 0.8 V. The spectrum clearly shows the vibration signal at 20 Hz with a signal-to-noise ratio (SNR) of about 40 dB. It then shows that the proposed scheme works very well at low frequency. Equation (4) shows that at the same level of acceleration, the response of the proposed scheme will be vibration frequency dependent due to the differential of the velocity. Therefore, the response of the proposed scheme will be higher at lower frequency and lower at higher frequency for the same level of acceleration.

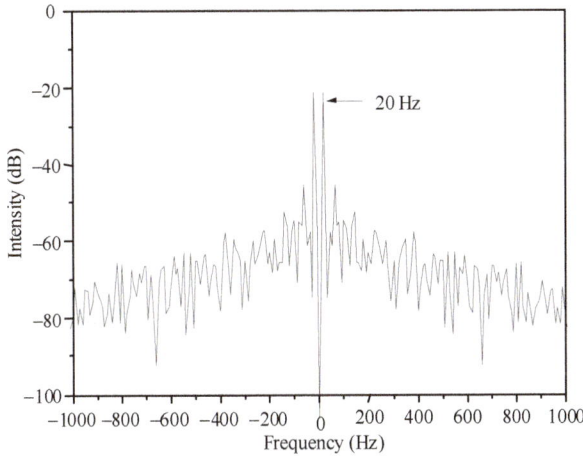

Fig. 2 Measured spectrum for a vibration at 20 H (the driving voltage of the vibrator is 0.8 V).

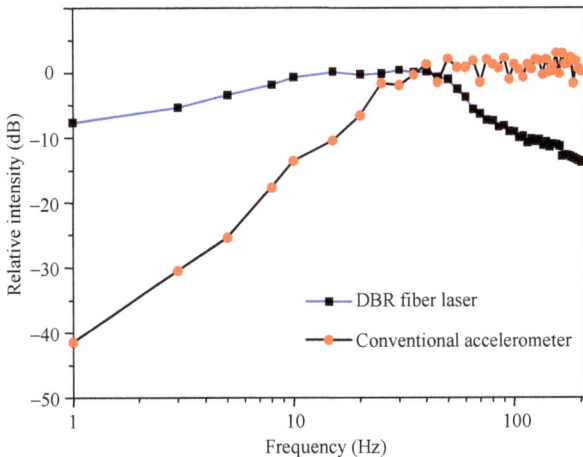

Fig. 3 Measured response at various vibrating frequencies for the DBR fiber laser based measurement and a conventional accelerometer based measurement.

Figure 3 shows the measured response at various frequencies. For comparison, the measured response of a conventional accelerometer is also plotted. The two response curves are normalized to their responses at 35 Hz, respectively. It then shows that the response of the proposed scheme by the DBR fiber laser is higher at lower frequency while the response of the conventional accelerometer degrades heavily at low frequency although this degradation may be the combined response degradation of the conventional accelerometer and the vibrator at low frequency. The proposed scheme works very well down to 1 Hz. Actually, the scheme can work at even lower vibrating frequency. However, the memory

size of the data acquisition unit limits the measurement to 1 Hz.

Figure 4 shows the response of the proposed scheme relative to that of the conventional accelerometer based scheme. The frequency dependent response of the vibrator is then removed from the curve. The curve shows a tendency basically proportional to $1/\omega^2$ which is shown as a linear curve in log-log plot. For the proposed scheme based on the Doppler effect, the direct measured is velocity, while for the accelerometer, it is acceleration. Because acceleration is the differential of velocity, the ratio between the velocity and acceleration is proportional to $1/\omega$ in frequency domain. With the vertical scale in decibels, $1/\omega^2$ is therefore resulted as shown by the dash curve for fitting according to $1/\omega^2$ in Fig. 4. The slope of the curve is the theoretical value, that is, −20 dB/decade. However, the best fitting of the measured results is by $1/\omega^{2.2}$ as shown by the solid curve with a slope of −22 dB/decade in Fig. 4. It then shows that the accelerometer has higher gain at high frequencies, which confirms again that the proposed scheme works better in low frequency in which the conventional accelerometer is quite inefficient.

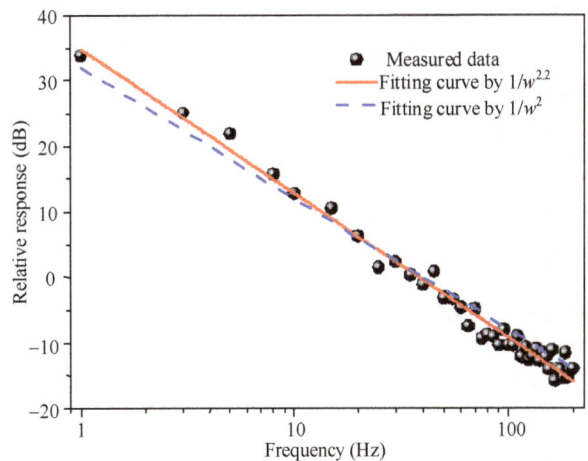

Fig. 4 Response of the proposed scheme relative to that of the conventional accelerometer based scheme.

Figure 5 shows the response of the proposed scheme at various driving levels of the vibrator. A quite linear curve is observed. Actually, the deviation of the measured data from the linear curve should be largely attributed to the non-uniform response of the vibrator at various driving voltages because the proposed scheme is a non-contact measurement.

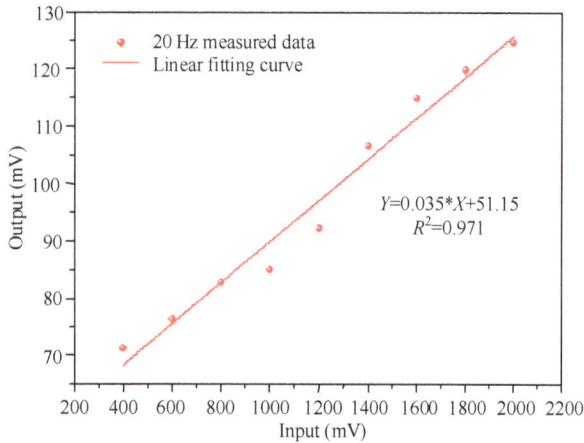

Fig. 5 Measured response at various driving voltages of the vibrator for a vibrating frequency of 20 Hz.

3. Conclusions

Based on the Doppler effect, a fiber-optic vibration measurement scheme is proposed and demonstrated by utilizing a dual-frequency DBR fiber laser, which results in a compact scheme due to the miniature dimensions of the fiber laser. The scheme demonstrates a linear response and high sensitivity at low frequency in which conventional accelerometers are quite inefficient. The scheme works very well in the band below 200 Hz and can works down to 1 Hz. Therefore, the proposed scheme is very useful for low-frequency vibration measurement.

Acknowledgment

This work was supported by the National Natural Science Foundation of China (Grant Nos. 11474133 and 61235005), Natural Science Foundation of Guangdong Province of China (No. 2014A030310419), and the Guangzhou Key Collaborative Innovation Foundation of China (No. 2016201604030084).

References

[1] E. Reithmeier, S. Mirzaei, and N. Kasyanenko, "Optical vibration and deviation measurement of rotating machine parts," *Optoelectronics Letters*, 2008, 4(1): 45–48.

[2] S. Wang, X. Fa, and Q. Liu, "Distributed fiber-optic vibration sensing based on phase extraction from time-gated digital OFDR," *Optics Express*, 2015, 23(26): 33301–33309.

[3] A. Wada, S. Tanaka, and N. Takahash, "Multipoint vibration sensing using fiber Bragg gratings and current-modulated laser diodes," *Journal of Lightwave Technology*, 2016, 34(19): 4610–4614.

[4] D. Jiang and H. Wei, "Review of applications for fiber Bragg grating sensor," *Journal of Optoelectronics Laser*, 2002, 13(4): 420–430.

[5] R. P. Linessio, K. D. Sousa, T. D. Silva, C. A. Bavastri, P. F. D. C. Antunes, and J. C. C. D. Silva, "Induction motors vibration monitoring using a biaxial optical fiber accelerometer," *IEEE Sensors Journal*, 2016, 16(22): 8075–8082.

[6] Y. Weng, X. Qiao, Z. Feng, M. Hu, J. Zhang, and Y. Yang, "Compact FBG diaphragm accelerometer based on L-shaped rigid cantilever beam," *Chinese Optics Letters*, 2011, 9(10): 22–25.

[7] C. R. Farrar, T. W. Darling, A. Migliori, and W. E. Baker, "Microwave interferometers for non-contact vibration measurements on large structures," *Mechanical Systems and Signal Processing*, 1999, 13(2): 241–253.

[8] T. Li, Y. Tan, Z. Zhou, L. Cai, S. Liu, Z. He, *et al.*, "Study on the non-contact FBG vibration sensor and its application," *Photonic Sensors*, 2015, 5(2):128–136.

[9] K. M. Singh and P. Sumathi, "Synchronization technique for Doppler signal extraction in ultrasonic vibration measurement systems," *IEEE Transactions on Instrumentation and Measurement*, 2015, 64(12): 3162–3172.

[10] J. H. Zhou, G. Chen, and Q. S. Cao, "Design of ultrasonic system for vibration measurement based on Doppler effect," *Instrument Technique and Sensor*, 2015, 1(7): 61–62.

[11] P. Castellini, M. Martarell, and E. P. Tomasini, "Laser doppler vibrometry: development of advanced solutions answering to technology's

needs," *Mechanical Systems and Signal Processing*, 2006, 20(6): 1265–1285.

[12] Z. Y. Kuang, L. H. Cheng, B. O. Guan, H. Liang, and B. Guan, "Dual-polarization fiber grating laser-based laser Doppler velocimeter," *Chinese Optical Letter*, 2016, 14(5): 050602–050605.

[13] Y. Zhang, B. O. Guan, and H. Y. Tam, "Ultra-short

distributed Bragg reflector fiber laser for sensing application," *Optics Express*, 2009, 17(12): 10050–10055.

[14] Y. Z. Liang, L. Jin, L. H. Cheng, and B. O. Guan, "Stabilization of microwave signal generated by a dual-polarization DBR fiber laser via optical feedback," *Optics Express*, 2014, 22(24): 29356–29362.

20

Research on Multi-Component Gas Optical Detection System Based on Conjugated Interferometer

Xin GUI[1*], Yuheng TONG[2], Honghai WANG[1], Haihu YU[1], and Zhengying LI[1,2]

[1]*National Engineering Laboratory for Fiber Optic Sensing Technology, Wuhan University of Technology, Wuhan, 430070, China*

[2]*Key Laboratory of Fiber Optic Sensing Technology and Information Processing, Ministry of Education, Wuhan University of Technology, Wuhan, 430070, China*

[*]Corresponding author: Xin GUI   E-mail: guixin@whut.edu.cn

Abstract: An optical multi-component gas detection system based on the conjugated interferometer (CI) is proposed and experimentally demonstrated. It can realize the concentration detection of mixture gas in the environment. The CI can transform the absorption spectrum of the target gases to a conjugated emission spectrum, when combining the CI with the broadband light source, the spectrum of output light matches well with the absorption spectrum of target gases. The CI design for different target gases can be achieved by replacing the kind of target absorbing gas in the CI filter. The traditional fiber gas sensor system requires multiple light sources for detection when there are several kinds of gases, and this problem has been solved by using the CI filter combined with the broadband light source. The experimental results show that the system can detect the concentration of multi-component gases, which are mixed with C_2H_2 and NH_3. Experimental results also show a good concentration sensing linearity.

Keywords: Gas conjugated interference filter; gas sensing; spectral absorption

1. Introduction

The multi-component gas detection has played a significant role in many fields, such as in the atmospheric condition detection, the safe operation of the transformer, industrial process control, and medical diagnosis [1–3]. The optical gas detection technology based on Beer-Lambert's law is widely applied in detecting different kind of gases such as carbon monoxide, carbon dioxide, methane, hydrogen, and acetylene. Tunable diode laser absorption spectroscopy (TDLAS) is one of the most commonly used gas detection methods with high sensitivity. The system based on the distributed feedback (DFB) laser [4–5] has narrow line-widths whose wavelength can be precisely tuned to match a gas absorption line, giving high spectral power density. However, DFB lasers have several disadvantages, for example, they are generally expensive and gas specific. Hence, a separate DFB laser is often required for each gas absorption line of interest. Besides, it cannot be applied in multiple-component gas due to the wide distribution of gas absorption lines.

The selection of a light source plays a vitally important role in the spectrum absorption gas sensor system. Worldwide researchers have proposed various methods to solve this problem. The bandwidth of a broadband source is typically 20 nm – 100 nm and can cover absorption lines of multiple types of gases, such as C_2H_2, CO, CO_2, H_2S, and NH_3, which have absorption lines in the range of 1 510 nm to 1 590 nm. For its stable performance, long service life, easy operation, and wide range, the broadband source is widely used in the gas detection. Broadband sources combined with optical filters to choose the gas absorption lines provide a good choose for the gas detection system. Applications of a broadband light source with filters of the gas sensor have been reported of which some are based on the fiber Bragg grating (FBG) [6], Fabry-Perot etalon [7–9], and the filter plate [10, 11]. In addition, using the method of the comb filter [12], the broadband light source can accurately match a plurality of CH_4 gas absorption lines. Fiber interferometers have been reported using as a filter, for example, the filters based on the Michelson interferometer principle [12] and Mach-Zehnder interference [13] principle combining with the FBG or other optical devices to achieve quality narrowband transmission filter can be used in the vicinity of the gas absorption 1 550 nm. Spectral matching with the gas absorption lines is the most challenging part of the filter approach.

Recently, we have proposed single gas detection method based on the conjugated interferometer (CI) filter [14], which transforms the absorption spectrum of the sample gas to a conjugated emission spectrum. Here, a comb light source which emits light at gas absorption lines is produced. However, for the previous case, the CI filter essentially acts as a single gas. In the present study, based on the broadband light source, the CI filter can be designed as the output of the wavelength light source matching the absorption lines of the multi-target gases so as to be used in the multi-gases detection system. By changing the gases in the CI filter, it can produce the optical output corresponding to multiple gases, and the output exactly corresponds to their intrinsic absorption lines of the gases, which greatly improves the low sensibility caused by the broadband light source. It provides a solution to the simplicity of gas selection in the light source. In the paper, the experimental results show that the system can detect multi-component gases concentrations, which are mixed with C_2H_2 and NH_3, and achieve a good concentration sensing linearity.

2. Principle of conjugated interferometer

2.1 Design of conjugated interferometer

The CI is the key component of the gas sensing system. As shown in Figs. 1(a) and 1(b), the light source of the multi-component gas detection system consists of a broadband light source cascade CI filter. The CI includes two collimators and a sandwiched etalon. The etalon has two units: reference cell and absorption cell. The absorption cell is filled with the high concentration absorptive multi-gases for target gases wavelength, and the reference cell contains gas which is non-absorptive for target gases and its refractive index is closed to absorptive gas. We choose NH_3 and C_2H_2 for garget gases in the paper.

The light from a broadband source is split into two parts by etalon and injected into two units. The amplitude powers can be expressed as

$$I_1(\lambda) = \sqrt{\eta_{in} r I(\lambda) \times [1 - \alpha(\lambda)]} \qquad (1)$$

$$I_2(\lambda) = \sqrt{\eta_{in}(1-r)I(\lambda)} \qquad (2)$$

where $I_1(\lambda)$ and $I_2(\lambda)$ are the amplitude powers of the absorption and reference units, η_{in} is the insertion loss, r is the splitting ratio of gas etalon, $I(\lambda)$ is the incident light intensity, and $\alpha(\lambda)$ is the gas absorption coefficient.

The output light intensity of the CI filter at the optical spectrum analyzer (OSA) is

$$I_0(\lambda) = \eta_{out}[I_1^2 + I_2^2 + 2I_1(\lambda)I_2(\lambda)\cos \Delta\varphi] \qquad (3)$$

where $\Delta\varphi$ is the phase difference between the two units, and η_{out} is the coupling coefficient from the   etalon to the other fiber coupled collimator.

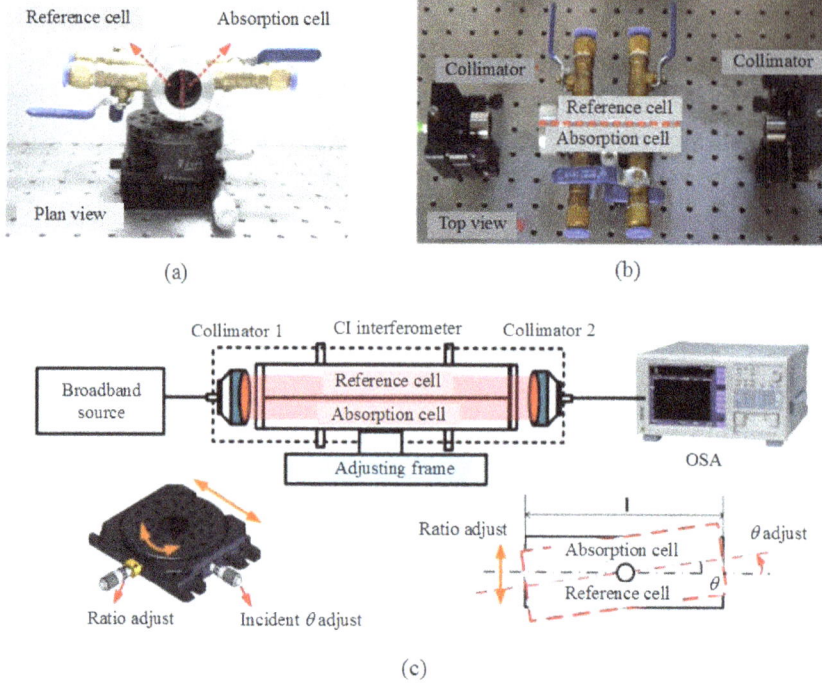

(a)

(b)

(c)

Fig. 1 Gas conjugated interferometer: (a) plan view of CI, (b) top view of CI, and (c) working principle of CI.

If $r = 50\%$, (3) can be simplified as follows:

$$I_0(\lambda) = \eta_{in}\eta_{out}\left[1 - \frac{1}{2}\alpha(\lambda) + [1 - \alpha(\lambda)]\cos\Delta\varphi\right]I(\lambda).$$

(4)

From (4), we can find that the transmission light intensity I_0 depends on the gas absorption coefficient α and the phase φ. When the phase different $\varphi = \pi$, (4) becomes

$$I_0(\lambda) = \frac{1}{2}\alpha(\lambda)\eta_{in}\eta_{out}I(\lambda).$$

(5)

The important role of the phase difference can be expressed as

$$\Delta\varphi = \frac{2\pi}{\lambda}(\Delta nL / \cos\theta)$$

(6)

where Δn is the refractive index difference of the gases filled in two air cells, and θ is the incident angle. As shown in Fig. 1(c), the optical path difference ΔL in two gas cells is resulted from different refractive indexes and the shift of the rotation angle θ. Different values of Δn correspond to different output values of ΔL, which also results in a change in φ accordingly. In this essay, the π phase shift condition can be achieved by altering the

optical path difference (OPD) of the reference cell and absorption cell (ΔL). From (6), we can find that when $\Delta L = (\lambda/2) \cdot (2n + 1)$, ($n = 0, 1, 2, \cdots$), the π phase shift condition is achieved. In order to make the absorption peak of the measuring spectral range flip at the same time, we must make the interference period greater, ΔL smaller, and the refractive indexes of two air cells as close as possible. Our targets gases are C_2H_2 and NH_3, whose gas refractive indexes are 1.00051 and 1.000376. Then we choose CH_4 with a refractive index of 1.000444, which has the most similar refractive index but at the same time whose absorption peak would not overlap with the target gas.

2.2 Achievement of multi-component optical output

The performance of the conjugated interferometer depends on the $\Delta\varphi$, α, and the splitting ratio. We simulate and compare their effects around two assumed gas absorption lines at $1\,529\,nm$ and $1\,531\,nm$. The results are shown in Fig. 2. In the design and adjustment of the CI, controlling the parameter $\Delta\varphi$ to interval $(0.9\pi, 1.1\pi)$ and the splitting ratio to $(0.3, 0.5)$ is better.

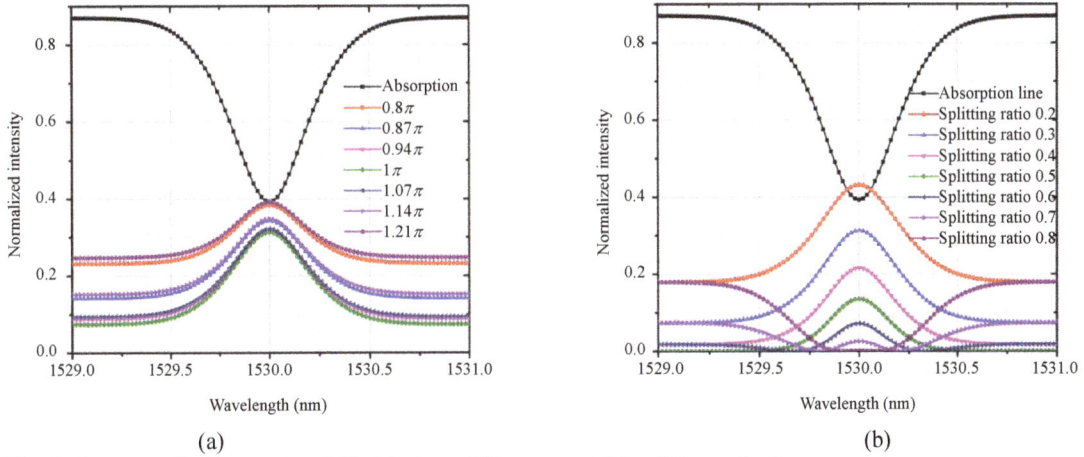

(a)

(b)

Fig. 2 Simulation transmission spectra of CI: (a) phase difference and (b) splitting ratio changes.

In a gas CI, we can simultaneously filter multi-component gases, so that among the outputs of the light source, we can obtain various light outputs of multi-component corresponding to gas absorption lines. As shown in Fig. 3, we use acetylene (C_2H_2) and ammonia (NH_3) as absorptive gases, which have absorptions in the wavelength range of $1\,510\,nm$ – $1\,535\,nm$ and methane (CH_4) as the reference gas which has no absorption in the range of $1\,510\,nm$ – $1\,535\,nm$. By adjusting the incident θ to change the OPD, the π phase shift condition is achieved. The gas absorption curve of the target gas is obtained by collecting the spectrum through the OSA. Figure 3 shows the transmission of the CI, and Figs. 3(a) and 3(b) show two target gases absorption peaks in the range of $1\,527\,nm$ – $1\,532\,nm$ and $1\,511\,nm$ – $1\,516\,nm$. In these ranges, two target gases have some strong absorption peaks and do not affect each other. Figures 3(c) and 3(d) show the wavelength peaks of the CI match well with two target gases absorption lines throughout the wavelength ranges of $1\,527\,nm$ – $1\,532\,nm$ and $1\,511\,nm$ – $1\,516\,nm$.

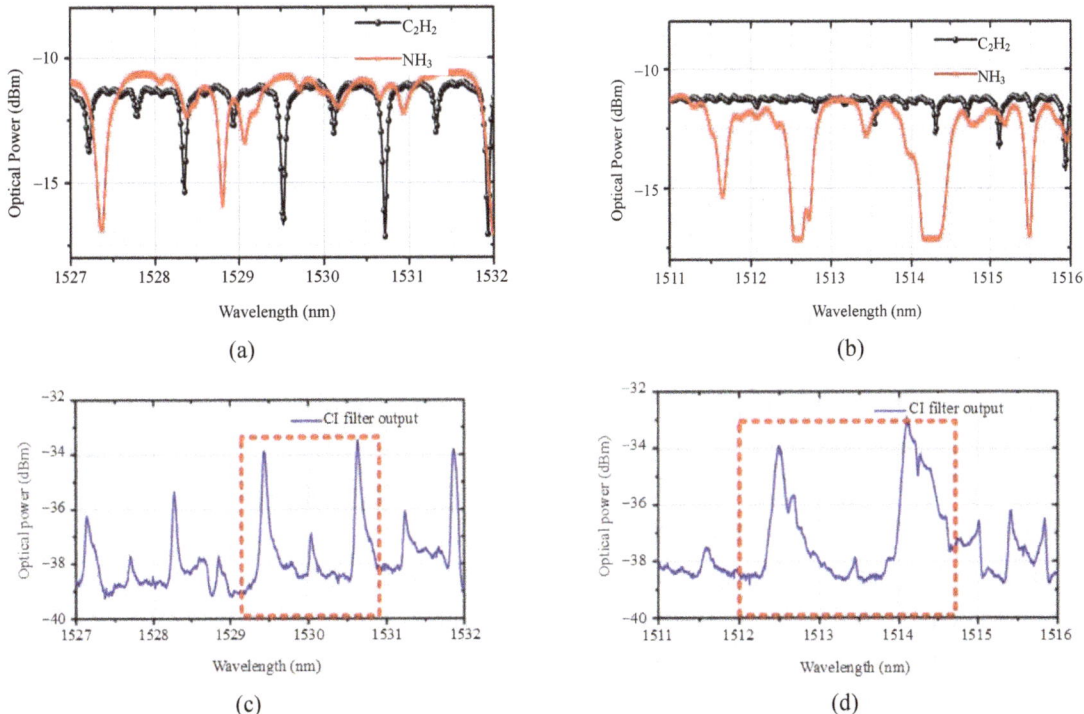

(a)

(b)

(c)

(d)

Fig. 3 Analysis of the CI filter: (a) absorption of C_2H_2 and NH_3 ($1\,527\,nm$ – $1\,532\,nm$), (b) absorption of C_2H_2 and NH_3 ($1\,512\,nm$ – $1\,516\,nm$), (c) CI filter output ($1\,527\,nm$ – $1\,532\,nm$), and (d) CI filter output ($1\,512\,nm$ – $1\,516\,nm$).

Figure 4(a) shows the CI transmission spectrum range of $1510\,nm - 1535\,nm$. The wavelength peaks of the CI match well with the gas absorption lines at the wavelength of $1510\,nm - 1535\,nm$ from Hitran database [15]. Only light close to the gas absorption lines can effectively pass the CI. The background

light, which usually limits the gas detection sensitivity, is minimized with our CI. After that, we test the stability of the CI at room temperature and record the output spectrum of the CI 2 hours by every 10 min. As shown in Fig. 4(b), the CI has a good stability.

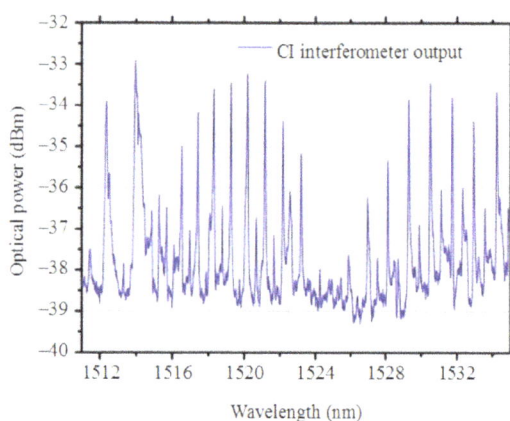

(a)

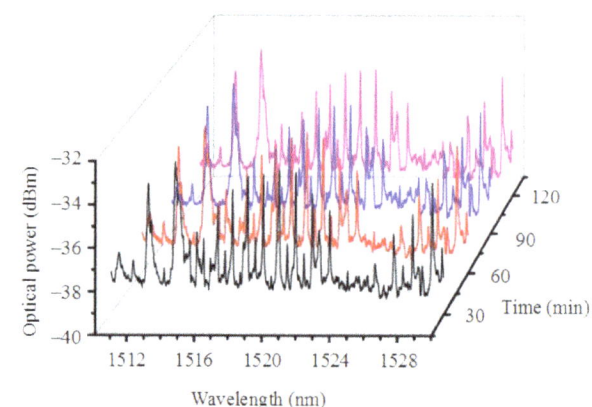

(b)

Fig. 4 Gas conjugated interferometer: (a) spectrum of CI and (b) stability of conjugated interferometer.

3. Optical gas sensor system for multi-component gas

We used C_2H_2 and NH_3 gases as sample gases to test the multi-component gas concentration detection system. The multi-gas detection system was analyzed by qualitative and quantitative analysis.

3.1 Multi-gases qualitative experiment

The gas cell with a long-length of 1 m was connected to the CI filter to verify the feasibility of gas concentration detection. The results are shown in Fig. 5, when the different concentrations of NH_3 and C_2H_2, respectively were filled in the gas cell, using the OSA on the spectrum acquisition. Due to the fact that the CI output contained the spectrum absorption information of C_2H_2 and NH_3, when there was only one type of gas in the testing gas cell, for instance, under the condition that one of the testing gases was C_2H_2 gas, the output would be acetylene gas concentration information. When the optical signal contained the C_2H_2 gas concentration subtracting the initial signal, the insertion loss of testing gas was

negligible, since the NH_3 concentration did not change, the acquired signal would contain the C_2H_2 concentration information only, but the NH_3 absorption spectrum information would be eliminated.

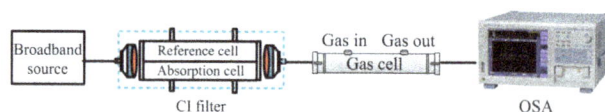

Fig. 5 Schematic of multi-gas basis verification test.

The two gases were analyzed by the above mentioned treatment method. Respectively, the 1 000 ppm and 2 000 ppm concentrations of C_2H_2, and the 0.5% and 1% concentrations of NH_3 were filled into the detected gas cell. The four signals were collected from the OSA subtracting the initial signal, respectively. The results obtained are shown in Figs. 6 and 7. The concentration information of the corresponding gas matches well with the absorption spectrum information, and with an increase in the concentration, the absorption is stronger, which is consistent with the theory deduction.

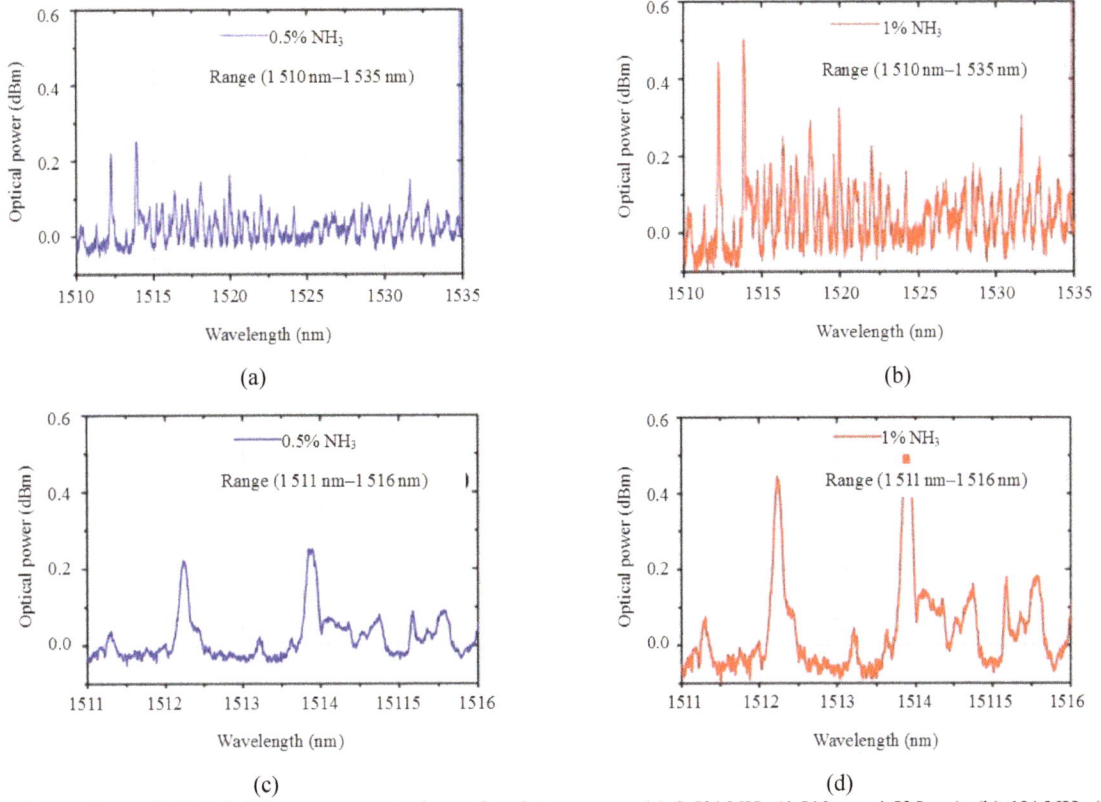

(a)

(b)

(c)

(d)

Fig. 6 Comparison of NH$_3$ of different concentrations after data process: (a) 0.5% NH$_3$ (1 510 nm–1 535 nm), (b) 1% NH$_3$ (1 510 nm – 1 535 nm), (c) 0.5% NH$_3$ (1 511 nm – 1 516 nm), and (d) 1% NH$_3$ (1 511 nm – 1 516 nm).

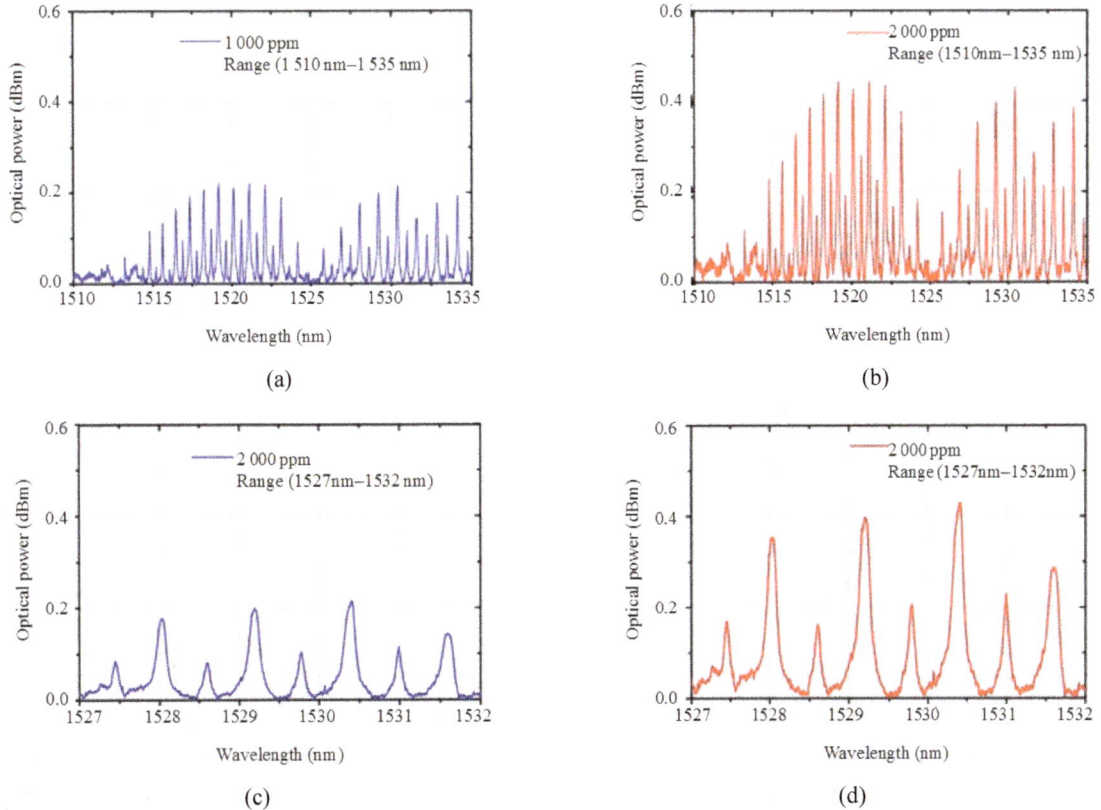

(a)

(b)

(c)

(d)

Fig. 7 Comparison of C$_2$H$_2$ of different concentrations after data process: (a) 1 000 ppm C$_2$H$_2$ (1 510 nm – 1 535 nm), (b) 2 000 ppm C$_2$H$_2$ (1 510 nm – 1 535 nm), (c) 1 000 ppm C$_2$H$_2$ (1 527 nm – 1 532 nm), and (d) 2 000 ppm C$_2$H$_2$ (1 527 nm – 1 532 nm).

3.2 Multi-gas experiment

The schematic setup of the sensor prototype and its testing system are shown in Fig. 8. The system selected a broadband light source with a wavelength range of 1 510 nm – 1540 nm and a 1-m-length detection gas cell, and the optic propagation loss in the entire cell was less than 0.5 dB. The light output of the CI was divided into two beams by a 90 : 10 coupler. The light beam with 10% power was detected by photo detector 1 (PD1) and served as a reference to compensate the source power fluctuation. Another light beam with 90% power passed the detection gas cell and an isolator, and then got detected by PD2. Data were recorded with a 16-bit data acquisition card.

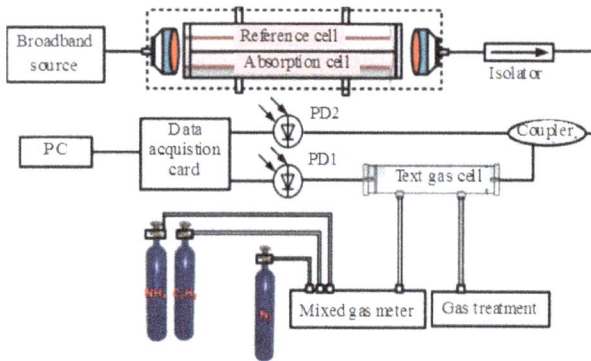

Fig. 8 Schematic of multiple gases detection system.

A reference channel was used to compensate for changes in the emission of the source, which were assumed to affect the reference and test channel wavelengths in the equal proportion. This relationship tended to result from the influence of the optoelectronic devices (such as coupler and PD). Thus, before the experiment, we should set a zero value calibration in the system to maintain that the concentration of the target gas was 0 (filled with pure N_2) in the test cell, so there was a fixed linear relationship between the two signals collected. The experiment was carried out by introducing the gas of different concentrations and collecting light intensities of the two signals. Then we processed the light intensity by the linear relationship obtained by the calibration value of zero, and we could obtain complete gas concentration information after doing the difference.

A mixture of NH_3, C_2H_2, and N_2 (99.99% concentration) was tested as the sample. The NH_3 concentration was altered at 0.1% step in the range of 0% – 1% by a mixing system (LFIX–2000), and the C_2H_2 concentration was altered at 200 ppm step in the range of 0 ppm – 2 000 ppm. For each step change in the concentration applied, a total of 2 min was allowed for passage of gas down the connecting pipe work, diffusion into the cell. At each concentration step, signals from the test and reference channels were recorded by data acquisition and the data were processed by the method mentioned above. The results are shown in Fig. 9. There were 10 concentrations of two target gases spectra information. Figure 9 reflects the result of subtraction between two channels in various gas concentrations. The main different fluctuations mainly concentrated on the range (1 511 nm–1 516 nm) of C_2H_2 and the range (1 527 nm–1 532 nm) of NH_3 absorption spectra, but no fluctuations in other wavelength ranges. The result could be determined accurately to reflect the concentration of the gas.

The ranges of the gas absorption peaks of 1 511 nm–1 516 nm and 1 527 nm–1 532 nm were selected corresponding to NH_3 gas and C_2H_2 gas, respectively, and the wavelength range was integrated in order to eliminate subtraction errors caused by a low concentration. The integral value was the change in the light intensity information corresponding to the gas concentration. According to the Lambert-Beer law, the concentrations of the C_2H_2 and NH_3 are in a linear relationship with the absorbance change ΔI. The calibration curve is shown in Fig. 10. Figure 10(a) is a linear fit chart with a scale from 200 ppm to 2 000 ppm C_2H_2, and Fig. 10(b) is a linear fit chart with a scale from 0.2% to 1% NH_3. The figure shows a good linearity of the two curves, $R^2 = 0.9987$ and $R^2 = 0.9968$. The lowest detections of the system are 20 ppm of C_2H_2 and 0.01% NH_3.

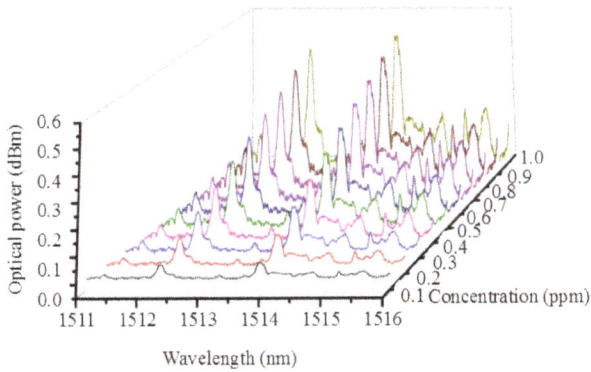

(a)

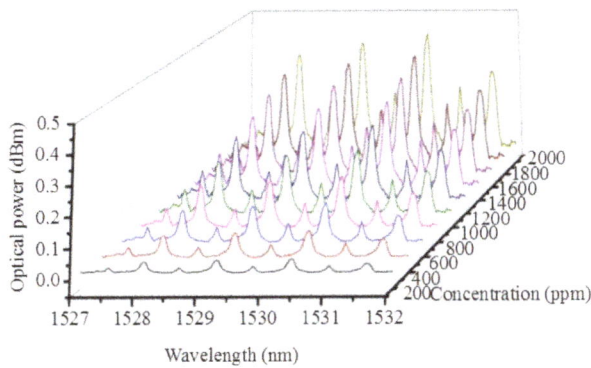

(b)

Fig. 9 Data process of the gas concentration test: (a) NH$_3$ and (b) C$_2$H$_2$.

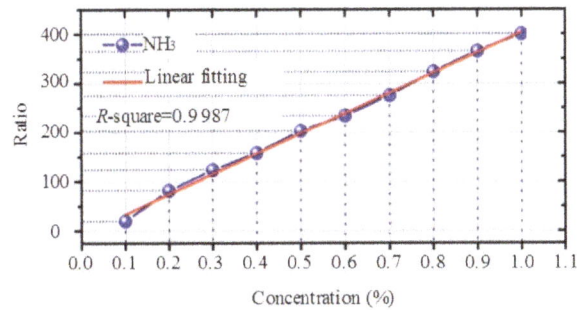

(a)

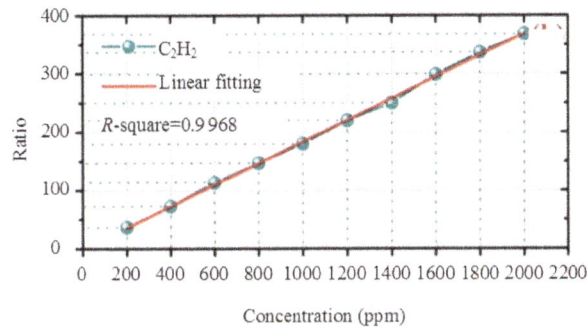

(b)

Fig. 10 Gas detection system concentration analysis: (a) NH$_3$ and (b) C$_2$H$_2$.

4. Conclusions

In this program, an optical multi-component gas sensor detection system based on the conjugated interferometer has been set up. The CI can transform the absorption spectrum of the target gases to a conjugated emission spectrum, and when combining the CI with the broadband light source, the spectrum of output light matches well with the absorption spectrum of target gases. The CI design for different target gases can be achieved by replacing the kind of target absorbing gases in the CI filter. Concerning the multi-component gas detection, the differential detection system based on the CI developed in this paper can be applied in the detection of multi-component gas detection, and the inversion of gas concentration with the output result of the system can be highly reliable as well. According to the experimental results, the system can be applicable in the multi-component gas concentration detection under the coexistence of various gases. The lowest detection of system is 20 ppm of C$_2$H$_2$ and 0.01% NH$_3$.

Acknowledgment

This research was supported by the Natural National Science Foundation of China, NSFC (Grant No. 61575149, 61290311), and the Major Project of Hubei Technological Innovation Special Fund (2016AAA008).

References

[1] R. Dhawan, M. M. Khan, N. Panwar, U. Tiwari, R. Bhatnagar, S. C. Jain, et al., "A low loss mechanical splice for gas sensing using hollow-core photonic crystal fibre," Optik – International Journal for Light and Electron Optics, 2013, 124(18): 3671–3673.

[2] S. Schilt, L. Thevenaz, M. Nikles, L. Emmenegger, and C. Huglin, "Ammonia monitoring at trace level using photoacoustic spectroscopy in industrial and environmental applications," Spectrochimica Acta Part A Molecular & Biomolecular Spectroscopy, 2004, 60(14): 3259–3268.

[3] D. Smith and P. Spanel, "The challenge of breath analysis for clinical diagnosis and therapeutic monitoring," *Analyst*, 2007, 132(5): 390–396.

[4] R. J. Lu, D. M. Shen, Q. Q. Du, B. Z. Huang, and J. S. Shi, "Tuning characteristics of DFB diode laser and its application to TDLAS gas sensor design," *Applied Mechanics & Materials*, 2014, 511: 173–177.

[5] M. Jahjah, R. Lewicki, K. F. Tittle, K. Krzempek, P. Stefanski, S. So, *et al.*, "CW DFB RT diode laser-based sensor for trace-gas detection of ethane using a novel compact multi-pass gas absorption cell," *SPIE*, 2013, 112(4): 461–465.

[6] J. Ye and Z. A. Li, "Method for the measurement of methane gas based on multi-beam interferometry," *Journal of the Optical Society of Korea*, 2013, 17(6): 481–485.

[7] H. Ding, J. Q. Liang, J. H. Cui, and X. N. Wu, "A novel fiber Fabry-Perot filter based mixed-gas sensing system," *Sensors & Actuators B: Chemical*, 2009, 138(1): 154–159.

[8] K. L. Chan, Z. Ning, D. Westerdahl, K. C .Wong, Y. W. Sun, A. Hartl, *et al.*, "Dispersive infrared spectroscopy measurements of atmospheric CO_2 using a Fabry-Perot interferometer sensor," *Science of the Total Environment*, 2014, 472: 27–35.

[9] W. Jin, S. Murray, D. Pinchbeck, G. Stewart, and B. Culshaw, "Absorption measurement of methane gas with a broadband light source and interferometric signal processing," *Optics Letters*, 1993, 18(16): 1364.

[10] J. Hodgkinson, R. Smith, W. O. Ho, J. R. Saffell, and R. P. Tatam, "Non-dispersive infra-red (NDIR) measurement of carbon dioxide at 4.2 µm in a compact and optically efficient sensor," *Sensors & Actuators B: Chemical*, 2013, 186(186): 580–588.

[11] Z. P. Zhu, Y. H. Xu, and B. G. Jiang, "A one ppm NDIR methane gas sensor with single frequency filter denoising algorithm," *Sensors*, 2012, 12(9): 12729–12740.

[12] F. Bilodeau, K. O. Hill, B. Malo, D. C. Johnson, and J. Albert, "High-return-loss narrowband all-fiber bandpass Bragg transmission filter," *IEEE Photonics Technology Letters*, 1994, 6(1): 80–82.

[13] P. Q. Liu, X. Wang, and C. F. Gmachl, "Single-mode quantum cascade lasers employing asymmetric Mach-Zehnder interferometer type cavities" *Applied Physics Letters*, 2012, 101(16): 219–221.

[14] Z. Y. Li, X. Gui, C. C. Hu, L. Zheng, H. H. Wang, and J. M. Gong, "Optical gas sensor based on gas conjugated interference light source," *IEEE Photonics Technology Letters*, 2015, 27(14): 1550–1552.

[15] HITRAN on the Web. Available online: http://hitran. iao.ru (accessed on 1 July 2012).

21

Design of a Porous Cored Hexagonal Photonic Crystal Fiber Based Optical Sensor With High Relative Sensitivity for Lower Operating Wavelength

Shuvo SEN[1], Sawrab CHOWDHURY[1], Kawsar AHMED[1, 2*], and Sayed ASADUZZAMAN[2]

[1]*Department of Information and Communication Technology, Mawlana Bhashani Science and Technology University, Santosh, Tangail-1902, Bangladesh*

[2]*Group of Biophotomatix, Bangladesh*

*Corresponding author: Kawsar AHMED   Email: kawsar.ict@mbstu.ac.bd and k.ahmed.bd@ieee.org

Abstract: In this article, highly sensitive and low confinement loss enriching micro structured photonic crystal fiber (PCF) has been suggested as an optical sensor. The proposed PCF is porous cored hexagonal (P-HPCF) where cladding contains five layers with circular air holes and core vicinity is formed by two layered elliptical air holes. Two fundamental propagation characteristics such as the relative sensitivity and confinement loss of the proposed P-HPCF have been numerically scrutinized by the full vectorial finite element method (FEM) simulation procedure. The optimized values are modified with different geometrical parameters like diameters of circular or elliptical air holes, pitches of the core, and cladding region over a spacious assortment of wavelength from 0.8 μm to 1.8 μm. All pretending results exhibit that the relative sensitivity is enlarged according to decrement of wavelength of the transmission band (O+E+S+C+L+U). In addition, all useable liquids reveal the maximum sensitivity of 57.00%, 57.18%, and 57.27% for n=1.33, 1.354, and 1.366 respectively by lower band. Moreover, effective area, nonlinear coefficient, frequency, propagation constant, total electric energy, total magnetic energy, and wave number in free space of the proposed P-HPCF have been reported recently.

Keywords: Optical sensor; relative sensitivity; porous core PCF; confinement loss; transmission band; effective area

1. Introduction

Photonic crystal fiber (PCF) is a novel invention of optical fiber technology which has ensured an immense advancement in telecom and nonlinear devices applications [1]. It consists of an arbitrary order [2] of tiny air holes which trip along the entire length of the fiber or periodic array [3]. Indexed guiding (IG) and photonic band gap (PBG) are two principal types of photonic crystal fibers according to propagation characteristics of light. In index guiding (IG), PCF has sustained superior refractive index than the cladding region with a solid core [4, 5]. In addition, light is instructed by photonic band gap principle with a large air core in PBG PCF [6]. In 1996, Knight *et al.* [7] first fabricated a hexagonal-PCF structure, and this procedure was modernized day by day to obtain better guiding properties. At the present time, researchers attained better guiding properties to utilize an octagonal [2], decagonal [8], honeycomb cladding [9], circular, and hybrid [10] designed PCF structure. Due to the promotion of fabrication technologies, high sensitivity [11], high birefringence [12], ultra-flattened dispersion [13], and high nonlinear effect [14] are achieved by scheming of PCF with altering the air hole diameters, pitch, and shape.

PCF can subdue many restrictions of conventional fibers by following a few supplemental features in design such as values of pitch, air-hole diameter, and number of rings both the core and cladding. Photonic crystal fibers become extremely admired due to low cost, small size, robustness, enhanced design freedom, and flexibility. It has been attracted much concentration for its improvable performance and enormous diversity of applications [15]. PCF can be accustomed like as switches [16], electro-optical modulators [17, 18], filters [19], and polarization converters [20]. PCFs are guaranteed a revolutionary improvement in spectroscopy [21], super continuum generation [22], and optical communication [1] applications due to its singular characteristic. Optical sensors are updated day by day with the advancement of technology and a unique geometrical structure.

Refractive index (RI) sensors [23], gas sensing [4], liquid sensors [24], pressure sensors [25], temperature sensors [26], mechanical sensors [27], chemical sensors, and P^H sensors [28] are the broad range implementation of PCF sensors. According to fascinate attraction in chemical and biomedical [28, 29] applications, the evanescent wave based PCF sensors are expanding quickly. The evanescent field of PCFs is prevalently entangled in gas sensing with chemical and bio sensing [28–30] and different index materials [23]. In addition, the bacteria sensor [30] is accustomed with the evanescent wave based PCF sensor. To subdue the safety problem in the

industrial adaptation [31] particularly for perception of toxic and flammable chemicals, noble sensitive chemical (liquid and gas) sensors are performed a significant rule. PCF based sensors are also attracted much concentration of researchers in environmental and safety monitoring [23, 31] applications.

In order to maximize the relative sensitivity and reduce confinement loss at adequate degree in liquid [24] or chemical sensing [28] applications, the researchers has been reported plenty of articles by replacing distinctive geometric parameters of the PCF. Lee and Asher [28] sensed echanical sensors [27], chemical sensors, and P^H level and ionic strength by using a PCF based chemical sensor. In 2001, Hansen *et al.* [32] reported a highly birefringence index guiding PCF which depicted a little bit higher birefringence contrast with conventional fibers. A hybrid-PCF structure is constituted by three rings cladding of circular air holes and micro structured core with elliptical air holes which depicts high birefringence, and sensitivity and low confinement loss was proposed by Asaduzzaman *et al.* [10]. In [33], it was recommended that lower confinement loss, dispersion, and nonlinear effect are pointed out by a nanostructure index guiding PCF.

In this article, a porous cored hexagonal photonic crystal fiber (P-HPCF) has been suggested which reveals high sensitivity and low confinement loss for three thermo optical coefficient like water (*n*=1.330), ethanol (*n*=1.354), and benzene (*n*=1.366). Our proposed P-HPCF is formed by five layered circular air holes based cladding and two layered porous cored with elliptical air holes. The geometrical parameters have been diverse with both the optimized structure core and cladding territory of proposed P-HPCF and improved the relative sensitivity with low confinement loss at the same time compared with prior PCFs [10, 34, 35].

2. Geometries of the proposed P-HPCF

The schematic cross sectional scenery of the proposed P-HPCF is presented in Fig. 1. In addition, this figure is plainly outlined the entire geometrical structure of the PCF. The proposed P-HPCF structure is hexagonal where the core is porous shaped. In the cladding region, the vertices of the attaching air cavities contain a 60° angle that forms hexagonal shape.

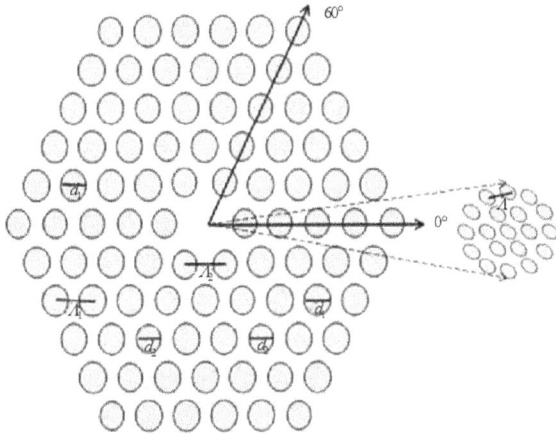

Fig. 1 Schematic cross sectional view of the proposed porous cored hexagonal photonic crystal fiber (P-HPCF) and enlarged view of its 2-layer porous core.

The first, second, third, fourth and fifth layers are restrained with 6, 12, 18, 24, and 30 air cavities, respectively. Pitch can be determined by the distance between the two air holes (at cladding) for entire types of fibers and procedures which are denoted by Λ_1 and Λ_2. The diameters of air cavities in every layer of cladding are presented by d_1 and d_2. At the core territory, the vertices of the adjoining 6 elliptical air holes of the first layer contain a 60° angle which is arranged in a porous shape. Due to the second layers, the vertices of the adjoining elliptical air holes contain a 30° angle, and the number of elliptical air holes is 12. All elliptical air holes in the core region are arranged in a 135° angle perspective to original position. As a result, the proposed PCF design cannot be collapsed. Utilizing the Sellmeier equation's [36] refractive index (n) and absolute silica as background, material has been chosen for all kinds of fibers. The diameter and pitch of the complementary elliptical air holes of porous cored are marked by d_{c1}, d_{c2}, and Λ, respectively. All supplementary elliptical air holes are filled with three different thermo optic coefficients like water

($n=1.33$), ethanol ($n=1.354$), and benzene ($n=1.366$) respectively at the core territory.

3. Synopsis of numerical method

All numerical results of the proposed P-HPCF can be examined by applying the full vectorial finite element method (FEM). To model and improve photonic ingredients or devices for engineers, perfectly matched layers (PML) boundary condition is referred as the most effective numerical approach [37–41]. According to the boundary condition, PML is fixed at 10% of the entire diameter of the proposed P-HPCF [42]. The thickness of circular PML is chosen as 1.1 μm. The propagation characteristics such as relative sensitivity and the confinement loss are obtained from Maxwell's equations [43].

The power of light can be infiltrated of the cladding territory from the core because the finite number of air cavities of a photonic crystal fiber is considered as the confinement loss which is denoted by L_c. Kaneshima et al. [44] expressed the confinement loss (dB/m) by

$$L_c = 8.68k_0\text{Im}\,[n_{\text{eff}}] \tag{1}$$

where $k_0=2\pi\lambda$; λ is the wavelength of light, and Im[n_{eff}] is known as the imaginary part of the refractive index. The relative sensitivity coefficient r can be computed by the mutual action between light and the analyte [45]. To estimate the relative sensitivity, coefficient r is enforced by the following equation:

$$r = \frac{n_r}{n_{\text{eff}}} f \tag{2}$$

where n_r and n_{eff} are marked as the refractive index of sensed material within the air cavities and the modal effective index. The fraction of air cavity power and the absolute power percentage can be calculated through (3) and represented by f:

$$f = \frac{\int_{\text{sample}} \text{Re}\left(E_x H_y - E_y H_x\right)dxdy}{\int_{\text{total}} \text{Re}\left(E_x H_y - E_y H_x\right)dxdy} \times 100 \tag{3}$$

where E_x and H_x represent the diagonal electric field

and magnetic field. E_y and H_y represent the longitudinal electric field and magnetic field. The FEM is implemented to procure the mode field pattern E_x, H_x, E_y, H_y, and formal effective index n_{eff} correspondingly. The effective area A_{eff} [46] of a PCF can be calculated by the following equation:

$$A_{eff} = \frac{\left(\iint |E|^2 \, dxdy\right)^2}{\iint |E|^4 \, dxdy} \qquad (4)$$

where optical power is denoted by E. A small effective area is caused for high optical power density which also indicates the nonlinear effect. Equation (5) is utilized to calculate the nonlinear coefficient (γ) [46]

$$\gamma = \frac{n_2 \omega}{C A_{eff}} = \frac{n_2 2\pi}{\lambda A_{eff}} \left[\omega = 2n\pi \text{ and } C = n\lambda\right] \qquad (5)$$

where C, n, and λ present the velocity, frequency, and wavelength of light, respectively. The nonlinear-index coefficient n_2 in the nonlinear part of the refractive index is $\delta_n = n_2 |E|^2$.

4. Numerical results and discussion

In this section, the guiding properties of the reported P-HPCF have been examined in agreement with assorted geometrical parameters. The complementary elliptical air cavities of core region is satiated by three thermo optic coefficients such as water ($n=1.330$), ethanol ($n=1.354$), and benzene ($n=1.366$). Numerical investigation into the P-HPCF has been carried out at a wide range of wavelength from $0.8\,\mu m$ to $1.8\,\mu m$. The extensive pretending procedure has been performed by COMSOL Multiphysics 4.2 versions. The convergence error of the proposed P-HPCF is very low which is approximately at $1.50 \times 10^{-8}\%$ for the optimum parameters.

The electric field distributions for (a) x-polarization and (b) y-polarization at an activation wavelength of $1.33\,\mu m$ have been presented in Fig. 2. The mode field is firmly restricted in the core vicinity which increases the sensitivity of the proposed P-HPCF. As a result, the leakage loss of the fiber is very low.

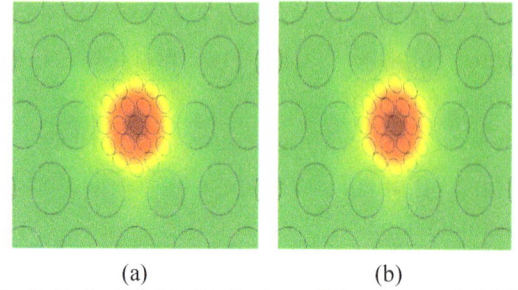

(a)   (b)

Fig. 2 Optical field distribution of the proposed P-HPCF: (a) x-polarization and (b) y-polarization for ethanol ($n=1.354$) at $\lambda=1.33\,\mu m$.

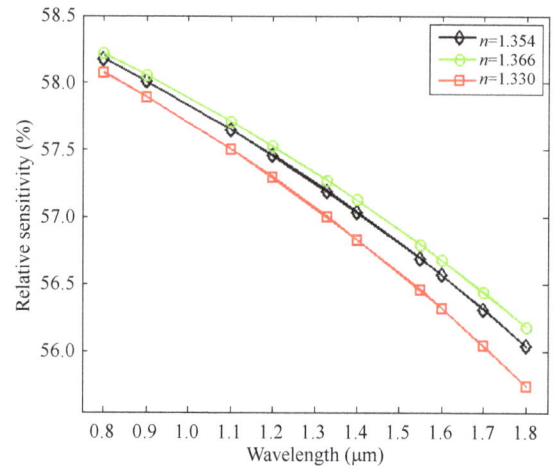

Fig. 3 Variation of the relative sensitivity versus wavelength of the proposed P-HPCF when all parameters are kept optimum.

Figure 3 reveals the influence on relative sensitivity of the proposed P-HPCF for the optimum values. This figure also indicates that the sensitivity is increased with the decrement of wavelength. The proposed P-HPCF shows conflict result compared with those shown in [10, 34, 35] due to the fact that those articles achieved high sensitivity with the increment of wavelength. The relative sensitivity of proposed P-HPCF is 57%, 57.18%, and 57.27%, respectively for three liquids like water ($n=1.330$), ethanol ($n=1.354$) and benzene ($n=1.366$) at $1.33\,\mu m$ wavelength. Besides, confinement loss of the fiber is 2.15×10^{-10} dB/m, 1.11×10^{-11} dB/m, and 1.97×10^{-11} dB/m, respectively in the same condition. In addition, the degree of freedom is $d_1=1.80\,\mu m$, $d_2=1.96\,\mu m$, $\Lambda_1=2.8\,\mu m$, $\Lambda_2=3\,\mu m$, $d_{c1}=0.40\,\mu m$, $d_{c2}=0.30\,\mu m$, and $\Lambda=0.80\,\mu m$ for the optimum structure. The optimization of the proposed P-HPCF has done following a simple technique [35]. The first cladding region is optimized.

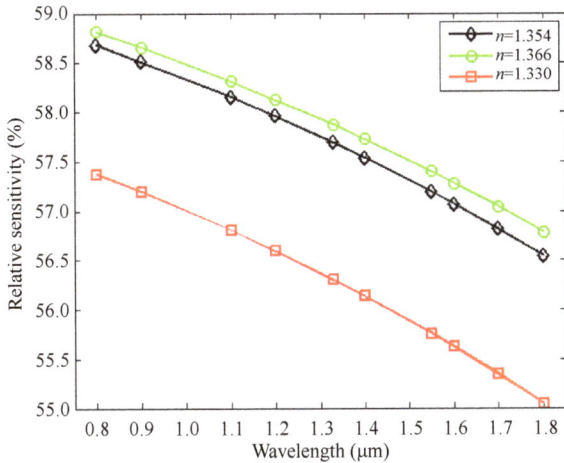

Fig. 4 Variation of the relative sensitivity versus wavelength of the P-HPCF changing the cladding diameter of +4% compared with the optimum value when other parameters are kept constant.

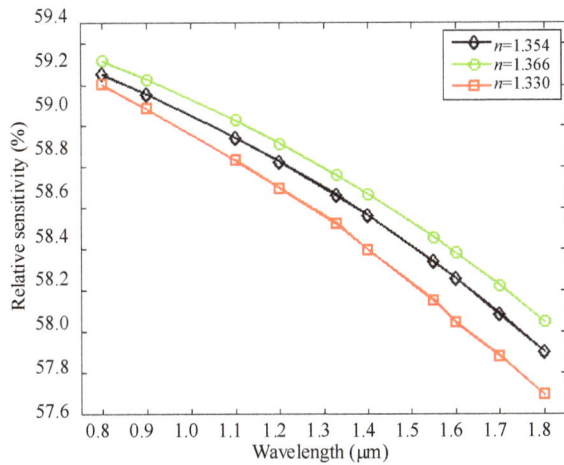

Fig. 5 Variation of the relative sensitivity versus wavelength of the P-HPCF changing the cladding diameter of +8% compared with the value when other parameters are kept constant.

Figure 4 demonstrates the effect on the relative sensitivity to fluctuate the cladding air-hole diameters (d_1 and d_2) of +4% with the optimum values when other parameters are kept same for the proposed P-HPCF. The sensitivity of 56.52%, 57.82%, and 57.96% and the confinement loss of 3.13×10^{-10} dB/m, 5.95×10^{-10} dB/m, and 2.5×10^{-10} dB/m correspondingly have gained at 1.33 μm wavelength. Now it is clearly visualized that sensitivity has increased for two liquids except water (n=1.330) compared with the optimum value.

Figure 5 shows the impact on the relative sensitivity, +8% variation with the optimum values of cladding air-hole diameters (d_1 and d_2) when

other parameters have remained constant for the proposed P-HPCF. The sensitivity of 58.37%, 58.51%, and 58.59% accordingly has been reported for three analytes at 1.33 μm wavelength which is little bit high compared with the optimum value. Besides, the confinement loss of the order of 1.92×10^{-9}dB/m, 5.80×10^{-11} dB/m, and 4.79×10^{-11} dB/m accordingly has also been achieved for desired liquids but it is also high compared with the optimum value. Same results have been found for +12% increment of cladding diameters which is shown in Fig. 6.

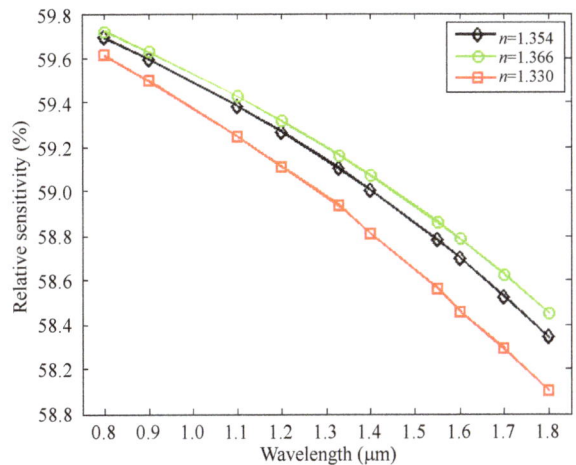

Fig. 6 Variation of the relative sensitivity versus wavelength of the P-HPCF changing the cladding diameter of +12% compared with the optimum value when other parameters are kept constant.

All examined results are shown in Figs. 4, 5, and 6. It has been signified that relative sensitivity of the proposed P-HPCF is enlarged when we increase the values of d_1 and d_2 with the optimum value. But to avoid fabrication complexity, d_1=1.80 μm and d_2= 1.96 μm have been selected as an optimum value in the cladding region.

Figure 7 demonstrates the influence on the relative sensitivity for decreasing the cladding air-hole diameters (d_1 and d_2) of –4% with the optimum values when other parameters are remained the same for the proposed P-HPCF. The sensitivity of 56.23%, 56.43%, and 56.52% and confinement loss of 1.92×10^{-9} dB/m, 5.80×10^{-11} dB/m, and 4.79×10^{-11} dB/m respectively have been achieved at 1.33 μm wavelength. Now it is clearly visualized that sensitivity has decreased due to –4% variations with the optimum value for three liquids. Similar results have been found for –8% decrement of cladding diameters in Fig. 8.

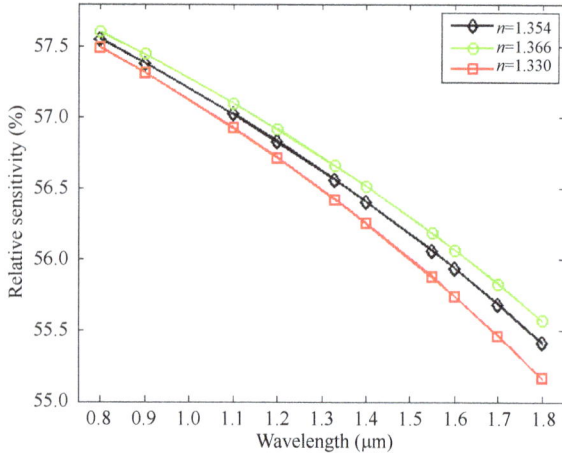

Fig. 7 Variation of the relative sensitivity versus wavelength of the P-HPCF changing the cladding diameter of –4% compared with the optimum value when other parameters are kept constant.

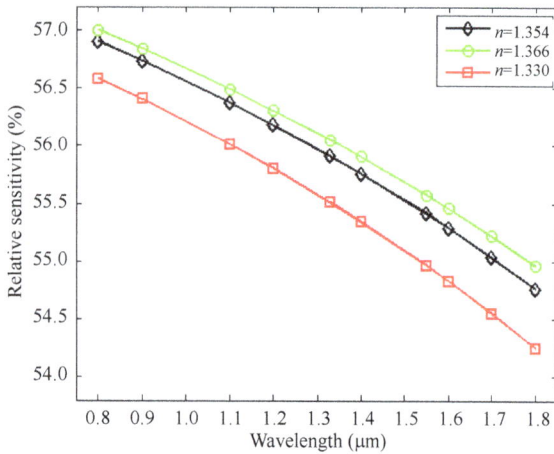

Fig. 8 Variation of the relative sensitivity versus wavelength of the P-HPCF changing the cladding diameter of –8% compared with the optimum value when other parameters are kept constant.

After completing all investigations from Figs. 7 and 8, it has been indicated that propagation characteristics like sensitivity of the reported P-HPCF is diminished due to a decrease in the optimum values of cladding air-hole diameters (d_1 and d_2). Next the core territory is optimized.

Figure 9 represents the effect on the relative sensitivity of the proposed P-HPCF for interchanging the air-hole diameters (d_{c1} and d_{c2}) of core vicinity with the optimum values. After completing this procedure, the sensitivity and confinement loss are turned into little bit changes. The sensitivity of 56.90%, 57.19%, and 57.14% and confinement loss of 2.79×10^{-10} dB/m, 4.39×10^{-10} dB/m, and 1.81×10^{-11} dB/m correspondingly are obtained at 1.33 µm wavelength for three thermo optic coefficients. Since the relative sensitivity is slightly decreased with exchanging core air-hole diameters, d_{c1}=0.40 µm and d_{c2}= 0.30 µm are chosen as the optimum values of air-hole diameters in the core region.

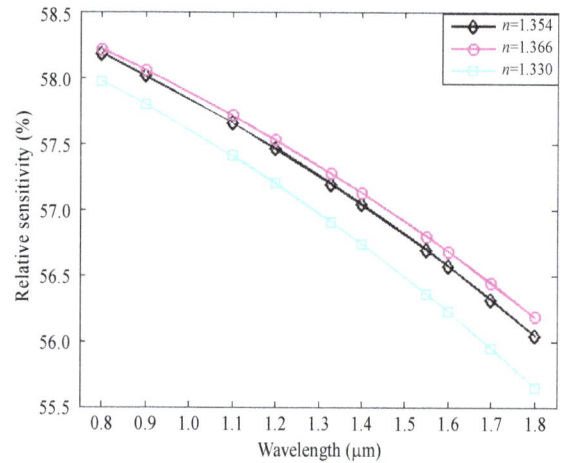

Fig. 9 Variation of the relative sensitivity versus wavelength of the P-HPCF interchanging the core vicinity air-hole diameters when other parameters are kept constant.

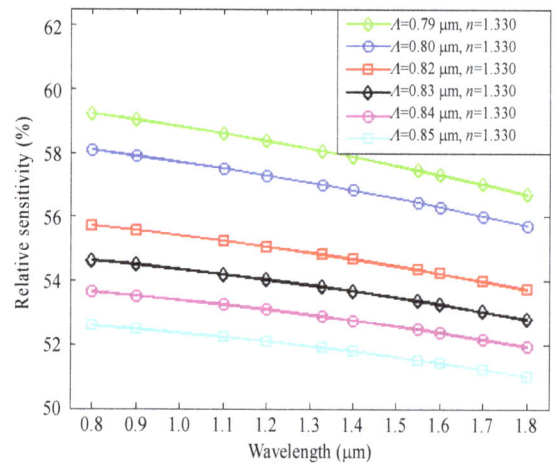

Fig. 10 Variation of the relative sensitivity versus wavelength of the P-HPCF due to different values of core pitch Λ when other parameters are remained same for n=1.33.

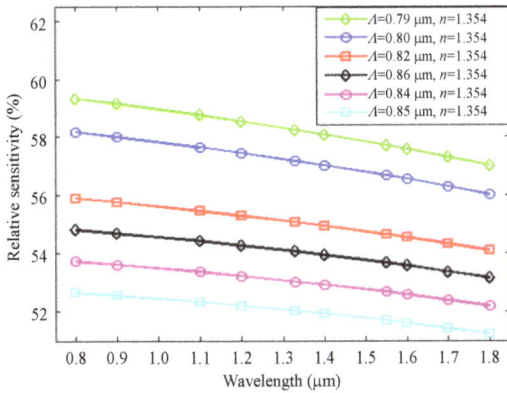

Fig. 11 Variation of the relative sensitivity versus wavelength of the P-HPCF due to different values of core pitch Λ when other parameters are remained same for n=1.354.

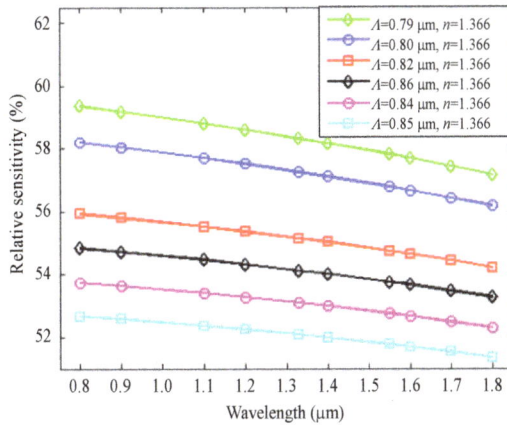

Fig. 12 Variation of the relative sensitivity versus wavelength of the P-HPCF due to different values of core pitch Λ when other parameters are remained the same for n=1.366.

Figure 10 indicates the impact on the relative sensitivity of proposed P-HPCF for distinctive values of core pitch Λ when applied liquid is water (n=1.330). The sensitivities of 58.06%, 57.00%, 54.82%, 53.80%, 52.90%, and 51.90% correspondingly have been attained due to Λ= 0.79 μm, 0.80 μm, 0.82 μm, 0.83 μm, 0.84 μm, and 0.85 μm at 1.33 μm wavelength. In addition, the confinement losses of 4.32×10^{-11} dB/m, 2.15×10^{-10} dB/m, 1.46×10^{-10} dB/m, 7.22×10^{-10} dB/m, 5.96×10^{-12} dB/m, and 3.12×10^{-11} dB/m accordingly have been gained for similar requirements.

After finishing all examinations from Figs. 10, 11, and 12, it has been revealed that the high relative sensitivity and low confinement loss are achieved for Λ=0.79 μm due to applying three chemicals like water (n=1.330), ethanol (n=1.354), and benzene

(n=1.366) at 1.33 μm wavelength. To evade cost and complexity during the fabrication process, Λ= 0.80 μm is selected as the optimum pitch value of the core vicinity.

In a selective manner, filling the air holes of photonic crystal fiber by analytes is referred as a dare job. The air cavities of PCF in either the core or cladding region can be satiated by distinctive analytes with the progress of nanotechnology. A number of technical skills have been reported by the researchers. In 2004, Huang et al. [47] proposed a unique technique to fill the micro structured air cavities of both the core and cladding vicinities with distinctive analytes which may be accustomed to reveal the functionality of the PCF applications. This filling adaptation takes place by compelling the ultraviolet-curable polymer within the PCF. Zhang et al. [48] suggested a method for practical exhibition and speculate pretending of a liquid filled core based PCF to improve the sensing application. All investigated results in Table 1 have been manifestly indicated that the relative sensitivity and confinement loss are slightly varied with the optimum parameters after all modifications. So we confirmed that no global effects can influence the reported structure afterwards the fabrication process.

We have also examined the optimum parameters of ±1% and ±2% replacement as global parameters of the proposed P-HPCF. Comparison of relative sensitivity and confinement loss among the optimum parameters and the change in global parameters for water (n=1.330), ethanol (n=1.354), and benzene (n=1.366) at λ=1.33 μm has been revealed in Table 1. Table 2 demonstrates the comparison of the effective area and nonlinear coefficient of the optimum P-HPCF structure for three applied liquids such as water, ethanol, and benzene at λ=1.33 μm.

Several numbers of guiding attributes like frequency, propagation constant, total electric energy, total magnetic energy, and wave number in the free space of the proposed P-HPCF have been shown in Table 3. After completing the calculation process, it has been seen that propagation constant, total electric energy, and total magnetic energy are changed with different refractive index chemicals such as water (n =1.330), ethanol (n=1.354), and benzene (n=1.366) except frequency and wave number in the free space.

Table 1 Comparison among the optimum parameters and the change in global parameters at λ=1.33 μm.

Change in global parameters (%)	Relative sensitivity (%)			Confinement loss (dB/m)		
	n=1.33	n=1.354	n=1.366	n=1.33	n=1.354	n=1.366
+2%	57.99	57.24	57.32	6.84×10^{-10}	2.01×10^{-10}	2.84×10^{-10}
+1%	56.93	57.21	58.16	1.21×10^{-9}	3.42×10^{-10}	8.16×10^{-11}
Optimum	57.00	57.18	57.27	2.15×10^{-10}	1.11×10^{-11}	1.97×10^{-11}
−1%	56.87	57.16	57.24	1.22×10^{-10}	8.31×10^{-10}	1.36×10^{-10}
−2%	57.79	57.13	47.79	3.98×10^{-9}	1.90×10^{-11}	3.98×10^{-9}

Table 2 Numerical values of effective area and nonlinear coefficient of the optimum P-HPCF structure at λ=1.33 μm.

Chemical (n)	Effective area (μm^2)	Nonlinear coefficient (W^{-1}km^{-1})
Water (n=1.330)	11.99	11.82
Ethanol (n=1.354)	11.85	11.95
Benzene (n=1.366)	11.78	12.02

Table 3 Different guiding values of the optimum P-HPCF structure at λ=1.33 μm.

Investigation parameters	Water	Ethanol	Benzene
Frequency (Hz)	2.25×10^{14}	2.25×10^{14}	2.25×10^{14}
Propagation constant (rad/m)	6.21×10^{6}	6.32×10^{6}	6.38×10^{6}
Total electric energy (J)	1.67×10^{-7}	1.74×10^{-7}	1.77×10^{-7}
Total magnetic energy (J)	1.67×10^{-7}	1.74×10^{-7}	1.77×10^{-7}
Wave number in free space (rad/m)	4.72×10^{6}	4.72×10^{6}	4.72×10^{6}

The comparative performance analysis between the prior PCFs and proposed P-HPCF has been listed in Table 4. Comparing with the prior PCFs, our proposed PCF depicts upward relative sensitivity and low confinement for ethanol (n=1.354) analytes. The proposed P-HPCF is nobly improved the relative sensitivity approximately 2.41, 1.30, and 1.16 times compared with the previous PCFs [34], [35], and [10] respectively as well as diminishing the confinement loss from [34] and [10].

Table 4 Comparison of simulated result and structural shape among proposed P-HPCF and prior PCFs at λ=1.33μm for n=1.354.

PCFs	Sen. (%)	Con. loss (dB/m)	No. of ring	Structural shape	
				Core	Cladding
Prior PCF$_1$ [34]	23.75	2.40×10^{-4}	3	Elliptical holes	Circular holes in octagonal
Prior PCF$_2$ [35]	46. 87	2.28×10^{-14}	5	Circular holes	Circular holes in octagonal
Prior PCF$_3$ [10]	49.17	2.75×10^{-10}	3	Elliptical holes	Circular holes in circular
Pro. PCF	57.18	1.11×10^{-11}	5	Elliptical holes in porous	Circular holes in hexagonal

The fabrication procedure is a significant topic for all proposed photonic crystal fibers. So the fabrication process of the proposed P-HPCF may be not comfortable. The proposed P-HPCF includes two different types of holes in cladding regions. Due to the technological progress in the fabrication of PCFs, our reported structure can be feasibly fabricated. Various fabrication techniques have been advanced for micro structured PCF like extrusion [49], stack and draw [50], drilling [51], and sol-gel casting [52]. Extrusion method contributes exemption in design. To utilize this method, material losses are extremely high for soft glasses, which is regarded as key drawback of this technique. The drilling technique allows fitting both the holes size and spacing. But circular shape fiber is restricted to drilling technique. In 2012, Hamzaoui *et al.* [52] fabricated an ionic copper-doped micro structured optical fiber by a new fabrication method which is renowned as a sol-gel technique. This method provides the freedom to modify air holes size, shape, and pitches. So, the proposed P-HPCF will be successfully fabricated by sol-gel casting technique.

5. Conclusions

A micro structured porous cored hexagonal photonic crystal fiber (P-HPCF) formed by five layered cladding embracing circular air holes and two layered porous coredwith elliptical air holes is reported in this paper. The proposed P-HPCF is accustomed as a chemical sensor. The FEM and PML circular boundary conditions are enforced for numerically investigated variation of effects on propagation characteristics of the proposed P-HPCF. All examined outcomes show that the proposed P-HPCF exposes decent sensitivity and lower confinement loss compared with the prior PCFs at a wide scope of wavelength from 0.8 μm to 1.8 μm. The proposed PCF reveals the relative sensitivity 57.00%, 57.18%, and 57.27% correspondingly for water (n=1.33), ethanol (n=1.354), and benzene (n=1.366) analytes. In addition, the confinement losses of 2.15×10^{-10} dB/m, 1.11×10^{-11} dB/m, and 1.97×10^{-11} dB/m are gained for the same analytes. So it is apparent that the proposed P-HPCF based chemical sensor exhibits novel execution in industrial security intention at the lower band.

References

[1] P. Russell, "Photonic crystal fibers," *Science*, 2003, 23(299): 358–362.

[2] H. Ademgil, "Highly sensitive octagonal photonic crystal fiber based sensor," *Optik–International Journal for Light and Electron Optics*, 2014, 125(20): 6274–6278.

[3] J. C. Knight, T. A. Birks, P. S. J. Russell, and D. M. Atkin, "All-silica single-mode optical fiber with photonic crystal cladding," *Optics Letters*, 1996, 21(19): 1547–1549.

[4] Y. L. Hoo, W. Jin, H. L. Ho, and D. N. Wang, "Measurement of gas diffusion coefficient using photonic crystal fiber," *IEEE Photonics Technology Letters*, 2003, 15(10): 1434–1436.

[5] M. Deng, C. Huang, D. Liu, W. Jin, and T. Zhu, "All fiber magnetic field sensor with ferrofluid-filled tapered microstructured optical fiber interferometer," *Optics Express*, 2015, 23(16): 20668–20674.

[6] J. M. Fini, "Microstructure fibres for optical sensing in gases and liquids," *Measurement Science and Technology*, 2004, 15(6): 1120–1128.

[7] J. C. Knight, T. A. Birks, P. S. J. Russell, and D. M. Atkin, "All-silica single-mode optical fiber with photonic crystal cladding," *Optics Letters*, 1996, 21(19): 1547–1549.

[8] S. M. A. Razzak, Y. Namihira, F. Begum, S. Kaijage, N. H. Hai, and N. Zou, "Design of a decagonal photonic crystal fiber for ultra-flattened chromatic dispersion," *IEICE Transactions on Electronics*, 2007, 90(11): 2141–2145.

[9] Y. Hou, F. Fan, Z. W. Jiang, X. H. Wang, and S. J. Chang, "Highly birefringent polymer terahertz fiber with honeycomb cladding," *Optik–International Journal for Light and Electron Optics*, 2013, 124(7): 3095–3098.

[10] S. Asaduzzaman, K. Ahmed, T. Bhuiyan, and T. Farah, "Hybrid photonic crystal fiber in chemical sensing," *SpringerPlus*, 2016, 5(1): 1–11.

[11] M. Morshed, M. I. Hasan, and S. M. A. Razzak, "Enhancement of the sensitivity of gas sensor based on microstructure optical fiber," *Photonic Sensors*, 2015, 5(4): 312–320.

[12] M. S. Habib, M. S. Habib, S. M. A. Razzak, and M. A. Hossain, "Proposal for highly birefringent broadband dispersion compensating octagonal photonic crystal fiber," *Optical Fiber Technology*, 2013, 19(5): 461–467.

[13] M. S. Habib, M. S. Habib, M. A. Hasan, and S. M. A. Razzak, "Single mode ultra-flat high negative residual dispersion compensating photonic crystal fiber," *Optical Fiber Technology*, 2014, 20(4): 328–332.

[14] F. Begum, Y. Namihira, S. M. A. Razzak, S. Kaijage, N. H. Hai, T. Kinjo, *et al.*, "Design and analysis of novel highly nonlinear photonic crystal fibers with ultra-flattened chromatic dispersion," *Optics Communications*, 2009, 282(7): 1416–1421.

[15] M. F. H. Arif, K. Ahmed, S. Asaduzzaman, and M. A. K. Azad, "Design and optimization of photonic crystal fiber for liquid sensing applications," *Photonic Sensors*, 2016, 6(3): 279–288.

[16] K. Nozaki, T. Tanabe, A. Shinya, S. Matsuo, T. Sato, H. Taniyama, *et al.*, "Sub-femtojoule all-optical switching using a photonic-crystal nanocavity," *Nature Photonics*, 2010, 4(7): 477–483.

[17] J. M. Brosi, C. Koos, L. C. Andreani, M. Waldow, J. Leuthold, and W. Freude, "High-speed low-voltage electro-optic modulator with a polymer-infiltrated silicon photonic crystal waveguide," *Optics Express*, 2008, 16(6): 4177–4191.

[18] Y. Gao, R. J. Shiue, X. Gan, L. Li, C. Peng, I. Meric, *et al.*, "High-speed electro-optic modulator integrated with graphene-boron nitride heterostructure and photonic crystal nanocavity," *Nano Letters*, 2015, 15(3): 2001–2005.

[19] D. Chen, "Stable multi-wavelength erbium-doped fiber laser based on a photonic crystal fiber Sagnac loop filter," *Laser Physics Letters*, 2007, 4(6): 437–439.

[20] H. Xuan, J. Ma, and W. Jin, "Polarization converters in highly birefringent microfibers," *Optics Express*, 2014, 22(3): 3648–3660.

[21] F. Benabid, F. Couny, J. C. Knight, T. A. Birks, and P. S. J. Russell, "Compact, stable and efficient all-fiber gas cells using hollow-core photonic crystal fibers," *Nature*, 2005, 434(7032): 488–491.

[22] J. M. Dudley, G. Genty, and S. Coen, "Supercontinuum generation in photonic crystal fiber," *Reviews of Modern Physics*, 2006, 78(4):

1135–1184.

[23] K. Mileńko, D. J. J. Hu, P. P. Shum, T. Zhang, J. L. Lim, Y. Wang, et al., "Photonic crystal fiber tip interferometer for refractive index sensing," Optics Letters, 2012, 37(8): 1373–1375.

[24] T. Larsen, A. Bjarklev, D. Hermann, and J. Broeng, "Optical devices based on liquid crystal photonic bandgap fibers," Optics Express, 2003, 11(20): 2589–2596.

[25] H. Y. Fu, H. Y. Tam, L. Y. Shao, X. Dong, P. K. A. Wai, C. Lu, et al., "Pressure sensor realized with polarization-maintaining photonic crystal fiber-based Sagnac interferometer," Applied Optics, 2008, 47(15): 2835–2839.

[26] C. Zhao, X. Yang, C. Lu, W. Jin, and M. S. Demokan, "Temperature-insensitive interferometer using a highly birefringent photonic crystal fiber loop mirror," IEEE Photonics Technology Letters, 2004, 16(11): 2535–2537.

[27] M. Eichenfield, J. Chan, R. M. Camacho, K. J. Vahala, and O. Painter, "Optomechanical crystals," Nature, 2009, 462(7269): 78–82.

[28] K. Lee and S. A. Asher, "Photonic crystal chemical sensors: PH and ionic strength," Journal of the American Chemical Society, 2000, 122(39): 9534–9537.

[29] N. Skivesen, A. Têtu, M. Kristensen, J. Kjems, L. H. Frandsen, and P. I. Borel, "Photonic-crystal waveguide biosensor," Optics Express, 2007, 15(6): 3169–3176.

[30] L. D. Bonifacio, D. P. Puzzo, S. Breslav, B. M. Willey, A. McGeer, and G. A. Ozin, "Towards the photonic nose: a novel platform for molecule and bacteria identification," Advanced Materials, 2009, 22(12): 1351–1354.

[31] A. M. R. Pinto and M. Lopez-Amo, "Photonic crystal fibers for sensing applications," Journal of Sensors, 2012, 2012: 1–21.

[32] T. P. Hansen, J. Broeng, S. E. Libori, E. Knudsen, A. Bjarklev, J. R. Jensen, et al., "Highly birefringent index-guiding photonic crystal fibers," IEEE Photonics Technology Letters, 2001, 13(6): 588–590.

[33] S. Olyaee, M. Seifouri, A. Nikoosohbat, and M. S. E. Abadi, "Low nonlinear effects index-guiding nanostructured photonic crystal fiber," International Journal of Chemical, Nuclear, Materials and Metallurgical Engineering, 2015, 9(2): 253–257.

[34] H. Ademgil and S. Haxha, "PCF based sensor with high sensitivity, high birefringence and low confinement losses for liquid analyte sensing applications," Sensors, 2015, 15(12): 31833–31842.

[35] K. Ahmed and M. Morshed, "Design and numerical analysis of microstructured-core octagonal photonic crystal fiber for sensing applications," Sensing and

Bio-Sensing Research, 2016, 7: 1–6.

[36] A. N. Z. Rashed, E. N. A. E. G. Mohamed, S. A. E. R. S. Hanafy, and M. H. Aly, "A comparative study of the performance of graded index perfluorinated plastic and alumino silicate optical fibers in internal optical interconnections," Optik–International Journal for Light and Electron Optics, 2016, 127(20): 9259–9263.

[37] S. E. Kim, B. H. Kim, C. G. Lee, S. Lee, K. Oh, and C. S. Kee, "Elliptical defected core photonic crystal fiber with high birefringence and negative flattened dispersion," Optics Express, 2012, 20(2): 1385–1391.

[38] N. Luan and J. Yao, "Surface plasmon resonance sensor based on exposed-core microstructured optical fiber placed with a silver wire," IEEE Photonics Journal, 2016, 8(1): 1–8.

[39] S. Haxha, A. Teyeb, F. A. Malek, E. K. Akowuah, and I. Dayoub, "Design of environmental biosensor based on photonic crystal fiber with bends using finite element method," Optics and Photonics Journal, 2015, 05(3): 69–78.

[40] B. T. Kuhlmey, B. J. Eggleton, and D. K. C. Wu, "Fluid-filled solid-core photonic bandgap fibers," Journal of Lightwave Technology, 2009, 27(11): 1617–1630.

[41] M. Vieweg, T. Gissibl, S. Pricking, B. T. Kuhlmey, D. C. Wu, B. J. Eggleton, et al., "Ultrafast nonlinear optofluidics in selectively liquid-filled photonic crystal fibers," Optics Express, 2010, 18(24): 25232–25240.

[42] S. Asaduzzaman and K. Ahmed, "Proposal of a gas sensor with high sensitivity, birefringence and nonlinearity for air pollution monitoring," Sensing and Bio-Sensing Research, 2016, 10: 20–26.

[43] F. Shi, G. Zhou, D. Li, L. Peng, Z. Hou, and C. Xia, "Surface plasmon mode coupling in photonic crystal fiber symmetrically filled with Ag/Au alloy wires," Plasmonics, 2014, 10(2): 335–340.

[44] K. Kaneshima, "Numerical investigation of octagonal Photonic crystal fibers with strong confinement field," IEICE Transactions on Electronics, 2006, 89(6): 830–837.

[45] T. Matsui, J. Zhou, K. Nakajima, and I. Sankawa, "Dispersion-flattened photonic crystal fiber with large effective area and low confinement loss," Journal of Lightwave Technology, 2005, 23(12): 4178–4183.

[46] T. Sato, S. Makino, Y. Ishizaka, T. Fujisawa, and K. Saitoh, "A rigorous definition of nonlinear parameter ¦Γ and effective area A_{eff} for photonic crystal optical waveguides," Journal of the Optical Society of America B, 2015, 32(6): 653–657.

[47] Y. Huang, Y. Xu, and A. Yariv, "Fabrication of

functional microstructured optical fibers through a selective-filling technique," *Applied Physics Letters*, 2004, 85(22): 5182–5184.

[48] Y. Zhang, C. Shi, C. Gu, L. Seballos, and J. Z. Zhang, "Liquid core photonic crystal fiber sensor based on surface enhanced Raman scattering," *Applied Physics Letters*, 2007, 90(19): 193504-1–193504-3.

[49] H. Ebendorff-Heidepriem, P. Petropoulos, S. Asimakis, V. Finazzi, R. Moore, K. Frampton, *et al.*, "Bismuth glass holey fibers with high nonlinearity," *Optics Express*, 2004, 12(21): 5082–5087.

[50] J. Broeng, D. Mogilevstev, S. E. Barkou, and A. Bjarklev, "Photonic crystal fibers: a new class of optical waveguides," *Optical Fiber Technology*, 1999, 5(3): 305–330.

[51] M. N. Petrovich, A. V. Brakel, F. Poletti, K. Mukasa, E. Austin, V. Finazzi, *et al.*, "Micro structured fibers for sensing applications," *Proc. SPIE*, 2005, 6005: 1–15.

[52] H. H. El, Y. Ouerdane, L. Bigot, G. Bouwmans, B. Capoen, A. Boukenter, *et al.*, "Sol-gel derived ionic copper-doped micro structured optical fiber: a potential selective ultraviolet radiation dosimeter," *Optics Express*, 2012, 20(28): 29751–29760.

22

Micro-Opto-Mechanical Disk for Inertia Sensing

Ghada H. DUSHAQ[*], Tadesse MULUGETA, and Mahmoud RASRAS

Department of Electrical Engineering and Computer Science, iMicro Center, MASDAR Institute, Building A1, PO Box 54224, Masdar City, United Arab Emirates

[*]Corresponding author: Ghada H. DUSHAQ   E-mail: gdushaq@masdar.ac.ae

Abstract: An optically enabled z-axis micro-disk inertia sensor is presented, which consists of a disk-shaped proof mass integrated on top of an optical waveguide. Numerical simulations show that the optical power of laser beam propagating in a narrow silicon nitride (Si_3N_4) waveguide located under the disk is attenuated in response to the vertical movement of the micro-disk. The high leakage power of the TM mode can effectively be used to detect a dynamic range of $1\,g$ – $10\,g$ (g=9.8 m/s^2). At lest, the waveguide is kept at a nominal gap of $1\,\mu m$ from the proof mass. It is adiabatically tapered to a narrow dimension of $W \times H = 350 \times 220\,nm^2$ in a region where the optical mode is intended to interact with the proof mass. Furthermore, the bottom cladding is completely etched away to suspend the waveguide and improve the optical interaction with the proof mass. The proposed optical inertia sensor has a high sensitivity of $3\,dB/g$ when a $50\,\mu m$-long waveguide is used (normalized sensitivity $0.5\,dB/\mu m^2$) for the vertical movement detection.

Keywords: Micro-opto-Mechanical system; photonic inertia sensor; hybrid integration

1. Introduction

Silicon micro-electro-mechanical systems (MEMS) devices are widely used for inertia and pressure sensing applications. Traditional MEMS inertia sensors employ large proof mass attached to springs which yield resonant frequency of few kilohertz [1, 2]. A variety of transduction mechanism has been used for sensing the proof mass displacement, which includes piezoresistive [3], tunneling [4], thermal [5], capacitive [6], and optical [7]. Optically enabled micro-accelerometers can offer high resolution detection and improve sensitivity. They are also immune to electromagnetic interference and have the potential for being integrated with electronics on the same silicon platform [1]. Such a platform can provide compact device size in addition to a low fabrication cost when produced in mass. They have been used in a wide range of applications including biomedical, industrial processes such as robotics, human-activities monitoring, and consumer electronics [8].

Typically, accelerometers quality can be specified by their sensitivity, maximum operation range, frequency response, resolution, off-axis sensitivity, and shock survival. In addition, a trade-off between the sensor's sensitivity and bandwidth should be attained. For example, low resonance frequency yields large displacements, hence resulting in a good sensor resolution but restricting its bandwidth. Capacitive accelerometer can eliminate the trade-off between the sensitivity and bandwidth by implementing a feedback circuit [9]. On the other hand, optically enabled inertia sensors are able to achieve sub nm/g resolution with smaller masses [1, 2]. However, those optical detection based systems employ optical resonators or photonics crystal cavities with a narrow transmission bandwidth. Therefore, they require tunable lasers with complex control of their resonance wavelength to align with that of the optical resonator. Consequently, these systems are complex in nature. In our system, we are addressing this issue by using only passive optical components as discussed in the optical configuration section of this paper.

In this work, we propose an optically-enabled micro-disk inertia sensor. The suspended disk shape proof mass has flexibility to move in three dimensions (3-axis) with the main focus placed on the out-of-plane movement. The movement is detected by placing birefringent waveguide under the proof mass separated by an air gap. The proof mass is designed using an inertial measurement unit (IMU) platform while the waveguide is in silicon photonics (SiPh). The proposed proof mass structure is suspended using four serpentine springs. The main focus of this work is placed on the detection of the out-of-plane displacement. Therefore, the serpentine springs are designed to provide a low spring constant and optimized to allow the maximum displacement in the out-of-plane direction. This movement is detected using birefringent suspended waveguides hybrid integrated under the proof mass. They are designed using a relatively low-index- contrast silicon nitride (Si$_3$N$_4$) waveguides where they can carry either transverse electric (TE) or transverse magnetic (TM) optical polarizations. Furthermore, the system is based on the detection of light intensity transmission modulation in a passive waveguide, therefore, it eliminates the tedious tuning of optical resonators which significantly simplify the detection method, and in addition low cost lasers can be used. The details of sensor mechanical design are described in Section 2.1. Section 2.2 explains the optical configuration optimized for the maximum sensitivity in the out of plane movement of the inertia disk.

2. Inertia sensor mechanical & optical configuration

Inertia sensors generally consist of proof mass suspended by beams which are anchored to a fixed frame and this system can be modeled by the second-order mass-damper-spring system. The out-of-plane (z) movement is detected by two sets of nano-photonic waveguides which are placed under the proof mass. Figure 1(a) shows a schematic of the waveguide designed using a relatively low index-contrast silicon nitride (Si$_3$N$_4$) platform [10]. In the proposed hybrid integrated platform, the silicon on insulator (SOI) wafer can be bonded to an IMU wafer where the initial gap between the two wafers is 1 μm as shown in Fig. 1(b).

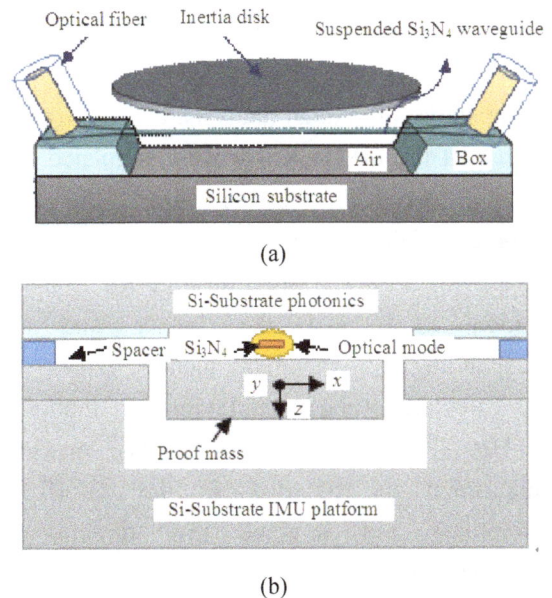

(a)

(b)

Fig. 1 Schematic representation of the proposed design of the inertia sensor: (a) a cartoon of the silicon nitrite suspended waveguide and (b) hybrid integration of SOI and IMU platform with initial gap of 1 μm.

2.1 Mechanical model

Figure 2 shows a schematic of the proposed inertia sensor. A classic serpentine springs are adopted in this design, which can offer a low spring constant and reasonable occupation area, furthermore they can be used for the in-plane as well as the out of plane displacements and they can behave like torsional spring. It is shown that the resonant frequency of the serpentine springs design is completely independent of residual stress value, while there is a large stress dependence for simple straight torsional rods with the same spring constant [11, 12]. The stiffness of serpentine springs and other beams shape can be calculated based on the standard small displacement beam theory as derived in [13, 14].

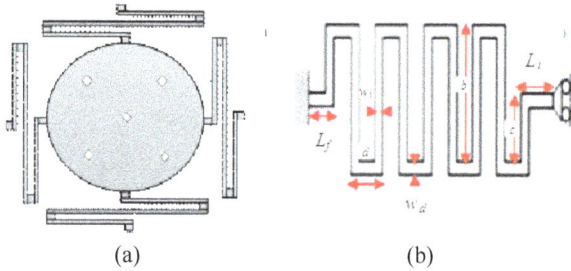

(a)   (b)

Fig. 2 Inertia sensor mechanical body: (a) schematic representation of the proposed inertia sensor and (b) serpentine spring schematic.

The static, the modal analysis, and the transient response of the inertia sensor were simulated using COMSOL Multi-Physics software. The design parameters of the sensor are summarized in Table 1. Figures 3(a), 3(b), and 3(c) illustrate the first three vibration modes of the sensor and their corresponding resonance frequency values. The three calculated fundamental vibration modes are 2.1 kHz, 3.6 kHz, and 3.6 kHz. The maximum displacement of the disk in the out-of-plane (z-direction) and in-plane (x-y direction) is calculated using a body load model ranging from $1\,g$ to $10\,g$. Figure 3(d) shows the maximum displacement of the sensor when a body force is acting on z-direction, the displacement values in the z-direction have the highest values which are consistent with the modal analysis results. This shows that the easiest energy of the system is in the z-direction, in addition it indicates a very tiny displacement in the in-plane direction under this z-loaded force recording ~1.3% cross axis sensitivity. Since the differential gap between the disk and the waveguides is restricted to 1 μm, by extrapolating the gap spacing to $16\,g$ in Fig. 3(d), the disk

displacement is $Z=1$ μm which corresponds to a zero spacing between the disk and waveguide. Therefore, $16\,g$ is the highest possible operational dynamic range of the sensor. From these numerical results, we choose the $1\,g - 10\,g$ as our sensor dynamic range to operate the system safely and avoid any collapse or stiction. Figure 3(e) shows the maximum displacement of the sensor when it is loaded by an in-plane force, clearly the displacement values are very small and consistent with the modal analysis which gives the z-direction the maximum displacement vibration value, hence this work focuses only on the z-displacement detection of an inertia sensor.

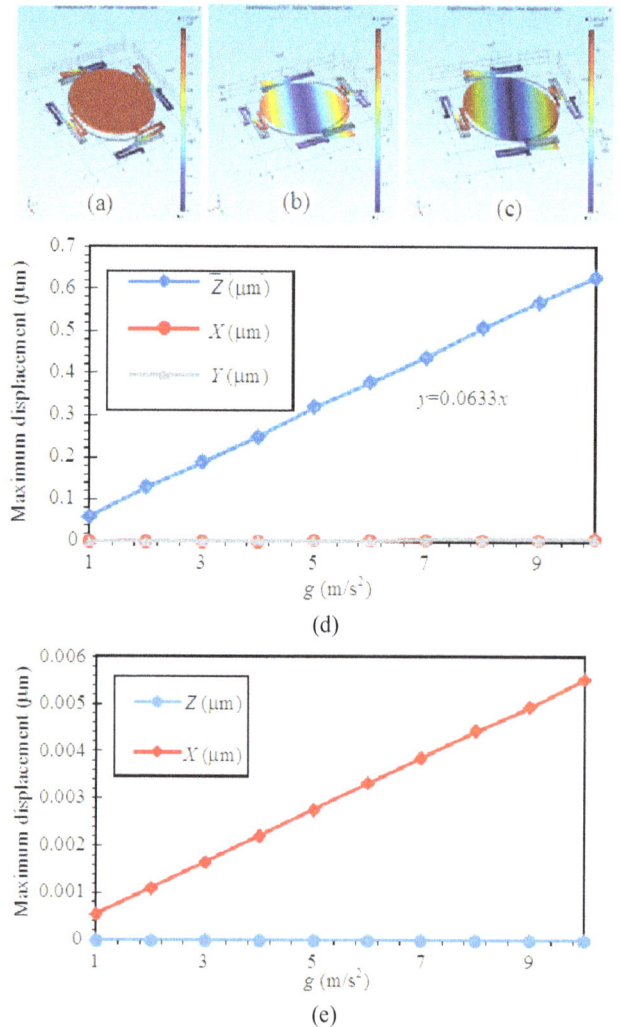

(a)   (b)   (c)

(d)

(e)

Fig. 3 Vibrational and steady state analysis of the sensor: (a) mode 1 with 2.1 kHz, (b) mode 2 with 3.6 kHz, (c) mode 3 with 3.6 kHz, (d) the maximum displacement of out-of-plane loaded inertia sensor, and (e) the maximum displacement of in-plane loaded inertia sensor.

Table 1 Design parameters of the proposed inertia sensor.

Design parameter	Value (µm)	Expression
L_i	100	The length of the initial part of the serpentine
w_i	100	The width of the initial part of the serpentine
L_f	100	The length of the final part of the serpentine
w_f	100	The width of the final part of the serpentine
w_l	80	The width of the beam
d	260	The turn length
b	2000	The beam length
c	1000	$c=b/2$
n	4	Number of turns
t	30	The thickness of the whole structure specified by the technology
R	1500	Proof mass radius

The modal analysis of the suspended Si_3N_4 beam is also carried out and the first three vibrational modes with their corresponding resonance frequencies are shown in Fig. 4. The beam has a vibrational mode in MHz range which makes it very rigid to perform any displacement in the in-plane (X-Y) or out of plane (Z) directions. Furthermore, under an external inertia force equivalent to $10\,g$, the calculated displacement in the Z-direction of the beam is ~4 pm, with a tiny displacement. Therefore, the suspended waveguide is mechanically stable. Furthermore, no fluctuation due to the optical forces can affect the system performance.

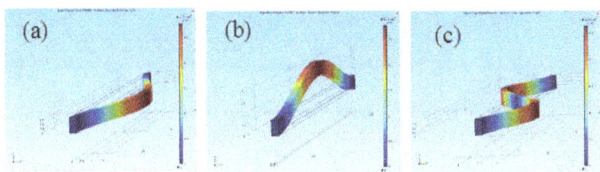

Fig. 4 The first three vibrational modes of the suspended beam (Si_3N_4 waveguide): (a) Mode 1 with 1 MHz, (b) Mode 2 with 2.2 MHz and (c) Mode 3 with 2.8 MHz.

2.2 Time response

Logarithmic decrement approach has been used to give an approximation of the sensor damping and quality factor. The approach depends on measuring the transient response of the structure when subjected to a sudden acceleration. Figure 5 shows the z-displacement of the proof mass center as a function of time when a rectangle pulse of 0.1 ms width is applied to the sensor under atmospheric condition (an air box surrounded with the structure is designed and fluid-mechanics interaction is detected in the sensor area). By measuring the ratio of any two successive amplitudes (X_1 and X_2 time difference) as shown in Fig. 5, the logarithmic decrement (δ) is calculated. Then, it can be shown that the damping ratio (ζ) is calculated as [15, 16]

$$\zeta = \frac{\delta}{\sqrt{\delta^2 + 4\pi^2}}. \tag{1}$$

By plotting $lg(X_j)$ vs j where j=1, 2, ⋯, the slope δ is calculated as 0.4. By substituting δ value in (1), this gives $\zeta = 0.07$, therefore the system is underdamped with a quality factor $Q = \dfrac{1}{2*\zeta} \sim 7$.

The proof mass settles after 0.25 ms which gives it a possibility to be implemented in vibrating analysis devices.

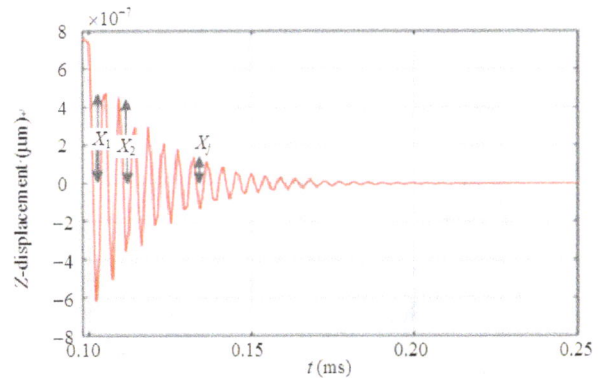

Fig. 5 Time response of inertia sensor under a sudden acceleration pulse of width 0.1 ms.

2.3 Optical model

The optical waveguides are designed using relatively low-index-contrast Si_3N_4 waveguides. This optical structure is flip-chipped on top of the IMU proof mass. In this configuration, the evanescent field of the optical waveguide interacts with the top surface of the proof mass. The larger the interaction of the optical fields with the proof mass is, the greater the scattering of the optical mode in the waveguide is, which will result in an attenuation of the optical signal.

As the mass vibrates in the out-of-plane dimensions, it will get closer or far away from the waveguide. This vibration can be detected as a modulation of the optical signal intensity. To maximize the interaction between the two platforms, the width of the waveguide is reduced, and the bottom SiO_2 cladding is completely etched away below the waveguide leaving a suspended Si_3N_4 with a top SiO_2 cladding structure. In this layout, Si_3N_4 waveguide has cross-section dimensions of $W{\times}H = 350{\times}220$ nm^2. The oxide box thickness is 2 μm. Figure 6 shows the mode shapes TE and TM of waveguide with W=0.35 μm.

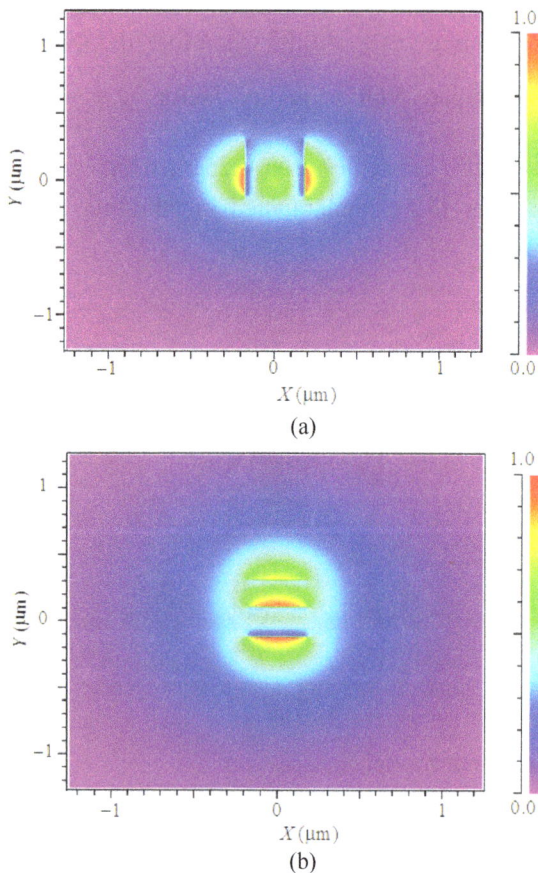

(a)

(b)

Fig. 6 Mode shapes of 0.35 μm width Si_3N_4/SiO_2 waveguide: (a) TE mode and (b) TM mode.

3. Results and discussion

Numerical simulations are used to compute the leakage of the TE and TM polarizations propagating in a 50 μm-long waveguide as a function of gap between the two wafers and for a scan of waveguide width. Simulation results of the sensitivity of the out-of-plane disk movement are shown in Fig. 7. The sensitivity of our device is defined as the attenuation

of light signal due to the mechanical movement of the disk. From Fig. 7, it can be seen that TE and TM modes have clearly distinct sensitivity behaviors, and a high detection capability up to 25 dB/μm (or normalized sensitivity 0.5 dB/μm^2) can be achieved by using a TM mode and a narrow waveguide with a width 0.25 μm. In both polarizations, the sensitivity has a low value for a gap more than 1 μm, however as the disk becomes closer to the waveguide with gap spacing below 1 μm, the TM mode becomes highly sensitive, which is also possible to calculate the gap spacing by monitoring the intensity of each polarization or the ratio between them.

Note that, in practice, a tap from the light source (~6%) can be used as a monitor of the actual optical power lunched from the laser. The variation of the signal at the output of the accelerometer waveguide due to the disk displacement is then compared to this reference monitor measurement.

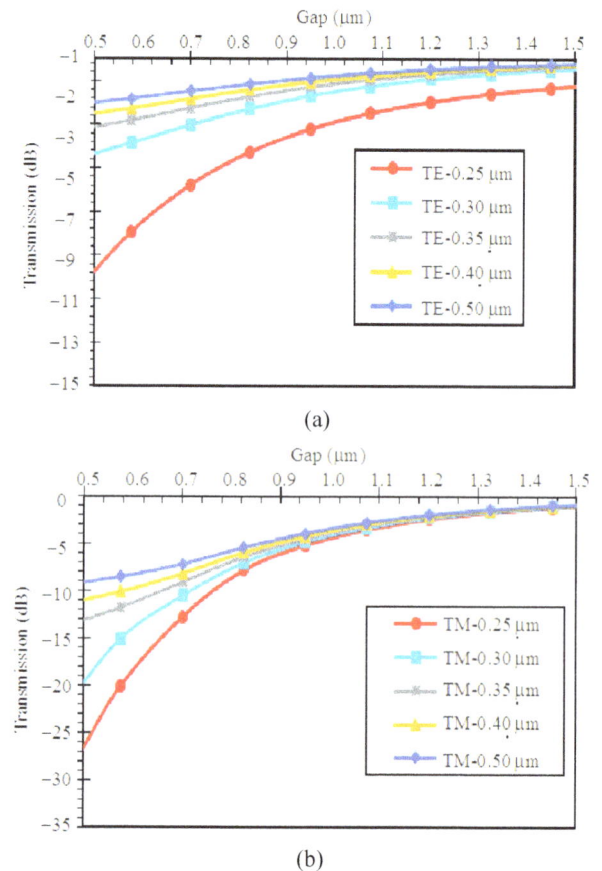

(a)

(b)

Fig. 7 Power leakage of Si_3N_4 waveguide modes at waveguide length of 50 μm: (a) TE (b) TM.

Figure 8 shows the TE and TM modes of 0.35 µm waveguide width, the gap scan is reduced to 1.1 µm to define the operational regime of the device where the sensitivity detection is the maximum. In addition, it can be seen that the range is approximately linear which makes it an optimum choice for the readout circuit in an experimental setup. From fabrication, perspective wider waveguide with 0.35 µm width has more mechanical stability and offers high sensitivity values.

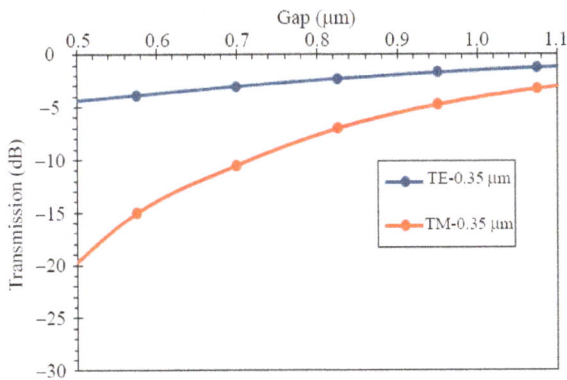

Fig. 8 Power leakage of TE and TM modes of 0.35 µm width Si$_3$N$_4$ waveguide.

TE mode can also be used for narrow gap detection where the waveguide is designed to be longer than 50 µm. The power leakage as a function of waveguide length for both 0.5 µm and 1 µm gaps of 0.35 µm waveguide width is shown in Fig. 9.

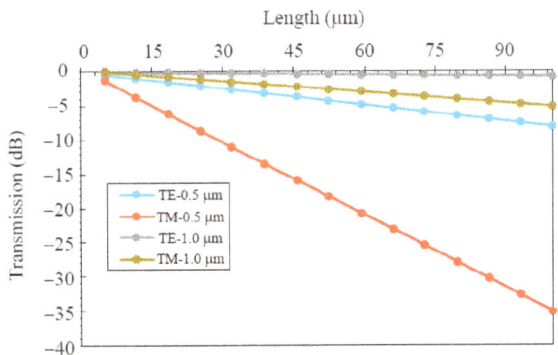

Fig. 9 Power leakage of 0.35 µm width Si$_3$N$_4$ waveguide as a function of length.

Based on the above numerical results, an optical integrated waveguide scheme for an accurate and yet a large dynamic range detection of the out-of-plane displacement is shown in Fig. 9. In this scheme, two sets of straight waveguides have been used, one of them with a TM component ($L \times W = 50 \times 0.35$ µm^2) is used to detect the course movement. On the other hand, the other straight waveguide structure ($L \times W = 100 \times 0.35$ µm^2) is dedicated to sense the fine movement of the disk. With this configuration, we can successfully measure a tiny displacement of <0.05 µm that corresponds to sub-g resolution over 10g range.

4. Conclusions

Optically enabled inertia sensor with a dynamic range up to 10 g of operation has been presented. The TE and TM light modes in relatively low-index-contrast Si$_3$N$_4$ suspended waveguide has been studied. The two light modes show different behaviors in the light intensity modulation, hence with this completely etched bottom cladding waveguide, TM mode is highly sensitive to any out of plane movement recording ~25 dB/µm change in light intensity for 0.25 µm Si$_3$N$_4$ width. The out of plane optical displacement detection and the time response behaviors of the proposed inertia sensor give the possibility to use it in the motion detection and smart user interface. In addition, low cost and high detection capability are offered in this design structure without the need to employ tunable optical resonators or photonics cavities.

Acknowledgment

This work is funded by Mubadala Development Company-Abu Dhabi, Economic Development Board-Singapore, and GLOBALFOUNDRIES-Singapore under the framework of 'Twinlab' project with participation of A*STAR Institute of Microelectronics-Singapore, Masdar Institute of Science and Technology-Abu Dhabi and GLOBALFOUNDRIES-Singapore.

References

[1] B. Dong, H. Cai, J. M. Tsai, D. L. Kwong, and A. Q. Liu, "An on-chip opto-mechanical accelerometer," in *Proceedings of IEEE Conference on Micro Electro Mechanical Systems (MEMS)*, Taipei, pp. 641–644, 2013.

[2] K. Zandi, J. A. Belanger, and Y. A. Peter, "Design and demonstration of an in-plane silicon-on-insulator optical MEMS Fabry-perot-based accelerometer integrated with channel waveguides," *Journal of Microelectromechanical Systems*, 2012, 21(6): 1464–1470.

[3] H. Seidel, U. Fritsch, R. Gottinger, J. Schalk, J. Walter, and K. Ambaum, "A peiezoresistive silicon accelerometer with monolithically integrated CMOS-circuitry," in *the 8th International Conference on Solid-State Sensors and Actuators, 1995 and Eurosensors IX.. Transducers '95*, Stockholm, Sweden, pp. 597–600, 1995.

[4] C. Yeh and K. Najafi, "A low-voltage bulk-silicon tunneling-based microaccelerometer," in *International Electron Devices Meeting, 1995*, Washington, DC, pp. 593–596, 1995.

[5] A .M. Leung, J. Jones, E. Czyzewska, J. Chen, and B. Woods, "Micromachened accelerometer based on convection heat transfer," in *Proceeding of IEEE Micro Electro Mechanical Systems Workshop (MEMS'98)*, Heidelberg, Germany, pp. 627–630, 1998 .

[6] C. Sun, C. Wang, and W. Fang, "On the sensitivity improvement of CMOS capacitive accelerometer," *Sensors & Actuators A Physical*, 2008, 14(12): 347–352.

[7] D. N. Hutchison and S. A. Bhave, "Z-axis optomechanical accelerometer," in *Proceeding of IEEE Conference on Micro Electro Mechanical Systems (MEMS)*, Paris, France, pp. 615–619, 2012.

[8] N. Yazdi, F. Ayazi, and K. Najafi, "Micromachined inertia sensors," *Proceeding of the IEEE*, 1998, 86(8): 1640–1659.

[9] B. E. Boser and R. T. Howe, "Surface micromachined accelerometer," *IEEE Journal of Solid-State Circuits*, 1996, 31(3): 366–375.

[10] J. F. Bauters, M. J. R. Heck, D. John, D. Dai, M. C. Tien, J. S. Barton, *et al.*, "Ultra-low-loss high-aspect-ratio Si_3N_4 waveguides," *Optics Express*, 2011, 19(4): 3163–3167.

[11] V. A. Aksyuk, M. E. Simon, F. Pardo, S. Arney, D. Lopez and A.Villanueva, A 2002 Optical MEMS design for telecommunications applications Solid-State Sensor, Actuator and Microsystem Workshop (Hilton Head Island, SC, 2002).

[12] G. Barillaro, A. Molfese, A. Nannini, and F. Pieri, "Analysis, simulation and relative performances of two kinds of serpentine spring," *Journal of Micromechanics & Microengineering*, 2005, 15(4): 736–746.

[13] L. D. Landau, L. P. Pitaevskii, A. M. Kosevich, and E. M. Lifshitz, *Theory of elasticity*. Massachusetts: Addison-Wesley, Inc. Reading, 1959.

[14] G. K. Fedder, "Simulation of micro-electromechanical systems," Ph.D. dissertation, Dept. University of California, Berkeley, 1994.

[15] M. I. Younis, *MEMS linear and nonlinear statics and dynamics*. Berlin: Springer, 2011.

[16] W. T. Thomoson and M. D Dahleh, *Theory of vibration with applications*. New Jersey: Prentice Hall, 1998.

23

Bridge Continuous Deformation Measurement Technology Based on Fiber Optic Gyro

Weibing GAN[1*], Wenbin HU[2], Fang LIU[2], Jianguang TANG[2], Sheng LI[2], and Yan YANG[1]

[1]*Key Laboratory of Fiber Optic Sensing Technology and Information Processing of Ministry of Education, Wuhan University of Technology, Wuhan, 430070, China;*

[2]*National Engineering Laboratory for Fiber Optic Sensing Technology, Wuhan University of Technology, Wuhan, 430070, China*

*Corresponding author: Weibing GAN   Email: ganweibing@whut.edu.cn; gwb_cxd@163.com

Abstract: Bridge is an important part of modern transportation systems and deformation is a key index for bridge's safety evaluation. To achieve the long span bridge curve measurement rapidly and timely and accurately locate the bridge maximum deformation, the continuous deformation measurement system (CDMS) based on inertial platform is presented and validated in this paper. Firstly, based on various bridge deformation measurement methods, the method of deformation measurement based on the fiber optic gyro (FOG) is introduced. Secondly, the basic measurement principle based on FOG is presented and the continuous curve trajectory is derived by the formula. Then the measurement accuracy is analyzed in theory and the relevant factors are presented to ensure the measurement accuracy. Finally, the deformation measurement experiments are conducted on a bridge across the Yangtze River. Experimental results show that the presented deformation measurement method is feasible, practical, and reliable; the system can accurately and quickly locate the maximum deformation and has extensive and broad application prospects.

Keywords: Long span bridge; continuous deformation measurement; FOG; structural safety

1. Introduction

Bridge is a very important part of modern transportation industry, and the safety monitoring is quite necessary for the bridge structure. As science and technology develops, an increasing number of devices are used in bridge monitoring. By analyzing these collected data, people can get knowledge of working conditions of bridges not only in the past, but also in the future. During this process,

continuous deformation is essential. Deformation is the key technical parameter which judges the vertical stiffness, load-bearing capacity, and integrality of the bridge. The structure of bridge will deform under the action of the external force, and various diseases such as cracks and pre-stress losses will also lead to the bridge deformation, so the deformation is one of the most important indexes for evaluating the health status of bridge structure [1].

In recent years, the number of methods of

measuring bridge deformation has been increasing. For example, the level gauge can only measure the deformation of discrete points, and the measurement process is more complex. Another method of measuring the deformation is to take advantages of the photoelectric level of communicating tubes. Although it has high precision, it is not easy to implement because the needed objective cursor is difficultly decided in the practical measurement. The third method is using the laser measurement system and photoelectric image measurement system, which capture the location of the spot or imaging changes to get the changes of relative position through the optical system, and this method must be fixed on the bridge with the equipment as a reference point. When the device is moved, the system can not get the initial measurement datum and meet the long-term measurement request, and different weather conditions have an effect on the measurement. The fourth method is the global position system (GPS), in which the precision is the rate of "cm". It can monitor the object in the storm and realize the automatic monitoring of three dimensional (3D) coordinates. However, the system is expensive, and it can not be applied in large scale. In summary, the traditional measurement methods have the characteristics of long detection period, high implementation cost, and discrete measurement points, which are easy to omit the potential disease parts of the structure, and it is difficult to meet the requirements of large engineering structures for the rapid, continuous, and accurate deformation measurement [2−9].

In this paper, according to the characteristics of the fiber optic gyro (FOG) sensitive angular velocity, the basic principle and method of FOG for measuring deformation are derived. The continuous deformation measurement system (CDMS) based on the inertial platform is designed, and it is applied to the scale bridge model in laboratory and the actual load of a large bridge.

2. Principle of measurement

2.1 Principle

Considering the FOG can accurately measure the attitude of the carrier motion, the CDMS can push out the moving track of the carrier by collecting the information of the rotational angular velocity and the linear acceleration of the relative inertia space of the gyroscope and accelerometer. The basic principle is described as shown in Fig. 1.

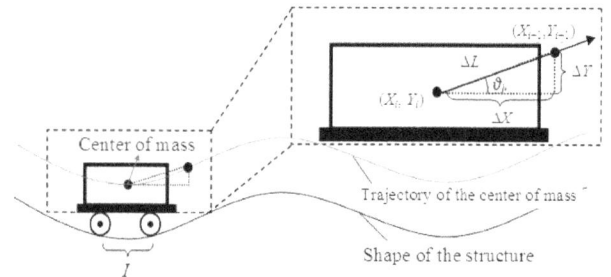

Fig. 1 Principle of CDMS based on FOG.

Here assume that the carrier moves from one point (X_i, Y_i) to another point (X_{i+1}, Y_{i+1}). According to the similarity theory of the integral limit, when the time interval t is very small, the coordinates of point (X_{i+1}, Y_{i+1}) can be approximated as [10]:

$$\begin{cases} X_{i+1} = X_i + \Delta X = X_i + \Delta L \cdot \cos\theta_{i+1} \\ \qquad = X_i + \Delta L \cdot \cos(\theta_i + \Delta\theta) \\ Y_{i+1} = Y_i + \Delta Y = Y_i + \Delta L \cdot \sin\theta_{i+1} \\ \qquad = Y_i + \Delta L \cdot \sin(\theta_i + \Delta\theta) \end{cases} \tag{1}$$

$$\Delta\theta = \int_{t_i}^{t_{i+1}} \omega_{i+1} dt, \quad \Delta L = \int_{t_i}^{t_{i+1}} v_{i+1} dt. \tag{2}$$

Among them, V_{i+1} is the line speed of the carrier and ΔL represents the minimum pulse interval. ω_i is the angular velocity, $\Delta\theta$ represents the angle change, and θ_i is the initial angle.

The CDMS algorithm assumes that the length of the carrier, L (shown in Fig. 1), is negligible in comparison with the fluctuations of the detected curve. The calculated trajectory, which is actually the trajectory of the system's center of mass, is regarded as the curve of the object. According to the established formulae (1, 2), the system can carry out the continuous curve trajectory calculation. We can get the continuous deformation curve of the bridge by comparing with the design data.

2.2 Constitution of the system

The CDMS consists of five parts, including the rigid four-wheel carrier, data acquisition system, mileage meter and interface circuit, storage battery, and fiber optic inertial navigation system (INS) as shown in Fig. 2.

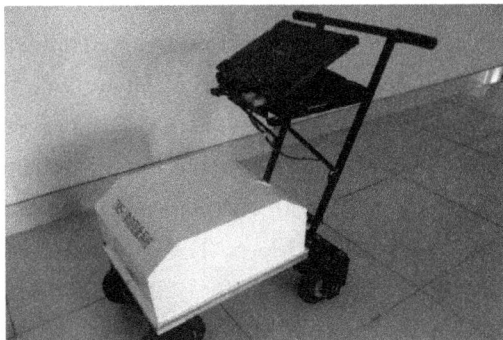

Fig. 2 CDMS based on FOG.

All devices are integrated in the rigid four-wheel carrier, and FOG completes the measurement of angular velocity when the carrier is moving. The mileage meter records the carrier's mileage in real time. In order to ease the bumps caused by uneven pavement, a shock absorber is mounted between the vehicle body and the wheel. The data acquisition system is used to complete the collection of all the devices, and the computer can finish the functions such as receiving data, analyzing the data, and displaying curves in real time by the serial port. By deducing the output information of the fiber optic INS and the mileage meter, we can get the curve trajectory of the carrier running. At the same time, the error of attitude angle and the error of mileage meter are compensated with the position information of the starting point, thus the measurement accuracy can be ensured.

2.3 Analysis of measurement accuracy

In this paper, the feasibility of FOG for measuring deformation of engineering structureis is discussed with a cable stayed bridge as an example. The distance between the two spans is D, and the height of the chord is δ. The curvature radius of the bridge is R, and the angle is θ, as shown in Fig. 3.

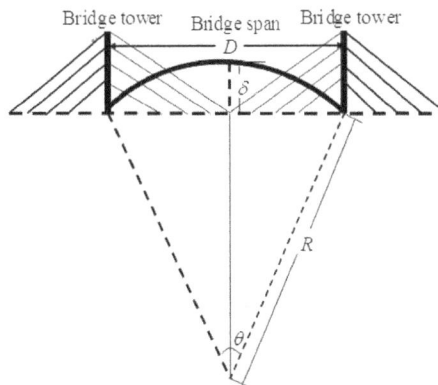

Fig. 3 Structure of cable stayed bridge.

When R is far greater than D, the minimum deformation which can be detected by FOG in theory at a given speed can be described by (3):

$$\delta_{min} = \Omega_{min} \cdot D^2 / 8V_{max}. \tag{3}$$

Among them, Ω_{min} is the zero deviation of FOG, and V_{max} is the maximum line speed of the detection carrier. δ_{min} represents the minimum deformation detected by FOG.

The selected FOG is a TXD6-A2 for a company in China, whose zero bias stability is 0.1°/h, and Ω_{min} is $4.845*10^{-7}$ rad/s. Assuming that the maximum speed of the carrier is 5 km/h and D is 500 m, then the minimum deformation value can be deduced by (3), which is 0.182 mm. The deformation value can satisfy the accuracy of the deformation measurement for the bridge structure.

2.4 Factors of ensuring measurement accuracy

Compared with the long-distance trajectory in the inertial navigation field, the CDMS has the following particularity in the detection process.

(a) The distance is short, and the test can be repeated many times.

(b) The time is short, and the system can neglect the influence of time drift and temperature shift.

(c) The size of the bridge structure is known, and the curve data can be corrected by boundary conditions.

As the FOG is very important in the measurement system, we should analyze some important parameters of the FOG, e.g. the maximum input angular rate, temperature effect, scale factor nonlinearity, and random walk coefficient. We just take a few minutes to detect a bridge, so the system can ignore the impact of time and temperature changes. In the process of carrier running, the angular velocity of FOG is between 1°/s and 5°/s, and the nonlinearity of scale factor is about 100 ppm, which has little influence on the deformation measurement of bridge. At the same time, considering the systematic error and the random walk of the gyro, the system has adopted the method of initial alignment and the calibration of the known points to ensure the accuracy of the system.

3. Test of bridge model

3.1 Test scenario

The bridge is a scale model of a Yangtze River Bridge, and the scale is 1:40. The length of the main span is 9.7 m, with a height of 3.46 m and a width of 0.55 m, and the bridge contains 56 cables. On the surface of the bridge model, two tracks are arranged to ensure that the track of the car is basically the same as the line at each time. In order to test the measurement accuracy of the system, the wireless measurement car is carried out on the bridge model for multiple load tests. At the same time, in order to compare with the deformation value between CDMS and leveling method, the dial indicators are mounted for recording deformation under various conditions of value in the three measuring points section (bridge span 1/4, center of span and bridge span 3/4). The test scenario is shown in Fig. 4.

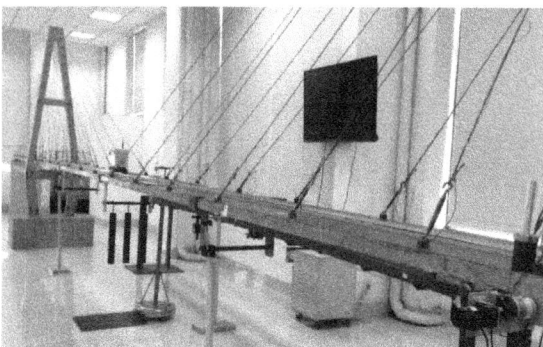

Fig. 4 Test scenario of bridge model.

3.2 Load experiment

At both ends of the bridge, the position sensors are installed for data calibration, and the total mileage of the car is 9.7 meters per time, respectively. Firstly, the car runs 3 times on the surface of the bridge model without external load, and by analyzing the data the repeatability of the system is shown in Fig. 5. Secondly, we simulate the load test by hanging 10 kg and 30 kg weights in the center of bridge span. The car runs 3 times in two cases repeatedly. After averaging the 3 trip data, we can get the curve of bridge structure before and after loading, as shown in Fig. 6. The horizontal axis is between the test point and the starting point of the distance, and the vertical coordinate is the bridge deck elevation for the relative starting point. The data comparison is shown in Table 1.

Table 1 Comparison of CDMS and indicator

Loading condition	Section position	Deformation/mm		
		CDMS	Indicator	Difference
10 kg	Span 1/4	−0.59	−0.55	−0.04
	Center of span	−1.12	−1.17	0.05
	Span 3/4	−0.36	−0.33	−0.03
30 kg	Span 1/4	−1.12	−0.19	−0.03
	Center of span	−2.70	−2.74	0.04
	Span 3/4	−1.03	−1.05	0.02

3.3 Summary

From Fig. 5, the system has good repeatability and the maximum deviation is 0.08 mm. From Fig. 6, it can be seen that the overall alignment of the bridge after loading is moved down as compared with the no-load. The deformation trend before loading is consistent with that of loading. The bridge deformation is the maximum near the load point, and it becomes smaller and smaller with the distance from the load point. This phenomenon is consistent with the load characteristics of the structure. The continuous curve also reflects that the actual maximum deformation is not located in the center of bridge span, which is related to the adjustment of cable force of the bridge model. In addition, the maximum deviation between the CDMS and the dial indicator does not exceed 0.05 mm, as shown in Table 1. The result shows that the CDMS can accurately reflect the scale bridge model loading section and the maximum deformation.

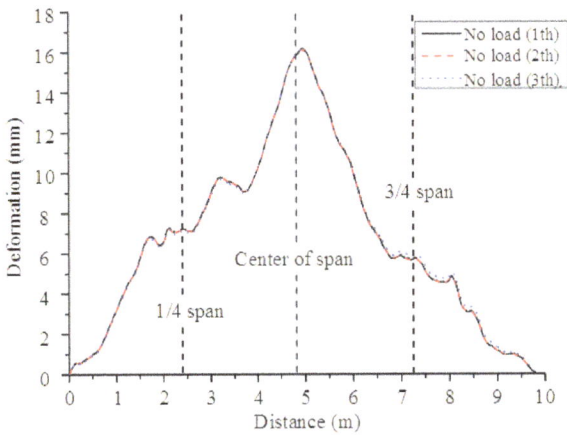

Fig. 5 Repeatability of the CDMS.

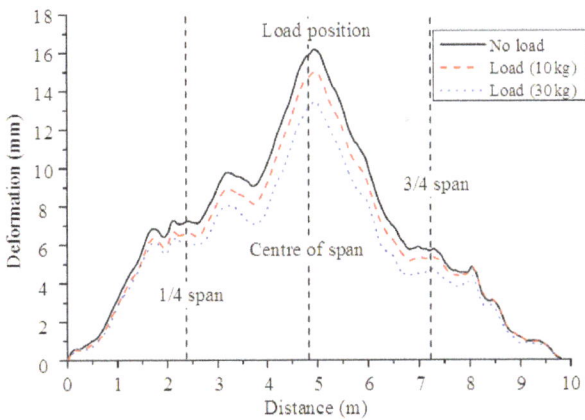

Fig. 6 Continuous curves of the bridge before and after loading.

4. Test of real bridge

4.1 Engineering Survey

A bridge across the Yangtze River is the world's longest and widest cable-stayed bridge, of which the total cost is 139 billion yuan, the length is 10 kilometers, and the width is 40.5 meters. It was designed with the standards of the eight-lane expressway, and the allowed maximum speed for vehicles is 120 km/h. The main bridge is about 2680 meters long, and it has the largest number of towers in the world. In order to ensure the safety of bridge, it is necessary to carry out the load test before the official opening. An overview of the bridge and load vehicle is shown in Fig. 7.

Fig. 7 Overview of a bridge across the river in China.

4.2 Load test

In order to further validate reliability and precision of the system in the practical application, the CDMS is used to test the continuous curve in the process of load test of a bridge, and the control points are checked by the level in partial conditions. Taking the 5#–6# pier loads as an example, there are 36 trains in the left and right load area of the working condition, and a train weighs 35 t. The whole process consists of 3 loading stages and 1 unloaded stage. The overall continuous curves and deformation values of the bridge are shown in Table 2 and Fig. 8.

Table 2 Z5#–Z6# load in the center of bridge.

Test position		Deformation (mm)			
		First stage	Second stage	Third stage	Un-load
Z4-Z5	Left	0.119	0.133	0.228	−0.023
Span	Right	0.107	0.175	0.223	−0.026
Z5-Z6	Left	−0.241	−0.493	−0.697	−0.006
Span	Right	−0.409	−0.622	−0.732	−0.002
Z6-Z7	Left	0.089	0.140	0.227	−0.051
Span	Right	0.103	0.184	0.260	−0.052

Note: the negative values in the table indicate the deformation of the beam.

From Fig. 8 and Table 2, with an increase in the vehicle load, the deformation of the cross section is gradually increasing, and the back arch is presented in the left and right sides of the loading area. The deformation of Z4-Z5 span and Z6-Z7 span is relatively small. After unloading, the overall alignment of the bridge is basically returned to the no-load state. The phenomenon is consistent with the structural deformation characteristics of the bridge.

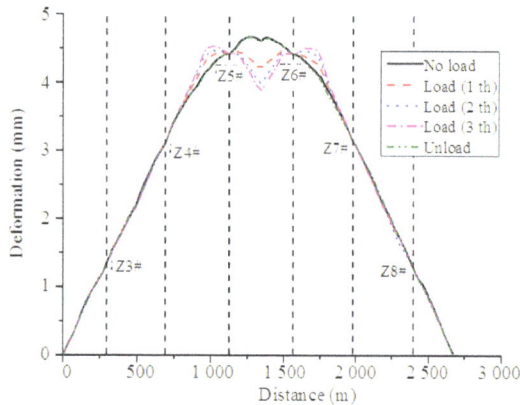

Fig. 8 Continuous curve and deformation of bridge.

At the same time, in order to test the accuracy of the system on the real bridge, the control points are checked by the level gauge in one case, and the comparison results are shown in Fig. 9 below.

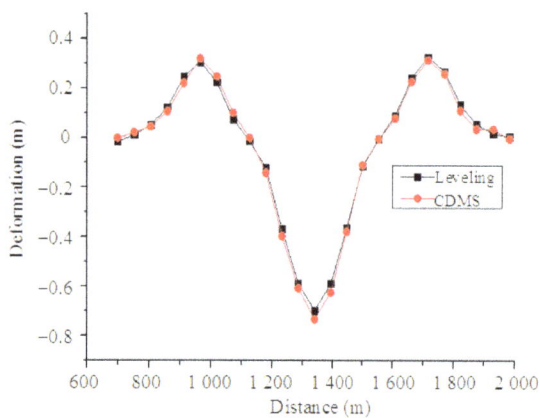

Fig. 9 Contrast curve of CDMS and level meter.

From Fig. 9, the dot represents the value of elevation obtained by the CDMS, and the square represents the value by the level meter. The measured values of the two methods are relatively close. It shows that the detection accuracy of the CDMS is almost the same as the level meter.

4.2 Summary

From Fig. 8, the deformation of the main girder is symmetric and consistent with the symmetric loading and the structure of the bridge. After unloading, the bridge alignment is basically recovered, which shows that the structure is of good elasticity. From Table 2, it can be known that the maximum deviation of the measured deformation is 0.0588 m.

From Fig. 9, the maximum deviation is 0.0688 meters between the measured deformation and the curve system measured by leveling. The deviation may be caused by random errors of bridge deck pavement and test. As for the bridge, when the load deformation reaches meter level, the measurement accuracy meets the engineering requirements.

5. Conclusions

The system detects the surface of bridge structure in the process of carrier movement, and no sensor is needed to be installed on the structure surface. In the detection process, there is no need to seal the road, without affecting the traffic, and the system has the features of convenient, fast, continuous measurement, and high accuracy. This system can accurately locate the maximum deformation of bridge span, and it has incomparable advantages especially in the linear detection of large span bridge.

Acknowledgement

This work was supported by the National Natural Science Foundation of China (NO. 61402345), National Science and Technology Support Program (2014BAG07B01), and Fundamental Research Funds for the Central Universities Special Fund (WUT: 2014-II-012).

References

[1] D. Jiang, D. Sun, and L. Liang, "Slab deflection measurement technique based on fiber optic gyro," in *Proc. SPIE*, vol. 4077, pp. 141–144, 2000.

[2] B. F. Spencer, "Opportunities and challenges for smart sensing technology," in *First International Conference on Structural Health Monitoring and Intelligent Infrastructure*, Japan, pp. 65–71, 2003.

[3] J. Ou and H. Li, "Wireless sensor information fusion for structural health monitoring," in *Proc. SPIE*, vol. 5099, pp. 356–362, 2003.

[4] D. V. Jáuregui, K. R. White, C. B. Woodward, and K. R. Leitch, "Noncontact photogrammetric measurement of vertical bridge deflection," *Journal of Bridge Engineering*, 2003, 8(4): 212–222.

[5] Y. Yu and J. Ou, "Design of wireless intelligent sensor for structural health monitoring," in *IEEE Sensor Networks and Information Processing Conference*, Hongkong, pp. 1–5, 2004.

[6] A. Nickitopoulou, K. Protopsalti, and S. Stiros, "Monitoring dynamic and quasi-static deformations of large flexible engineering structures with GPS: accuracy, limitations and promises," *Engineering Structures*, 2006, 28(10): 1471–1482.

[7] G. Liu, "Research about low frequency dynamic characteristics of deflection testing system for LianTongGuan type bridge," Dissertation for the Degree of Master of Chongqing University, China, 2007.

[8] S. Jang, H. Jo, S. Cho, K. Mechitov, J. A. Rice, S. Sim, *et al.*, "Structural health monitoring of a cable-stayed bridge using smart sensor technology: deployment and evaluation," *Smart Structures and Systems*, 2010, 6(5–6): 439–459.

[9] Y. Yu, J. Ou, and H. Li, "Design, calibration and application of wireless sensors for structural global and local monitoring of civil infrastructures," *Smart Structures and Systems*, 2010, 6(5): 641–659.

[10] S. Li, W. Hu, Y. Yang, F. Liu, and W. Gan, "Research of FOG-based measurement technique for continuous curve modes of long span bridge," *Bridge Construction*, 2014, 44(5): 69–74.

Permissions

All chapters in this book were first published in PS, by Springer International Publishing AG.; hereby published with permission under the Creative Commons Attribution License or equivalent. Every chapter published in this book has been scrutinized by our experts. Their significance has been extensively debated. The topics covered herein carry significant findings which will fuel the growth of the discipline. They may even be implemented as practical applications or may be referred to as a beginning point for another development.

The contributors of this book come from diverse backgrounds, making this book a truly international effort. This book will bring forth new frontiers with its revolutionizing research information and detailed analysis of the nascent developments around the world.

We would like to thank all the contributing authors for lending their expertise to make the book truly unique. They have played a crucial role in the development of this book. Without their invaluable contributions this book wouldn't have been possible. They have made vital efforts to compile up to date information on the varied aspects of this subject to make this book a valuable addition to the collection of many professionals and students.

This book was conceptualized with the vision of imparting up-to-date information and advanced data in this field. To ensure the same, a matchless editorial board was set up. Every individual on the board went through rigorous rounds of assessment to prove their worth. After which they invested a large part of their time researching and compiling the most relevant data for our readers.

The editorial board has been involved in producing this book since its inception. They have spent rigorous hours researching and exploring the diverse topics which have resulted in the successful publishing of this book. They have passed on their knowledge of decades through this book. To expedite this challenging task, the publisher supported the team at every step. A small team of assistant editors was also appointed to further simplify the editing procedure and attain best results for the readers.

Apart from the editorial board, the designing team has also invested a significant amount of their time in understanding the subject and creating the most relevant covers. They scrutinized every image to scout for the most suitable representation of the subject and create an appropriate cover for the book.

The publishing team has been an ardent support to the editorial, designing and production team. Their endless efforts to recruit the best for this project, has resulted in the accomplishment of this book. They are a veteran in the field of academics and their pool of knowledge is as vast as their experience in printing. Their expertise and guidance has proved useful at every step. Their uncompromising quality standards have made this book an exceptional effort. Their encouragement from time to time has been an inspiration for everyone.

The publisher and the editorial board hope that this book will prove to be a valuable piece of knowledge for researchers, students, practitioners and scholars across the globe.

List of Contributors

Ekaterina V. Loginova and Tatyana V. Zhidkova
M.V. Lomonosov Moscow State University, Chemistry Department, Moscow, 119991, Russia

Mikhail A. Proskurnin
M.V. Lomonosov Moscow State University, Chemistry Department, Moscow, 119991, Russia
Agilent Technologies Partner Laboratory / M.V. Lomonosov Moscow State University, Analytical Center, Moscow, 119991, Russia

Vladimir P. Zharov
Philips Classic Laser Laboratories, University of Arkansas for Medical Sciences, Little Rock, Arkansas, 72205, USA

Sihem Ben Zakour
Higher Institute of Management Tunis, University of Tunis, Tunisia

Hassen Taleb
Higher institute of Business and Accounting Bizerte, University of Carthage, Tunisia

Jun Chang, Jie Lian, Qiang Wang and Wei Wei
School of Information Science and Engineering and Shandong Provincial Key Laboratory of Laser Technology and Application, Shandong University, Jinan, 250100, China

Yubin Wei
School of Information Science and Engineering and Shandong Provincial Key Laboratory of Laser Technology and Application, Shandong University, Jinan, 250100, China
Laser Research Institute of Shandong Sciences Academy, Jinan, 250014, China

Wenxue Chen, Zhibin Ni, Xinhan HU and Xiaofeng Lu
Equipment Academy of the Rocket Force, Beijing, 100094, China

Fukun Bi, Tong Zheng and Hongquan Qu
College of Information Engineering, North China University of Technology, Beijing, 100144, China

Liping Pang
School of Aeronautic Science and Engineering, Beijing University of Aeronautics and Astronautics, Beijing, 100191, China

Susana Silva, Regina Magalhães, M. B. Marques, and O. Frazão
1INESC TEC, Rua do Campo Alegre 687, 4169-007 Porto, Portugal and Dept. de Física e Astronomia da Faculdade de Ciências da Universidade do Porto, Rua do Campo Alegre 687, 4169-007 Porto, Portugal

Rosa Ana Pérez-Herrera and Manuel Lopez-Amo
Public University of Navarra, Dept. of Electric and Electronic Engineering, Campus de Arrosadía, Pamplona, Spain

Fereshteh Mohammadi Nafchi
Department of Electrical Engineering, Majlesi Branch, Islamic Azad University, Majlesi, Isfahan, Iran

Sharifeh Shahi, Hossein Noormohammadi and Mohammad Kanani
Dental Biophotonics and Laser Research Center (DBLRS), Khorasgan (Isfahan) Branch, Islamic Azad University, Arghavanieh, Isfahan, Iran

Mohammad Taha Shaffaatifar
Department of Computer Engineering, Najafabad Branch, Islamic Azad University, Najafabad, Isfahan, Iran

Qian Cao, Long Jin, Yizhi Liang, Linghao Cheng, and Bai-Ou Guan
Guangdong Provincial Key Laboratory of Optical Fiber Sensing and Communications, Institute of Photonics Technology, Jinan University, Guangzhou, 510632, China

Md. Faizul Huq Arif, Kawsar Ahmed and Sayed Asaduzzaman
Department of Information and Communication Technology (ICT), Mawlana Bhashani Science and Technology University (MBSTU), Tangail-1902, Bangladesh

Md. Abul Kalam Azad
Department of Material and Metallurgical Engineering (MME), Bangladesh University of Engineering and Technology University (BUET), Dhaka-1000, Bangladesh

Xi Chen, Jun Chang, Fupeng Wang, Wei Wei, Yuanyuan Liu and Zengguang Qin
School of Information Science and Engineering and Shandong Provincial Key Laboratory of Laser Technology and Application, Shandong University, Jinan, 250100, China

Zongliang Wang
School of Physics Science and Information Technology and Shandong Key Laboratory of Optical Communication Science and Technology, Liaocheng University, Liaocheng, 252059, China

Fukun Bi, Chong Feng, Hongquan Qu and Tong Zheng
School of Electrical and Information Engineering, North China University of Technology, Beijing, 100144, China

Chonglei Wang
Beijing Institute of Technology Department of Information and Electronic, Beijing, 100081, China

Deming Liu, Qizhen Sun, Ping Lu, Li Xia, and Chaotan Sima
School of Optical and Electronic Information, Huazhong University of Science and Technology; National Engineering Laboratory for Next Generation Internet Access System, Wuhan, 430074, China

Chen Wang, Ying Shang, Xiaohui Liu and Chang Wang
Shandong Provincial Key Laboratory of Optical Fiber Sensing Technologies, Laser Institute of Shandong Academy of Sciences, Jinan, 250014, China

Hongzhong Wang
Shengli Oilfield Xinsheng Geophysical Technology Co. Ltd., No. 23 Xingfu Road, Dongying, 257086, China

Gangding Peng
School of Electrical Engineering & Telecommunications, The University of New South Wales, NSW, 2052, Australia

Ruquan Xu, Huiyong Guo, and Lei Liang
National Engineering Laboratory for Fiber Optic Sensing Technology, Wuhan University of Technology, Wuhan, 430070, China

Pinggang Jia
Science and Technology on Electronic Test & Measurement Laboratory, North University of China, Taiyuan, 030051, China

Qianyu Ren Yingping Hong and Jijun Xiong
Science and Technology on Electronic Test & Measurement Laboratory, North University of China, Taiyuan, 030051, China
Key Laboratory Instrumentation Science & Dynamic Measurement, Ministry of Education, North University of China, Taiyuan, 030051, China

Junhong LI
Department of Automation, Shanxi University, 030006, China

Chen Wu, Yongjun Wang, Lina Wang, and Fu Wang
School of Electronic Engineering, Beijing University of Posts and Telecommunications, Beijing, 100876, China

Jun YE, Jiangming Xu, Hanwei Zhang and Pu Zhou
College of Optoelectronic Science and Engineering, National University of Defense Technology, Changsha, 410073, China
Hunan Provincial Collaborative Innovation Center of High Power Fiber Laser, Changsha, 410073, China

Minghong Yang, Wei Bai, Huiyong Guo, Hongqiao Wen, Haihu Yu, and Desheng Jiang
National Engineering Laboratory for Fiber Optic Sensing Technology, Wuhan University of Technology, Wuhan, 430070, China

Bing Zhang, Linghao Cheng, Yizhi Liang, Long Jin, Tuan Guo, and Bai-Ou Guan
Guangdong Provincial Key Laboratory of Optical Fiber Sensing and Communications, Institute of Photonics Technology, Jinan University, Guangzhou, 510632, China

Xin Gui, Honghai Wang and Haihu Yu
National Engineering Laboratory for Fiber Optic Sensing Technology, Wuhan University of Technology, Wuhan, 430070, China

Yuheng Tong
Key Laboratory of Fiber Optic Sensing Technology and Information Processing, Ministry of Education, Wuhan University of Technology, Wuhan, 430070, China

Zhengying Li
National Engineering Laboratory for Fiber Optic Sensing Technology, Wuhan University of Technology, Wuhan, 430070, China
Key Laboratory of Fiber Optic Sensing Technology and Information Processing, Ministry of Education, Wuhan University of Technology, Wuhan, 430070, China

Shuvo Sen and Sawrab Chowdhury
Department of Information and Communication Technology, Mawlana Bhashani Science and Technology University, Santosh, Tangail-1902, Bangladesh

Sayed Asaduzzaman
Group of Biophotomatix, Bangladesh

Kawsar Ahmed
Department of Information and Communication Technology, Mawlana Bhashani Science and Technology University, Santosh, Tangail-1902, Bangladesh
Group of Biophotomatix, Bangladesh

Ghada H. Dushaq, Tadesse Mulugeta, and Mahmoud Rasras
Department of Electrical Engineering and Computer Science, iMicro Center, MASDAR Institute, Building A1, PO Box 54224, Masdar City, United Arab Emirates

Weibing Gan and Yan Yang
Key Laboratory of Fiber Optic Sensing Technology and Information Processing of Ministry of Education, Wuhan University of Technology, Wuhan, 430070, China;

Wenbin Hu, Fang Liu, Jianguang Tang and Sheng Li
National Engineering Laboratory for Fiber Optic Sensing Technology, Wuhan University of Technology, Wuhan, 430070, China

Index

www.ingramcontent.com/pod-product-compliance
Lightning Source LLC
Chambersburg PA
CBHW082033190326
41458CB00010B/3347